Bernadette Bouchon-Meunier, Luis Magdalena, Manuel Ojeda-Aciego,
José-Luis Verdegay, and Ronald R. Yager (Eds.)

Foundations of Reasoning under Uncertainty

Studies in Fuzziness and Soft Computing, Volume 249

Editor-in-Chief

Prof. Janusz Kacprzyk
Systems Research Institute
Polish Academy of Sciences
ul. Newelska 6
01-447 Warsaw
Poland
E-mail: kacprzyk@ibspan.waw.pl

Further volumes of this series can be found on our homepage: springer.com

Vol. 232. Eduardo Massad, Neli Regina Siqueira Ortega, Laécio Carvalho de Barros, Claudio José Struchiner
Fuzzy Logic in Action: Applications in Epidemiology and Beyond, 2008
ISBN 978-3-540-69092-4

Vol. 233. Cengiz Kahraman (Ed.)
Fuzzy Engineering Economics with Applications, 2008
ISBN 978-3-540-70809-4

Vol. 234. Eyal Kolman, Michael Margaliot
Knowledge-Based Neurocomputing: A Fuzzy Logic Approach, 2009
ISBN 978-3-540-88076-9

Vol. 235. Kofi Kissi Dompere
Fuzzy Rationality, 2009
ISBN 978-3-540-88082-0

Vol. 236. Kofi Kissi Dompere
Epistemic Foundations of Fuzziness, 2009
ISBN 978-3-540-88084-4

Vol. 237. Kofi Kissi Dompere
Fuzziness and Approximate Reasoning, 2009
ISBN 978-3-540-88086-8

Vol. 238. Atanu Sengupta, Tapan Kumar Pal
Fuzzy Preference Ordering of Interval Numbers in Decision Problems, 2009
ISBN 978-3-540-89914-3

Vol. 239. Baoding Liu
Theory and Practice of Uncertain Programming, 2009
ISBN 978-3-540-89483-4

Vol. 240. Asli Celikyilmaz, I. Burhan Türksen
Modeling Uncertainty with Fuzzy Logic, 2009
ISBN 978-3-540-89923-5

Vol. 241. Jacek Kluska
Analytical Methods in Fuzzy Modeling and Control, 2009
ISBN 978-3-540-89926-6

Vol. 242. Yaochu Jin, Lipo Wang
Fuzzy Systems in Bioinformatics and Computational Biology, 2009
ISBN 978-3-540-89967-9

Vol. 243. Rudolf Seising (Ed.)
Views on Fuzzy Sets and Systems from Different Perspectives, 2009
ISBN 978-3-540-93801-9

Vol. 244. Xiaodong Liu and Witold Pedrycz
Axiomatic Fuzzy Set Theory and Its Applications, 2009
ISBN 978-3-642-00401-8

Vol. 245. Xuzhu Wang, Da Ruan, Etienne E. Kerre
Mathematics of Fuzziness – Basic Issues, 2009
ISBN 978-3-540-78310-7

Vol. 246. Piedad Brox, Iluminada Castillo, Santiago Sánchez Solano
Fuzzy Logic-Based Algorithms for Video De-Interlacing, 2010
ISBN 978-3-642-10694-1

Vol. 247. Michael Glykas
Fuzzy Cognitive Maps, 2010
ISBN 978-3-642-03219-6

Vol. 248. Bing-Yuan Cao
Optimal Models and Methods with Fuzzy Quantities, 2010
ISBN 978-3-642-10710-8

Vol. 249. Bernadette Bouchon-Meunier, Luis Magdalena, Manuel Ojeda-Aciego, José-Luis Verdegay, Ronald R. Yager (Eds.)
Foundations of Reasoning under Uncertainty, 2010
ISBN 978-3-642-10726-9

Bernadette Bouchon-Meunier,
Luis Magdalena, Manuel Ojeda-Aciego,
José-Luis Verdegay, and
Ronald R. Yager (Eds.)

Foundations of Reasoning under Uncertainty

Springer

Author

Prof. Bernadette Bouchon-Meunier
LIP6
Université Pierre et Marie Curie
104 avenue du président Kennedy
75016 Paris, France
E-Mail: Bernadette.Bouchon-Meunier@lip6.fr

Prof. Luis Magdalena
European Centre for Soft Computing
Edificio Científico-Tecnológico.
3â Planta.
C Gonzalo Gutiérrez Quirós S/N
33600 Mieres, Asturias, Spain
E-Mail: luis.magdalena@softcomputing.es

Prof. Manuel Ojeda-Aciego
Computer Science Faculty
University of Málaga
Blv Louis Pasteur s/n
29071 Málaga, Spain
E-Mail: aciego@uma.es

Prof. José Luis Verdegay
Computer Science Faculty
University of Granada
C/Periodista Daniel
Saucedo Aranda s/n
18071 Granada, Spain
E-mail: verdegay@decsai.ugr.es

Prof. Ronald R. Yager
Machine Intelligence Institute
Iona College
New Rochelle, NY 10801, USA
E-mail: yager@panix.com

ISBN 978-3-642-26235-7 e-ISBN 978-3-642-10728-3

DOI 10.1007/978-3-642-10728-3

Studies in Fuzziness and Soft Computing ISSN 1434-9922

Softcover reprint of the hardcover 1st edition 2010

Typeset & Cover Design: Scientific Publishing Services Pvt. Ltd., Chennai, India.

Printed in acid-free paper

9 8 7 6 5 4 3 2 1

springer.com

Preface

Uncertainty exists almost everywhere, except in the most idealized situations; it is not only an inevitable and ubiquitous phenomenon, but also a fundamental scientific principle. Furthermore, uncertainty is an attribute of information and, usually, decision-relevant information is uncertain and/or imprecise, therefore the abilities to handle uncertain information and to reason from incomplete knowledge are crucial features of intelligent behaviour in complex and dynamic environments. By carefully exploiting our tolerance for imprecision and approximation we can often achieve tractability, robustness, and better descriptions of reality than traditional deductive methods would allow us to obtain. In conclusion, as we move further into the age of machine intelligence, the problem of reasoning under uncertainty, in other words, drawing conclusions from partial knowledge, has become a major research theme.

Not surprisingly, the rigorous treatment of uncertainty requires sophisticated machinery, and the present volume is conceived as a contribution to a better understanding of the foundations of information processing and decision-making in an environment of uncertainty, imprecision and partiality of truth.

This volume draws on papers presented at the 2008 Conference on Information Processing and Management of Uncertainty (IPMU), held in Málaga, Spain, organized by the University of Málaga. The conference brought together some of the world's leading experts in the study of uncertainty.

Since its first edition, held in 1986, the focus of IPMU conferences has been on the development of foundations and technology needed for the construction of intelligent systems. Over the years, IPMU has grown steadily in visibility and importance, and has evolved into a leading conference in its field, embracing a wide variety of methodologies for dealing with uncertainty and imprecision, and this explains the unusually wide variety of concepts, methods and techniques which are discussed in the book. The growth in importance of IPMU reflects the fact that as we move further into the age of machine intelligence and mechanized decision-making, the issue of how to deal with uncertain information becomes an issue of paramount concern.

The book starts with a revisited approach for possibilistic fuzzy regression methods proposed by Bisserier *et al.*, in which the identification problem is reformulated according to a new criterion that assesses the model fuzziness independently of the collected data. Later, Bonissone *et al.* propose the fundamentals to design and construct a "forest" of randomly generated fuzzy decision trees in an approach which combines the robustness of multi-classifiers, the construction efficiency of decision trees, the power of the randomness to increase the diversity of the trees in the forest, and the flexibility of fuzzy logic and the fuzzy sets for data managing. The third contribution, by Delgado *et al.*, is related to the well-known framework of mining association rules for extracting useful knowledge from databases; they introduce so-called double rules as a new type of rules which in conjunction with exception rules will describe in more detail the relationship between two sets of items. Next, Dubois discusses ignorance and contradiction, and argues that they cannot be viewed either as additional truth-values or processed in a truth-functional manner, and that doing it leads to weak or debatable uncertainty handling approaches.

The volume continues with Grzegorzewski's work, which introduces new algorithms for calculating the proper approximation by trapezoidal fuzzy numbers which preserves the expected interval. Next, Jenhani *et al.* investigate the problem of measuring the similarity degree between two normalized possibility distributions encoding preferences or uncertain knowledge. Later, Julián *et al.* propose an improved fuzzy query answering procedure for multi-adjoint logic programming which avoids the re-evaluation of goals and the generation of useless computations, thanks to the combined use of tabulation with thresholding techniques. Then, Kacprzyk and Wilbik focus on an extension of linguistic summarization of time series; in addition to the basic criterion of a degree of truth (validity), they also use a degree of imprecision, specificity, fuzziness and focus as an additional criteria.

In the final part of the volume, Kalina *et al.* discuss the possibility of applying the modified level-dependent Choquet integral to a monopersonal multicriterial decision-making problem; they propose an algorithm which produces an outranking of objects taking into account an interaction between criteria. Next, Llamazares and Marques Pereira consider mixture operators to aggregate individual preferences and characterize those that allow to extend some majorities rules, such as simple, Pareto and absolute special majorities, to the field of gradual preferences. Later, Mercier *et al.* concentrate on the links between the different operations that can be used in the theory of belief functions to correct the information provided by a source, given meta-knowledge about that source. Then, Miranda compares the different notions of conditional coherence within the behavioural theory of imprecise probabilities when all the referential spaces are finite. Finally, Soubaras focuses on evidential Markov chains as a suitable generalization of classical Markov chains to the Dempster-Shafer theory, replacing the involved states by sets of states.

Last, but not least, we would like to thank the following institutions for their help with the organization of the 12th IPMU Conference: Ministerio de Educación y Ciencia, grant TIN2007-30838-E, Junta de Andalucía, grant Res. 2/07-OC, Universidad de Málaga, Diputación Provincial de Málaga, Patronato de Turismo de la

Costa del Sol, Ayuntamiento de Málaga, Ayuntamiento de Torremolinos, European Society for Fuzzy Logic and Technology, EUSFLAT, IEEE Computational Intelligence Society.

Paris, Mieres, Málaga, Granada, New York
September 2009

Bernadette Bouchon-Meunier
Luis Magdalena
Manuel Ojeda-Aciego
José Luis Verdegay
Ronald R. Yager

Contents

List of Contributors

Salem Benferhat
CRIL, Université d'Artois,
Lens, France
benferhat@
cril.univ-artois.fr

Amory Bisserier
LISTIC, Université de
Savoie BP 80439 74941
Annecy-le-Vieux Cedex France
amory.bisserier@
univ-savoie.fr

Piero P. Bonissone
GE Global Research.
One Research Circle.
Niskayuna, NY 12309. USA
bonissone@crd.ge.com

Reda Boukezzoula
LISTIC, Université de
Savoie BP 80439 74941
Annecy-le-Vieux Cedex France
reda.boukezzoula@
univ-savoie.fr

José Manuel Cadenas
Dept. Ing. de la
Información y las Comunicaciones.
University of Murcia,
Spain
jcadenas@um.es

Miguel Delgado
Dept. Computer Science and
Artificial Intelligence,
University of Granada,
Spain
mdelgado@ugr.es

Thierry Denœux
Université de Technologie
de Compiègne,
UMR CNRS 6599
Heudiasyc, France
tdenoeux@hds.utc.fr

R. Andrés Díaz-Valladares
Dept. Ciencias Computacionales.
Universidad de Montemorelos, Mexico
rdiaz@um.edu.mx

Didier Dubois
IRIT-CNRS, Université de
Toulouse, France
dubois@irit.fr

Zied Elouedi
LARODEC, Institut Supérieur
de Gestion de Tunis, Tunisia
zied.elouedi@gmx.fr

Sylvie Galichet
LISTIC, Université de
Savoie BP 80439 74941
Annecy-le-Vieux Cedex France
sylvie.galichet@
univ-savoie.fr

María del Carmen Garrido
Dept. Ing. de la
Información y
las Comunicaciones.
University of Murcia, Spain
carmengarrido@um.es

Przemysław Grzegorzewski
Systems Research Institute,
Polish Academy of Sciences,
Newelska 6, 01-447 Warsaw,
Poland
pgrzeg@ibspan.waw.pl

and
Faculty of Mathematics
and Information Science,
Warsaw University
of Technology,
Plac Politechniki 1,
00-661 Warsaw, Poland
pgrzeg@mini.pw.edu.pl

Dana Hliněná
Dept. of Mathematics,
FEEC Brno Uni. of Technology
Technická 8,
616 00 Brno,
Czech Republic
hlinena@feec.vutbr.cz

Ilyes Jenhani
LARODEC, Institut Supérieur de
Gestion de Tunis, Tunisia
CRIL, Université d'Artois,
Lens, France
ilyes.j@lycos.com

Pascual Julián
University of Castilla-La Mancha,
Dept of Information
Technologies and Systems,
Ciudad Real (Spain)
Pascual.Julian@uclm.es

Janusz Kacprzyk
Systems Research Institute,
Polish Academy of Sciences,
ul. Newelska 6,
01-447 Warsaw, Poland

and
PIAP – Industrial Research
Institute for Automation
and Measurements,
Al. Jerozolimskie 202,
02-486 Warsaw, Poland
kacprzyk@ibspan.waw.pl

Martin Kalina
Dept. of Mathematics,
Slovak Uni. of Technology,
Radlinského 11,
813 68 Bratislava,
Slovakia
kalina@math.sk

Pavol Král'
Institute of Mathematics
and Computer Science,
UMB and MÚ SAV,
Ďumbierska 1,
974 11 Banská Bystrica,
Slovakia
pavol.kral@umb.sk

Bonifacio Llamazares
Department of
Applied Economics,
Avda. Valle Esgueva 6,
E-47010 Valladolid,
Spain
boni@eco.uva.es

Marie-Hélène Masson
Université de
Picardie Jules Verne,
UMR CNRS 6599
Heudiasyc, France
mmasson@hds.utc.fr

Ricardo Alberto Marques Pereira
Department of Computer
and Management Sciences,
Via Inama 5,
TN 38100 Trento,
Italy
ricalb.marper@unitn.it

Jesús Medina
University of Cádiz,
Department of Mathematics,
CASEM-Campus
Río San Pedro,
11510 Puerto Real,
Cádiz (Spain)
Jesus.Medina@uca.es

David Mercier
Univ. Lille Nord
de France,
UArtois, EA
3926 LGI2A, France
david.mercier@
univ-artois.fr

Enrique Miranda
University of Oviedo,
Dep. of Statistics and
Operations Research.
C-Calvo Sotelo,
s/n 33007 Oviedo (Spain)
mirandaenrique@uniovi.es

Ginés Moreno
University of Castilla-La Mancha,
Department of Computing Systems,
Campus Universitario,
02071 Albacete (Spain)
Gines.Moreno@uclm.es

Manuel Ojeda-Aciego
Department of Applied Mathematics,
University of Málaga,
Málaga (Spain)
aciego@uma.es

María Dolores Ruiz
Dept. Computer Science and
Artificial Intelligence,
University of Granada, Spain
mdruiz@decsai.ugr.es

Daniel Sánchez
Dept. Computer Science and
Artificial Intelligence,
University of Granada,
Spain
daniel@decsai.ugr.es

Hélène Soubaras
Thales R&T.
Campus Polytechnique,
1. av. A.
Fresnel - F91767 Palaiseau
helene.soubaras@
thalesgroup.com

Anna Wilbik
Systems Research Institute,
Polish Academy of Sciences,
Newelska 6,
01–447 Warsaw, Poland
wilbik@ibspan.waw.pl

Linear Fuzzy Regression Using Trapezoidal Fuzzy Intervals

Amory Bisserier, Reda Boukezzoula, and Sylvie Galichet

Abstract. In this paper, a revisited approach for possibilistic fuzzy regression methods is proposed. Indeed, a new modified fuzzy linear model form is introduced where the identified model output can envelop all the observed data and ensure a total inclusion property. Moreover, this model output can have any kind of spread tendency. In this framework, the identification problem is reformulated according to a new criterion that assesses the model fuzziness independently of the collected data. The proposed concepts are used in a global identification process in charge of building a piecewise model able to represent every kinds of output evolution.

1 Introduction

Model identification is based on a general principle which consists in determining, among candidate models, the one that best explains the behavior of the system to be modeled. Assuming a particular class of models such as linear functions, splines, rule-based systems, neural networks, ..., the best candidate is determined from the available information, usually a set of observations of the input and output variables. Classical identification techniques assume perfect knowledge of input and output values. It means that the observations are supposed to be both precise (point-valued) and certain. However, there are situations in which this assumption is not realistic, especially when the information about the output value is obtained through

Amory Bisserier
LISTIC, Université de Savoie BP 80439 74941 Annecy-le-Vieux Cedex France
e-mail: amory.bisserier@univ-savoie.fr

Reda Boukezzoula
LISTIC, Université de Savoie BP 80439 74941 Annecy-le-Vieux Cedex France
e-mail: reda.boukezzoula@univ-savoie.fr

Sylvie Galichet
LISTIC, Université de Savoie BP 80439 74941 Annecy-le-Vieux Cedex France
e-mail: sylvie.galichet@univ-savoie.fr

B. Bouchon-Meunier et al. (Eds.) Found. of Reas. under Uncert., STUDFUZZ 249, pp. 1–22.
springerlink.com

measuring devices with limited precision. In the framework of fuzzy modeling in which it is possible to handle imprecise representations using fuzzy set theory as proposed by Zadeh [19], the assumption of perfect data becomes even paradoxical. Nevertheless, most fuzzy model identification techniques used in practice, especially in fuzzy control, still consider crisp data.

In this context, we are interested in developing modeling techniques for building fuzzy models from fuzzy observations. For this purpose, it is proposed to focus on works about fuzzy regression techniques. Fuzzy regression, a fuzzy type of conventional regression analysis, has been proposed to evaluate the functional relationship between input and output variables in a fuzzy environment. This approach [14] is well adapted for situations where data are imprecise and/or partially available, and where human estimation is influential. It can be applied in many non quantitative fields, like social, health or biological sciences for example. According to [6], fuzzy regression techniques can be classified into two distinct categories. The first, initially proposed by Tanaka known as possibilistic regression aims at minimizing the total spread of the model output under data inclusion constraints. In this case, the problem is viewed as finding fuzzy coefficients of a regression model according to a mathematical programming problem. The second approach, developed by Diamond [5], is based on the minimization of the total square error between data and model outputs using a fuzzy least square method.

In this study, we adopt the possibilistic regression approach. Since its introduction by Tanaka and al. [15, 17] in a linear context, several improved methods have been proposed. For example, Tanaka, Hayashi and Watada [15] propose different expressions of the criterion to be optimized and different formulations of the constraints to be satisfied for possibility and necessity estimation models. Still in a linear context, Tanaka and Ishibushi [16] extends their approach for dealing with interactive fuzzy parameters. Furthermore, the complete specification of regression problems highly depends on the nature of input-output data [6]. Some works are thus devoted to crisp input - crisp output data [12, 14] while others [13] consider fuzzy input - fuzzy output data. Most commonly, a mixed approach (crisp input - fuzzy output) is chosen [7, 8, 9, 10, 15, 17]. That is the formalism we adopt here in a linear context with the idea of keeping a simple model, possibly invertible [2, 3].

Unfortunately, three types of problems emerge from most of above cited methods:

- The assumption of symmetrical triangular fuzzy parameters is most frequently used. However, such parameters have some limitations, especially when total inclusion of the observed data in the model output must be ensured.
- The identification is made at a chosen level α considered as a degree of fitting of the obtained model to the observed data. If this way of doing allows to simplify the problem by using interval arithmetics to express the inclusion problem, after reconstruction of the parameters, inclusion is no more guaranteed at any level α.
- The obtained models are not able to represent any tendency of the output spread. It follows that the identified model may become more imprecise than necessary in some situations.

The main objective of this paper is to revisit some theoretical works about fuzzy regression techniques [6] and to propose some slight improvements for suppressing the limitations mentioned previously. This paper is organized as follows. In Sect. 2, the concepts of intervals and fuzzy intervals are introduced. Sect. 3 is devoted to the conventional fuzzy linear regression. A revisited approach of the latter is detailed in Sect. 4. Sect. 5 and Sect. 6 present the identification process and its application in the identification of a piecewise model. A generalization to multi-input model identification is described in Sect. 7. Applications on several examples are shown in Sect. 8. Finally, conclusions and perspectives are presented in Sect. 9.

2 Intervals and Fuzzy Intervals

This section introduces some relevant concepts and notations about conventional intervals and fuzzy ones.

2.1 Conventional Intervals

An interval is defined by the set of elements lying between its lower and upper limits as:

$$a = \{x | a^- \le x \le a^+, x \in \mathbb{R}\} \tag{1}$$

Given an interval a, its Midpoint $M(a)$ and its Radius $R(a)$ are defined by:

$$M(a) = (a^- + a^+)/2 \text{ and } R(a) = (a^+ - a^-)/2 \tag{2}$$

For two intervals a and b, an inclusion relation of a in b is defined as follows [4] (see Fig. 1):

$$a \subseteq b \Leftrightarrow \begin{cases} b^- \le a^- \\ a^+ \le b^+ \end{cases} \Leftrightarrow \begin{cases} M(b) - R(b) \le M(a) - R(a) \\ M(a) + R(a) \le M(b) + R(b) \end{cases}$$

$$\Leftrightarrow \begin{cases} M(b) - M(a) \le R(b) - R(a) \\ M(a) - M(b) \le R(b) - R(a) \end{cases} \tag{3}$$

From Eq. 3 it follows:

$$a \subseteq b \Leftrightarrow |M(b) - M(a)| \le R(b) - R(a) \tag{4}$$

When a is a scalar value, the relation defined in Eq. 4 becomes:

$$a \in b \Leftrightarrow |M(b) - a| \le R(b) \tag{5}$$

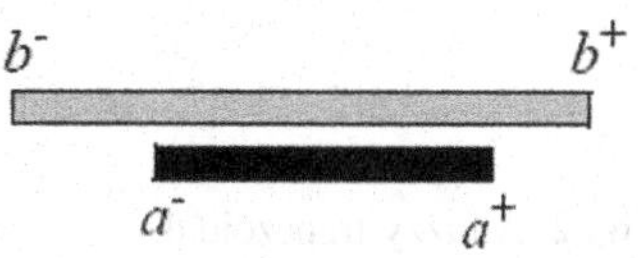

Fig. 1 Inclusion of two conventional intervals

2.2 Fuzzy Intervals

An interval a can be viewed as a special fuzzy number whose membership function $\mu_a(x)$ takes the value 1 over the interval and 0 anywhere else. A fuzzy interval A is represented by its membership function μ_A. In order to specify the fuzzy interval shape, one has to consider two dimensions. The first one (horizontal dimension) is similar to that used in interval representation, that is the real line $\Re$. The second one (vertical dimension) is related to the handling of the membership degrees and thus restricted to the interval $[0,1]$. In this context, two kinds of information are required for defining a fuzzy interval. Both pieces of information, called support and kernel intervals, are defined on the horizontal dimension, but are associated to two different levels (level 0 and level 1) on the vertical dimension (see Fig. 2). For a fuzzy trapezoidal interval A we have:

$$\text{Support: } S_A = [S_A^-, S_A^+], \text{ Kernel: } K_A = [K_A^-, K_A^+] \tag{6}$$

To completely define the fuzzy interval, two additional functions are used to link the support and the kernel:

$$\begin{cases} (A^-)_\alpha = \inf\{x \mid \mu_A(x) \geq \alpha \, ; x \geq S_A^-\} \\ (A^+)_\alpha = \sup\{x \mid \mu_A(x) \geq \alpha \, ; x \leq S_A^+\} \end{cases} \tag{7}$$

where $\alpha \in [0,1]$ represents the vertical dimension. In this case, for a given α-cut on the fuzzy interval A, a conventional interval is obtained:

$$[A]_\alpha = [(A^-)_\alpha, (A^+)_\alpha] \tag{8}$$

Finally, in the same way that the conventional interval a is denoted $[a^-, a^+]$, the fuzzy interval A will be defined by its support and kernel bounds:

$$A = (K_A, S_A) = ([K_A^-, K_A^+], [S_A^-, S_A^+]) \tag{9}$$

A particular case of trapezoidal fuzzy intervals are triangular symmetrical ones. In this case, the fuzzy number can be defined by its kernel (modal value) K_A and the radius of its support R_A, i.e. $A = (K_A, R_A)$, that is:

$$K_A^- = K_A^+ = K_A, \; S_A^+ = K_A + R_A, \; S_A^- = K_A - R_A \tag{10}$$

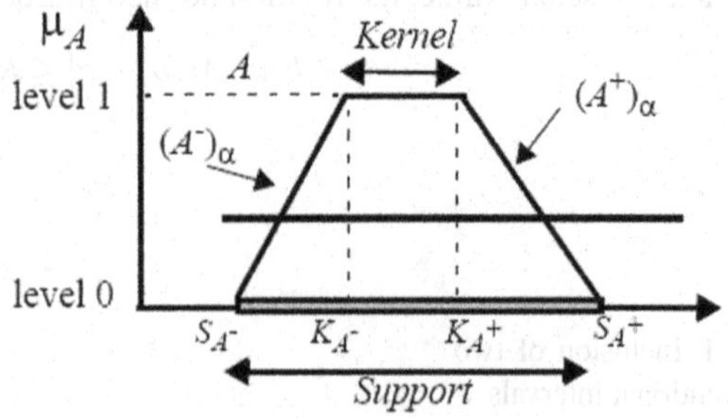

Fig. 2 A fuzzy trapezoidal interval

3 Fuzzy Linear Regression

Let us consider a set of N observed data samples with single input defined on the interval $D = [x_{min}, x_{max}]$. Let the jth sample be represented by the couple (x_j, Y_j), $j = 1, ..., N$ where x_j are crisp inputs sorted in increasing order and Y_j the corresponding fuzzy outputs which are assumed to be triangular and symmetrical fuzzy intervals. In this case, the jth fuzzy interval Y_j is completely defined by its modal value K_{Y_j} and its radius R_{Y_j}, that is:

$$Y_j = (K_{Y_j}, R_{Y_j}) \tag{11}$$

Like for any regression technique, the fuzzy regression objective is to determine a predicted functional relationship $\hat{Y} = h(x)$ between input x and output Y. In this paper, the function h is assumed to be linear and given by the following expression:

$$\hat{Y}(x) = A_0 \oplus A_1.x \tag{12}$$

defined on the domain D.

In order to consider the fuzziness of the observed outputs, the parameters A_0 and A_1 are fuzzy coefficients. The latter are assumed to be triangular and symmetrical, represented by:

$$A_0 = (K_{A_0}, R_{A_0}) \text{ and } A_1 = (K_{A_1}, R_{A_1}) \tag{13}$$

3.1 Inclusion Problem Statement

A fuzzy interval is a standard normal fuzzy set defined on the set of real numbers, whose α-cuts, are closed intervals of real numbers with bounded supports. Using an α-cut representation, a fuzzy interval is viewed as a weighted family of nested intervals. By doing so, for a specified α-cut, the fuzzy interval becomes a conventional interval, which states that a fuzzy interval representation is a generalization of a conventional one. Moreover, this strategy has the advantage to reduce the fuzzy computational complexity and makes easier its implementation, especially in optimization and identification problems. That is the approach proposed by Tanaka for the identification of a fuzzy model in the form of Eq. 12 in references [15, 17] where a possibilistic regression methodology is developed according to the α-cut representation principle.

Indeed, for a set of observed data, the authors try to identify the fuzzy model parameters A_0 and A_1 so that all observed data are included in the predicted ones for some α-cut, i.e.,

$$[Y_j]_\alpha \subseteq [\hat{Y}_j]_\alpha \tag{14}$$

Eq. 14 is viewed as a constraint in the identification procedure. The latter is based on the minimization of a criterion which exhibits the spreads of the predicted intervals, that is:

$$min_{K_{A_0}, K_{A_1}, R_{A_0}, R_{A_1}} \; N.R_{A_0} + R_{A_1}.\sum_{j=1}^{N} |x_j| \tag{15}$$

After optimization is performed, the obtained parameters computed for a given α-cut are assumed to be defined for all $\alpha \in [0,1]$.

Let us give a simple example used by Tanaka in [17] to illustrate this method (see Table 1). In this example, the pessimistic case (maximum of uncertainty) is adopted, i.e. $\alpha = 0$. In this case, the constraints are the following ones:

$$\begin{cases} K_{A_0} + K_{A_1}.x_j + R_{A_0} + R_{A_1}.|x_j| \geq K_{Y_j} + R_{Y_j} \\ K_{A_0} + K_{A_1}.x_j - (R_{A_0} + R_{A_1}.|x_j|) \leq K_{Y_j} - R_{Y_j} \end{cases} \quad (16)$$

The identification method gives the fuzzy symmetrical triangular coefficients $A_0 = (3.85, 3.85), A_1 = (2.1, 0)$, and the predicted intervals represented in Table 1.

Table 1 Observed and predicted intervals

j	x_j	observed intervals	predicted intervals
1	1	$[6.2, 9.8]$	$[2.1, 9.8]$
2	2	$[4.2, 8.6]$	$[4.2, 11.9]$
3	3	$[6.9, 12.1]$	$[6.3, 14]$
4	4	$[10.9, 16.1]$	$[8.4, 16.1]$
5	5	$[10.6, 15.4]$	$[10.5, 18.2]$

For example, when $j = 1$, the observed and the predicted output are respectively $Y_1 = [6.2, 9.8]$ and $\hat{Y}_1 = [2.1, 9.8]$. It can be stated that the inclusion constraint is respected for $\alpha = 0$, i.e.,

$$[Y_1]_{\alpha=0} \subseteq [\hat{Y}_1]_{\alpha=0} \quad (17)$$

According to Fig. 3, it is obvious that although the inclusion is respected for $\alpha = 0$, it is not respected for any $\alpha \in [0,1]$.

From a general point of view, if the fuzzy model parameters are identified for a chosen level α under the constraint of Eq. 14, the inclusion of all observed outputs in the predicted ones is not guaranted. Indeed, the inclusion relation between α-cuts is not sufficient to guarantee the total inclusion of the fuzzy intervals. For example, when the inclusion is ensured for $\alpha = 0$ (support inclusion), according to the kernel

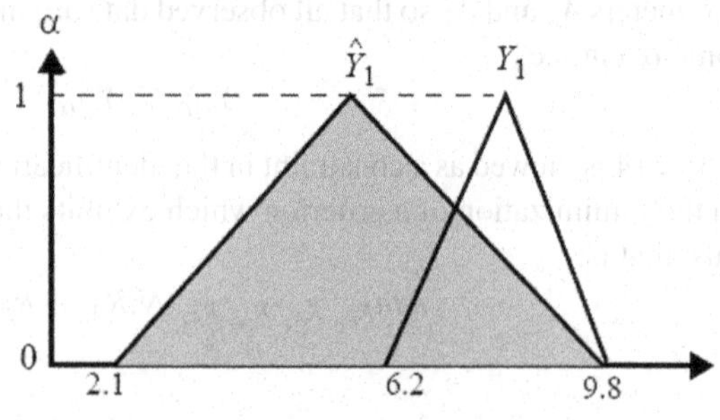

Fig. 3 Observed and predicted outputs for $j = 1$

value positions in the fuzzy interval, three cases can be obtained (see Fig. 4 for two fuzzy intervals A and B). It is then clear that the total inclusion of fuzzy intervals is respected if and only if modal values are equal. More generally, whatever the identification level α, total inclusion requires modal value equality. It follows that total inclusion for all observations (or equivalently identification at $\alpha = 1$) is achievable only if observed modal values strictly fit a straight line [6]. Moreover, the higher the α considered for identification, the wider the support of the predicted fuzzy number is [13]. These drawbacks weaken the potential use of this method, especially in real identification problems.

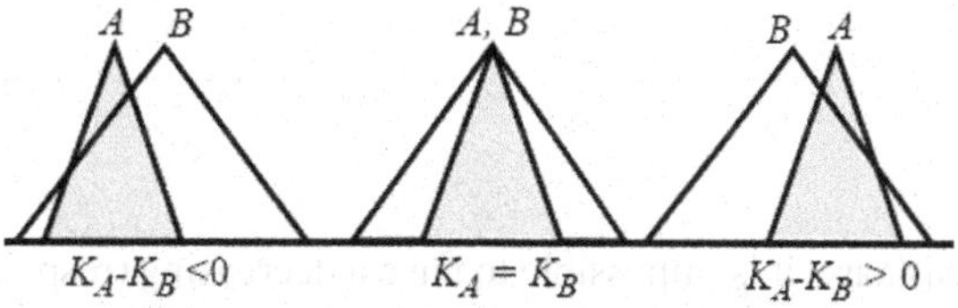

Fig. 4 Three cases of support inclusion for triangular fuzzy intervals

3.2 Tendency Problem Statement

Let us now apply the Tanaka identification method for another example presented in Table 2. In this case, it can be stated that the observed outputs have a spread which is globally decreasing. Applying Tanaka identification method for $\alpha = 0$, the predicted intervals given in Table 2 are obtained. The identified fuzzy parameters are $A_0 = (2.574, 4)$ and $A_1 = (2.43, 0)$. A representation of the model output is given in Fig. 5.

Table 2 Observed and predicted intervals

j	x_j	observed intervals	predicted intervals
1	1	[1,9]	[1,9]
2	2	[5.4, 10.6]	[3.43, 11.43]
3	3	[8, 12]	[5.85, 13.85]
4	4	[10, 12]	[8.28, 16.28]
5	5	[13.5, 14.5]	[10.71, 18.71]

According to Table 2 and Fig. 5, it can be observed that the identified model output spread is constant. Obviously, it would be better if it was decreasing, i.e. if the identified model presented the same spread variation than the observed data.

More generally, one weakness of this method is the fact that the fuzziness of the model output varies in the same way than the absolute value of the inputs. In

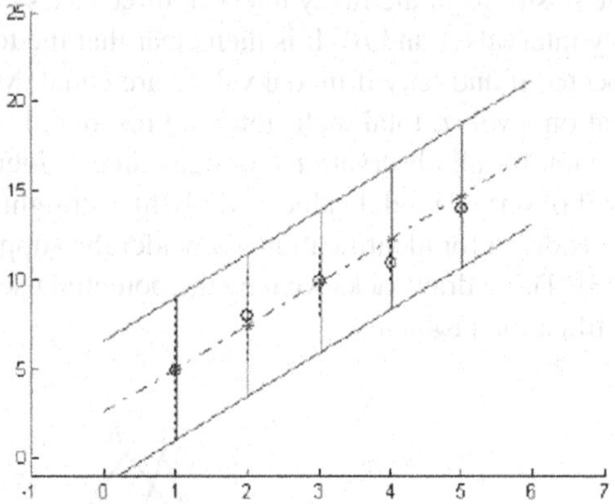

Fig. 5 Representation of the identified model

this case, it is impossible to have a decreasing (resp. increasing) spread of the model output for positive (resp. negative) inputs. This restriction is acceptable in a measurement context where it is usual to express percentage relative errors. However, when fuzziness is considered as an intrinsic characteristic of the system to be modeled, the assumption that the higher the input, the higher the fuzziness attached to the model output, is open to criticism. Finally, as classical fuzzy regression models are not able to represent any tendency of output spread, they become more imprecise than necessary in some situations. As a consequence, in piecewise fuzzy regression problems, in which collected data can have any kind of spread tendency, actual identification methods are clearly insufficient.

Let us now study the tendency output problem in order to release a suitable solution. From the model of Eq. 12, the output modal value and spread can be determined. Indeed, as A_0 and A_1 are symmetrical triangular fuzzy intervals, and x a crisp input, $\hat{Y}(x)$ is also a symmetrical triangular fuzzy interval. In this case, the modal value $M(\hat{Y}(x))$ and the spread $R(\hat{Y}(x))$ are given by:

$$\begin{cases} M(\hat{Y}(x)) = K_{\hat{Y}(x)} = K_{A_0} + K_{A_1}.x \\ R(\hat{Y}(x)) = R_{\hat{Y}(x)} = R_{A_0} + R_{A_1}.|x| \end{cases} \tag{18}$$

As x is varying on D, the variation of Eq. 18 needs to be analyzed according to the sign of x.

From Eq. 18, it follows that the variation of $M(\hat{Y}(x))$ depends on the sign of K_{A_1} and can be increasing or decreasing for any value of the input x. On the other hand, we see that the variation of $R(\hat{Y}(x))$ depends on the sign of the input. As R_{A_1} is always positive, it can be stated that when x is positive, the output radius is increasing, whereas when x is negative, the output radius is decreasing.

In short, it is possible to have any kind of variation of the output modal value, with an appropriate sign of K_{A_1}. However, the radius output variation is limited by the sign of the input x.

4 Revisited Fuzzy Linear Regression

In order to deal with the two drawbacks discussed in the previous section (inclusion and tendency problems), two evolutionary concepts are introduced into the conventional fuzzy regression model identification problems.

4.1 Inclusion Problem Solution

In order to overcome the inclusion problem relating to the α-cut specification, the fuzzy model parameters A_0 and A_1 are assumed to be trapezoidal. In this case, it is ensured that total inclusion of all observed inputs in the predicted ones at each level α can be respected. As the fuzzy parameters are trapezoidal, the model output $\hat{Y}(x)$ is also a trapezoidal interval.

By using trapezoidal fuzzy intervals, whose kernel values are not reduced to point values, inclusion can be guaranteed. Moreover, the membership function being linear, they are completely defined by only two α-cuts, and so easily expressible. This is an advantage in the parametric regression framework.

In order to extend the Tanaka interval method and solving the inclusion problem, two inclusion constraints must be taken into account in the identification method:

$$[Y_j]_{\alpha=0} \subseteq [\hat{Y}_j]_{\alpha=0} \text{ and } [Y_j]_{\alpha=1} \subseteq [\hat{Y}_j]_{\alpha=1} \tag{19}$$

In this case, as a trapezoidal fuzzy interval shape is assumed, it is obvious that if relations of Eq. 19 are respected, then the total inclusion is guaranteed for each level $\alpha \in [0,1]$, i.e.:

$$\forall \alpha \in [0,1],\ [Y_j]_\alpha \subseteq [\hat{Y}_j]_\alpha \tag{20}$$

Let us consider the jth observed data, whose output is the triangular symmetrical fuzzy interval $Y_j = (K_{Y_j}, R_{Y_j})$. The corresponding predicted output is the trapezoidal fuzzy interval given by:

$$\hat{Y}_j = (K_{\hat{Y}_j}, R_{\hat{Y}_j}) = ([K^-_{\hat{Y}_j}, K^+_{\hat{Y}_j}], [S^-_{\hat{Y}_j}, S^+_{\hat{Y}_j}]) \tag{21}$$

In this case, the constraints in Eq. 19 can be written as:

- for $\alpha = 1$:

$$[Y_j]_{\alpha=1} \subseteq [\hat{Y}_j]_{\alpha=1} \Leftrightarrow K_{Y_j} \in [K^-_{\hat{Y}_j}, K^+_{\hat{Y}_j}] \tag{22}$$

- for $\alpha = 0$:

$$[Y_j]_{\alpha=0} \subseteq [\hat{Y}_j]_{\alpha=0} \Leftrightarrow [K_{Y_j} - R_{Y_j}, K_{Y_j} + R_{Y_j}] \subseteq [S^-_{\hat{Y}_j}, S^+_{\hat{Y}_j}] \tag{23}$$

4.2 Tendency Problem Solution

As stated previously, the output model tendencies are not taken into account in the conventional method. In order to solve this problem, a modified model expression is proposed. Actually, the model output can have any kind of spread variation for any sign of x by introducing a shift on the original model input. Doing so, it is possible to obtain the desired sign for the shifted input variable, and so to influence the spread variation of the output.

In this case, the fuzzy linear model of Eq. 12 defined on its domain D, becomes:

$$\hat{Y}(x) = A_0 \oplus A_1.(x - shift) \tag{24}$$

where A_0 and A_1 are trapezoidal parameters.

In the model of Eq. 24, the output spread is given by the support radius, i.e.:

$$\forall x \in D,\ R(S_{\hat{Y}}) = R(A_0) + R(A_1)|x - shift| \tag{25}$$

According to Eq. 25 and by tuning the value of $shift$, the model output can have any spread variation on D. Indeed,

- if $x - shift \geq 0\ \forall x \in D$, i.e. $shift \leq x_{min}$, then the model output has an increasing spread on D.
- if $x - shift \leq 0\ \forall x \in D$, i.e. $shift \geq x_{max}$, then the model output has a decreasing spread on D.

For the sake of simplicity, the value $shift = x_{min}$ is chosen for a model whose output has an increasing radius. On the contrary, for decreasing radius output, $shift = x_{max}$ is taken (see Table 3).

Table 3 The two models

output spread variation	↗	↘
Used model	$A_0 \oplus A_1(x - x_{min})$	$A_0 \oplus A_1(x - x_{max})$

5 The Identification Process

In this section, a modified identification methodology for linear regression models is proposed. The latter exploits the concepts of inclusion and tendency discussed previously for determining the parameters of a fuzzy model in the form of Eq. 24.

When considering:

- a set of N observed data (x_j, Y_j), where x_j are crisp inputs, sorted in increasing order, and Y_j the corresponding fuzzy triangular outputs,
- a fuzzy model in the form of Eq. 24, where its output is defined on the domain D,

the identification statement lies in the answers given to the following questions:

1. In order to ensure the inclusion of all observed data in the predicted ones for any $\alpha \in [0,1]$, is it possible to identify the fuzzy trapezoidal parameters A_0 and A_1? In other words, what are the constraints to be taken into account for the optimization problem?
2. For a better representation of the observed data tendencies, is it possible to determine the parameter $shift$ which allows the integration of any kind of spread in the model?

So, two main steps have to be discussed: the choice of the value of $shift$ and the model parameter identification.

5.1 The $shift$ Value Determination

The first step of the identification concerns the choice of the $shift$ value according to the output radius tendency. The most appropriate tendency is determined from observed data, comparing the initial output radius R_{init} attached to minimal inputs with the final output radius R_{fin} attached to maximal inputs. If $R_{init} < R_{fin}$, an increasing tendency is chosen, otherwise a decreasing tendency is preferred.

The corresponding $shift$ value is defined as:

- If $R_{init} > R_{fin}$ then $shift = x_{max}$
- If $R_{init} \leq R_{fin}$ then $shift = x_{min}$

The R_{init} and R_{fin} values are estimated by computing mean values from k data, that is $R_{init} = mean(R_1, R_2, ..., R_k)$ and $R_{fin} = mean(R_{N-k+1}, ..., R_{N-1}, R_N)$. The next step of the identification concerns the optimization of the fuzzy coefficients A_0 and A_1.

5.2 The Identification Method

Like all linear regression identification methods, the proposed one is based on the minimization of a criterion under some constraints.

5.2.1 The Used Criterion

In the sequel, for the clarity and the simplicity of notations we take: $w_j = (x_j - shift)$. So w_j can be either always positive or always negative, depending on the chosen value of $shift$ for the considered domain. Indeed, the choice $shift = x_{min}$ leads to $w_{min} = 0$ and $w_{max} > 0$. On the contrary, when $shift = x_{max}$, $w_{min} < 0$ and $w_{max} = 0$.

According to the model expression, the output of the fuzzy model is a trapezoidal interval given by:

$$\forall w \in D: \begin{cases} K_{\hat{Y}}^- = K_{A_0}^- + (M(K_{A_1}) - R(K_{A_1}).\Delta).w \\ K_{\hat{Y}}^+ = K_{A_0}^+ + (M(K_{A_1}) + R(K_{A_1}).\Delta).w \\ S_{\hat{Y}}^- = S_{A_0}^- + (M(S_{A_1}) - R(S_{A_1}).\Delta).w \\ S_{\hat{Y}}^+ = S_{A_0}^+ + (M(S_{A_1}) + R(S_{A_1}).\Delta).w \end{cases} \tag{26}$$

where:

$$\Delta = sign(w_{min} + w_{max}) \tag{27}$$

The choice of the criterion to be minimized is also an important issue. In conventional methods [15], the used criteria are only based on the available data, their minimization does not guarantee that the identified model has the least global fuzziness that could be achieved on the whole domain D. If the identified model is to be used on the whole domain D, it may be more judicious to prefer a model with a lower global fuzziness, i.e. a less imprecise model. It has been shown in [1], that it is possible to minimize the whole spread of the identified model for fuzzy triangular output. The same approach is here proposed for dealing with trapezoidal fuzzy models. In this case, the global fuzziness of the model is the "volume" covered by its output on D, i.e. the integration of the area of the fuzzy output $\hat{Y}(w)$ on D. For a given input w, the area of $\hat{Y}(w)$ takes into account all possible α levels from 0 to 1 (vertical dimension).

It can be stated that the output area represented by a trapezoidal fuzzy number [18] is given by the following expression:

$$area(\hat{Y}(w)) = \frac{K_{\hat{Y}}^+ + S_{\hat{Y}}^+}{2} - \frac{K_{\hat{Y}}^- + S_{\hat{Y}}^-}{2} \tag{28}$$

In this case, the volume delimited by the model output on its whole domain D is given by:

$$volume = \int_{w_{min}}^{w_{max}} area(\hat{Y}(w))dw \tag{29}$$

Substitution of Eq. 28 in Eq. 29 yields:

$$volume = (w_{max} - w_{min})(R(K_{A_0}) + R(S_{A_0})) + \frac{1}{2}(w_{max}^2 - w_{min}^2)(R(K_{A_1}) + R(S_{A_1})).\Delta \quad (30)$$

It is clear that this criterion is independent from the data. The optimization is performed on the whole definition domain of the model, and not only at the learning points. So, the learning data distribution doesn't affect the model identification. This property which guarantees some kind of "robustness" of the proposed criterion allows the identification of models whose fuzziness is possibly lower than usually.

5.2.2 Assumed Constraints

In the optimization procedure, the constraints of Eq. 22 and Eq. 23 must be respected.

- For $\alpha = 1$:

$$K_{Y_j} \in [K_{\hat{Y}_j}^-, K_{\hat{Y}_j}^+] \Leftrightarrow |M(K_{\hat{Y}_j}) - K_{Y_j}| \leq R(K_{\hat{Y}_j}) \quad (31)$$

where:

$$\begin{cases} M(K_{\hat{Y}_j}) = M(K_{A_0}) + M(K_{A_1}).w_j \\ R(K_{\hat{Y}_j}) = R(K_{A_0}) + R(K_{A_1}).w_j.\Delta \end{cases} \quad (32)$$

- For $\alpha = 0$:

$$[K_{Y_j} - R_{Y_j}, K_{Y_j} + R_{Y_j}] \subseteq [S_{\hat{Y}_j}^-, S_{\hat{Y}_j}^+] \Leftrightarrow |M(S_{\hat{Y}_j}) - K_{Y_j}| \leq R(S_{\hat{Y}_j}) - R_{Y_j} \quad (33)$$

where:

$$\begin{cases} M(S_{\hat{Y}_j}) = M(S_{A_0}) + M(S_{A_1}).w_j \\ R(S_{\hat{Y}_j}) = R(S_{A_0}) + R(S_{A_1}).w_j.\Delta \end{cases} \quad (34)$$

- In order to obtain a fuzzy interval, another inclusion constraint must be verified, i.e., the inclusion of the kernel into the support:

$$[K_{\hat{Y}_j}^-, K_{\hat{Y}_j}^+] \subseteq [S_{\hat{Y}_j}^-, S_{\hat{Y}_j}^+] \Leftrightarrow |M(S_{\hat{Y}_j}) - M(K_{\hat{Y}_j})| \leq R(S_{\hat{Y}_j}) - R(K_{\hat{Y}_j}) \quad (35)$$

To sum up, the proposed identification method is performed by minimizing the linear criterion defined by Eq.30 under the inclusion constraints of Eq.31, Eq. 33 and Eq. 35.

6 Piecewise Linear Regression Problem

In this section, the previous identification method is used to identify a piecewise fuzzy linear model of the form presented in Eq. 36:

$$\hat{Y}(x) = \sum_{k=1}^{S \oplus} [A_{k0} \oplus A_{k1}.(x - shift_k)].1_{[x^k_{min}, x^k_{max}]} \quad (36)$$

where $shift_k \in \{x^k_{min}, x^k_{max}\}$, S is the number of segments which will compose the global model, $\Sigma^{\oplus}$ represents the sum of several fuzzy intervals. The coefficients A_{k0} and A_{k1} are fuzzy trapezoidal intervals. The function $1_{[x^k_{min}, x^k_{max}]}$ is equal to 1 on $[x^k_{min}, x^k_{max}]$ and to 0 otherwise.

At the beginning of the process, it is necessary to find a good segmentation of the data set, i.e. to determine on which domains we have to identify the different sub-models. As for a given data set, output modal value tendency and output radius one have to be considered, the segmentation is made on both.

So, in order to finally get the different $[x^k_{min}, x^k_{max}]$, $k = 1, ..., S$, we apply on collected data the following method:

- First, we make a segmentation on observed output modal values;
- On each interval got, we make another segmentation on the corresponding observed outputs radius values.

Then, we apply the identification method presented in Section 5 on each interval given by segmentation process, in order to determine the best model on this domain according to the volume criterion. An advantage of this approach is that the sub-models are independent, as well for their identification as for their potential use in prediction.

7 Generalization to Multi-input Models

This section is devoted to the identification of multi-input models. This case can be viewed as a straightforward generalization of the previous case, the used concepts being the same.

Let's consider a set of N observed data $(x_{1j}, x_{2j}, ..., x_{Mj}, Y_j), j = 1, ..., N$, where $\mathbf{x_j} = [x_{1j}, x_{2j}, ..., x_{Mj}]$ is the input vector of M components, and Y_j the corresponding fuzzy triangular output. Each component x_i of the input vector is defined on its domain $D_i = [x_i^{min}, x_i^{max}]$.

The objective is to identify a model of the form:

$$\hat{Y}(x) = A_0 \oplus A_1.(x_1 - shift_1) \oplus ... \oplus A_M.(x_M - shift_M) \quad (37)$$

The first step of the identification concerns the choice of $shift_i$ values. As stated in subsection 5.1, this choice is made according to the output radius tendency. In the case of multi-input models, a suitable value is chosen for each component

x_i of the input vector, considering the output radius tendency according to this component.

Once appropriate values of $shift_i$ are chosen, the parameters identification is lead by minimizing a linear criterion under constraints. In the sequel, for the simplicity of the notations, we take: $w_i = x_i - shift_i, i = 1, ..., M$. So, the model output is given by:

$$\begin{cases} K_{\hat{Y}}^- = K_{A_0}^- + \Sigma_{i=1}^M (M(K_{A_i}) - R(K_{A_i}).\Delta_i).w_i \\ K_{\hat{Y}}^+ = K_{A_0}^+ + \Sigma_{i=1}^M (M(K_{A_i}) + R(K_{A_i}).\Delta_i).w_i \\ S_{\hat{Y}}^- = S_{A_0}^- + \Sigma_{i=1}^M (M(S_{A_i}) - R(S_{A_i}).\Delta_i).w_i \\ S_{\hat{Y}}^+ = S_{A_0}^+ + \Sigma_{i=1}^M (M(S_{A_i}) + R(S_{A_i}).\Delta_i).w_i \end{cases} \tag{38}$$

where $\Delta_i = sign(w_{min}^i + w_{max}^i)$.

The considered criterion in the case of multiple inputs is a generalization of the one presented in Eq. 30:

$$volume = \int_{w_{min}^1}^{w_{max}^1} \dots \int_{w_{min}^M}^{w_{max}^M} area(\hat{Y}(w_1, ..., w_M))dw_M...dw_1 \tag{39}$$

By introducing Eq. 38 and Eq. 28 in Eq. 39, the criterion to be optimized to identify multi-input model is defined by:

$$volume = R(K_{A_0}) + R(S_{A_0}) + \sum_{i=1}^{M} (R(K_{A_i}) + R(S_{A_i})) \cdot M(D_i) \cdot \Delta_i \tag{40}$$

The optimization must be achieved under the inclusion constraints obtained by the generalization of those presented in Eq. 31, Eq. 33 and Eq. 35 to multi-input case.

Several points can be underlined here:

- The extension to multi-input model identification is still based on a linear criterion optimization under linear inclusion constraints.
- As stated prevouisly for single-input models, the considered criterion is a representation of the model output uncertainty on the whole definition domain, which allows some kind of "robustness" and the identification of potentially less uncertain models.
- From computationnal point of view, the identification complexity increases with the number of input vector components. However, the increase is limited by the fact that the criterion is independent of the observed data. Only the number of considered constraints is affected.
- This multi-input approach can be extended to piecewise models identification. In this case, segmentation has to be made on each component of the input vector, and appropriate values of $shift_i$ and parameters are determined for each submodel.

8 Application on Several Examples

The proposed identification method is applied on the first example presented in Table 1. The model given in Eq. 41, defined on $D = [1,5]$, is obtained:

$$\hat{Y}(x) = A_0 \oplus A_1.(x-1) \tag{41}$$

with:

$$\begin{cases} A_0 = ([4.45,8],[2.25,9.8]) \\ A_1 = ([1.95,1.95],[1.95,2.1]) \end{cases} \tag{42}$$

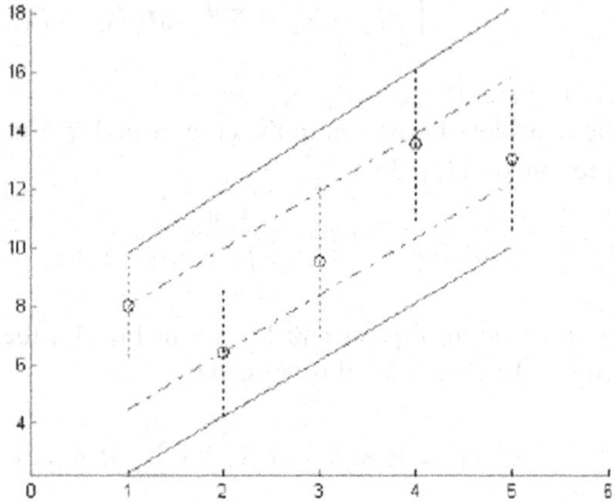

Fig. 6 Trapezoidal identified model

The model representation is illustrated in Fig. 6. The optimal volume computed on $D = [1,5]$ is 22.8.

In this case, it can be stated that all observed data and included in the predicted ones. For example, when $j = 1$, the observed and the predicted output are respectively $Y_1 = [6.2,9.8]$ and $\hat{Y}_1 = ([4.45,8],[2.25,9.8])$ which illustrates that the inclusion of the observed output into the predicted one is ensured $\forall\alpha \in [0,1]$ (see Fig. 7).

Fig. 7 Trapezoidal observed and predicted outputs for $j = 1$

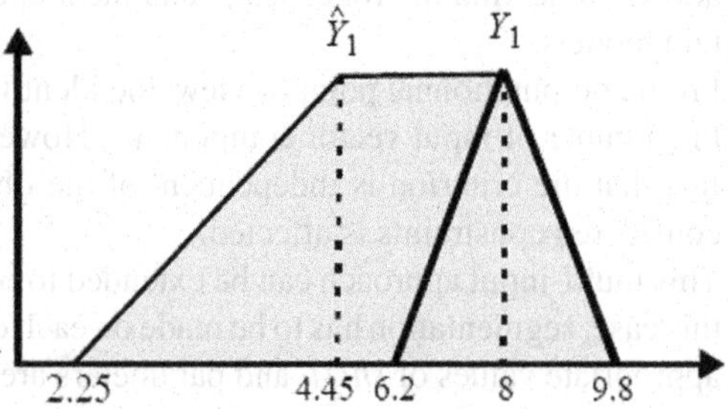

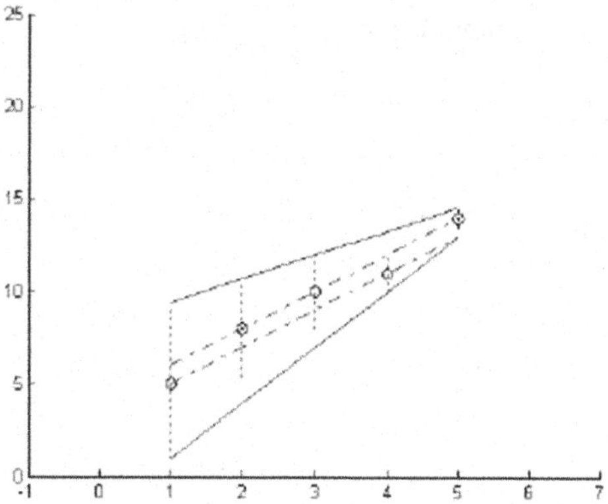

Fig. 8 Trapezoidal identified model representation

The identification method is also applied on the second example, presented in Table 2. It leads to the following model:

$$\hat{Y}(x) = A_0 \oplus A_1.(x-5) \tag{43}$$

with $D = [1,5]$ and:

$$\begin{cases} A_0 = ([13,14],[13,14.6]) \\ A_1 = ([2,2],[1.3,3]) \end{cases} \tag{44}$$

A representation of the model is given in Fig. 8.

The model output has a decreasing spread, and so it well represents the data tendency. So, the obtained model is less fuzzy than the one presented in Fig. 5.

Then, the piecewise identification process is applied on data proposed by Tanaka and Ishibuchi ([16], example 2). Two different segments can be distinguished where the change point is $x = 11$. The observed outputs on first segment present a globally decreasing spread. On the second segment, the spread is globally increasing. Identification method leads to the following piecewise model (Eq. 45) (see Fig. 9):

$$\begin{aligned} \hat{Y}(x) &= (A_{01} \oplus A_{11}(x-11))1_{[5,11]} \\ &+ (A_{02} \oplus A_{12}(x-11))1_{[11,17]} \end{aligned} \tag{45}$$

with:

$$\begin{cases} A_{01} = ([10,10.18],[9,11]) \\ A_{11} = ([0.392,0.5],[0,1]) \\ A_{02} = ([9.84,10],[9,11]) \\ A_{12} = ([0.385,0.5],[-0.83,1.83]) \end{cases} \tag{46}$$

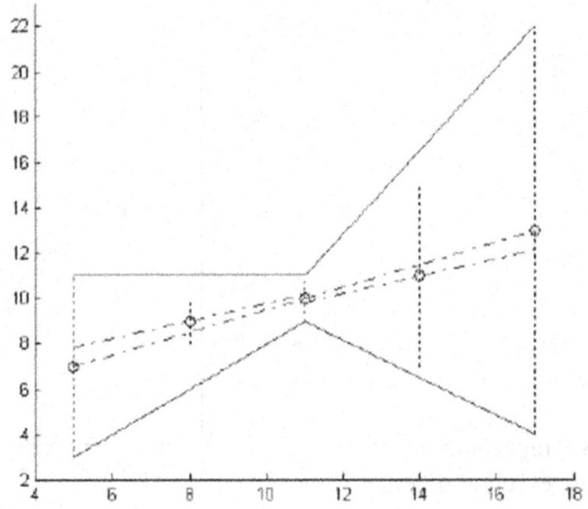

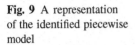

Fig. 9 A representation of the identified piecewise model

So, with a piecewise model and shifted inputs, a good data representation can be achieved, without using interactive coefficients as proposed by Tanaka and Ishibuchi in [16].

Then, the proposed identification method is applied on a data set presented in [6], concerning prefabricated houses prices variation according to several criterion (see Table 4). Observed crisp data are structured as follows. Input x_1 is the quality of construction material ($x_1 = 1$ for low grade, $x_1 = 2$ for medium grade, and $x_1 = 3$ for high grade), x_2 the area of the first floor, x_3 the area of the second floor, x_4 the total number of rooms and x_5 the number of Japanese rooms. The output y is the sale price, in million of yen.

Table 4 Data related to prefabricated houses

j	x_1	x_2	x_3	x_4	x_5	y_j
1	1	38.09	36.43	5	1	606
2	1	62.10	26.50	6	1	710
3	1	63.73	44.71	7	1	808
4	1	74.52	38.09	8	1	826
5	1	75.38	41.1	7	2	865
6	2	52.99	26.49	4	2	852
7	2	62.93	26.49	5	2	917
8	2	72.04	33.12	6	3	1031
9	2	76.12	43.06	7	2	1092
10	2	90.26	42.64	7	2	1203
11	3	85.70	31.33	6	3	1394
12	3	95.27	27.64	6	3	1420
13	3	105.98	27.64	6	3	1601
14	3	79.25	66.81	6	3	1632
15	3	120.5	32.25	6	3	1699

As stated in [6] and illustrated in Table 4, the price y globally decreases when the total number of rooms x_4 increases. Conventionnal methods fail to represent this trend. As a consequence, identified model in [6] does not consider this input x_4. Here, by setting $shift_4 = 8$, i.e. a *shift* value allowing a decreasing tendency according to this input, the identified model global uncertainty (optimal value of the identification criterion) is 83.58. It is better than for a conventional model without any *shift* value, whose global uncertainty is 88.58 (see Fig. 10).

Finally, the proposed method for multi-input model identification is applied on a data set presented in the Appendix section. In this example, the input vector has two components x_1 and x_2. The observed outputs present several cases of radius variation according to these components, and so a piecewise model is identified. Our identification method leads to a global model composed of 12 submodels, as illustrated in Fig. 11.

If we consider the submodel corresponding to $x_1 \in [-3,0]$ and $x_2 \in [-4,0]$, the identified submodel is the following:

$$\hat{Y}(x) = (A_0 \oplus A_1(x_1+3)) \oplus A_2 x_2) \cdot 1_{([-3,0]\times[-4,0])} \tag{47}$$

with:

$$\begin{cases} A_0 = ([5.4, 6.05], [1.4, 10.375]) \\ A_1 = ([-1.8, -1.8], [-3.8, -0.125]) \\ A_2 = ([-1.25, -0.8], [-1.25, -0.8]) \end{cases} \tag{48}$$

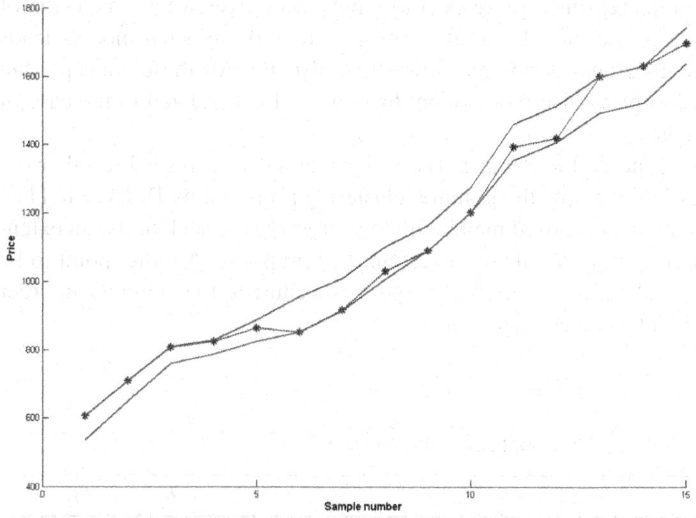

Fig. 10 A representation of the houses prices estimation

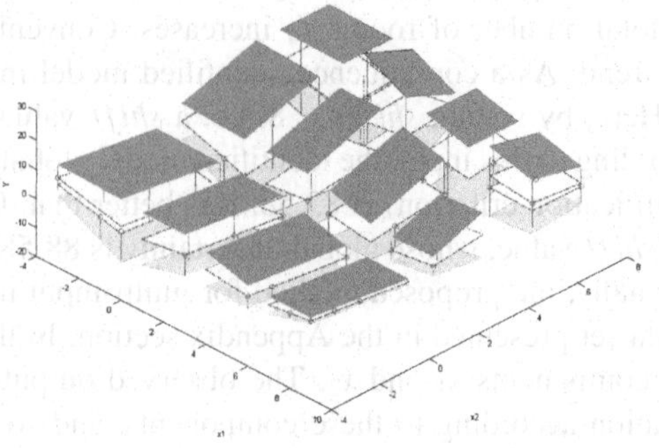

Fig. 11 A representation of the identified piecewise multi-input model

In this case, the chosen values of $shift_i$ (i.e. $shift_1 = x_1^{min} = -3$ and $shift_2 = x_2^{max} = 0$) allows the representation of an increasing output radius according to x_1, and of a decreasing one according to x_2.

By considering all submodels (see Fig. 11) with appropriate values of $shift_i$, a suitable representation of observed data tendencies is got on each subdomain, according to both x_1 and x_2. Obviously, as the parameters are trapezoidal fuzzy intervals, the global inclusion is respected.

9 Conclusion

The proposed methodology is based on the use of shifted models with trapezoidal fuzzy parameters. In this case, it becomes possible to represent output spreads either increasing or decreasing with respect to inputs. Moreover, a total inclusion of the observed data in the model output is ensured. Identifying such models leads to models whose fuzziness is possibly lower than usually. All these concepts can easily be applied to piecewise model identification, and generalized to the case of multi-input models.

Further works concern the comparison of our method with regard to other existing techniques, for example the granular clustering proposed by Pedrycz in [11]. Moreover, although the proposed method doesn't preserve the scalability, an extension can be found in order to take into account this propriety. Another point to be studied is the generalization of the identification procedure to fuzzy inputs, in order to manage uncertainties on collected data.

Appendix

Table 5 Multi-input data with several cases of tendencies

j	x_{1j}	x_{2j}	K_{Y_j}	R_{Y_j}	j	x_{1j}	x_{2j}	K_{Y_j}	R_{Y_j}
1	-3	-4	11	4	5	-3	4	11	12
2	-3	-2	7	4	6	-3	7	12	9
3	-3	0	6	4	7	-2	-3	8	6
4	-3	3	10	10	8	-2	-1	5.5	6

Table 5 *(continued)*

j	x_{1j}	x_{2j}	K_{Y_j}	R_{Y_j}	j	x_{1j}	x_{2j}	K_{Y_j}	R_{Y_j}
9	-2	0	4	6	42	4	1	3	8
10	-2	2	7	10	43	4	2	4	10
11	-2	5	9.5	13	44	4	6	8	12
12	-2	6	10.2	12	45	4	7	9	11
13	-1	-4	5	8	46	5	-2	3	5
14	-1	-2	3.4	8	47	5	-1	2	5
15	-1	1	3	10	48	5	0	1	5
16	-1	3	5	14	49	5	3	4	11
17	-1	7	9	13	50	5	4	5	13
18	0	-4	4	10	51	5	5	6	12
19	0	-1	1	10	52	6	-4	4	4
20	0	0	0	10	53	6	-3	3	4
21	0	1	1	12	54	6	0	0	4
22	0	4	4	18	55	6	2	2	8
23	0	6	6	16	56	6	4	4	12
24	1	-2	3	9	57	6	6	6	10
25	1	0	1	9	58	7	-3	4	3
26	1	2	3	13	59	7	-2	3	3
27	1	3	4	15	60	7	1	2	3
28	1	5	6	16	61	7	5	6	10
29	1	7	8	14	62	7	7	8	8
30	2	-3	5	8	63	8	-4	6	2
31	2	-1	3	8	64	8	-1	3	2
32	2	2	4	12	65	8	0	2	2
33	2	4	6	16	66	8	2	4	6
34	2	6	8	14	67	8	3	5	8
35	3	-4	7	7	68	8	6	8	8
36	3	-2	5	7	69	9	-4	7	1
37	3	0	3	7	70	9	-3	6	1
38	3	1	4	9	71	9	-2	5	1
39	3	5	8	14	72	9	1	4	3
40	4	-3	5	6	73	9	5	8	8
41	4	-1	3	6	74	9	7	5	6

References

1. Bisserier, A., Galichet, S., Boukezzoula, R.: Fuzzy piecewise linear regression. In: IEEE International Conference on Fuzzy Systems, FUZZ-IEEE 2008 (IEEE World Congress on Computational Intelligence), pp. 2089–2094 (2008)
2. Boukezzoula, R., Foulloy, L., Galichet, S.: Inverse controller design for fuzzy interval systems. IEEE Transactions on Fuzzy Systems 14(1), 111–124 (2006)
3. Boukezzoula, R., Galichet, S., Foulloy, L.: Nonlinear internal model control: Application of inverse model based fuzzy control. IEEE Transactions on Fuzzy Systems 11(6), 814–829 (2003)
4. Boukezzoula, R., Galichet, S., Foulloy, L.: MIN and MAX Operators for Fuzzy Intervals and Their Potential Use in Aggregation Operators. IEEE Transactions on Fuzzy Systems 15(6), 1135–1144 (2007)
5. Diamond, P.: Fuzzy least squares. Information Sciences: an International Journal 46(3), 141–157 (1988)
6. Diamond, P., Tanaka, H.: Fuzzy regression analysis. Kluwer Handbooks of Fuzzy Sets Series, pp. 349–387 (1999)

7. Ge, H., Wang, S.: Dependency between degree of fit and input noise in fuzzy linear regression using non-symmetric fuzzy triangular coefficients. Fuzzy Sets and Systems 158(19), 2189–2202 (2007)
8. Guo, P., Tanaka, H.: Dual models for possibilistic regression analysis. Computational Statistics and Data Analysis 51(1), 253–266 (2006)
9. Hao, P., Chiang, J.: Fuzzy Regression Analysis by Support Vector Learning Approach. IEEE Transactions on Fuzzy Systems 16(2), 428–441 (2008)
10. Hung, W., Yang, M.: An omission approach for detecting outliers in fuzzy regression models. Fuzzy Sets and Systems 157(23), 3109–3122 (2006)
11. Pedrycz, W., Bargiela, A.: Granular clustering: a granular signature of data. IEEE Transactions on Systems, Man, and Cybernetics, Part B 32(2), 212–224 (2002)
12. Roychowdhury, S., Pedrycz, W.: Modeling temporal functions with granular regression and fuzzy rules. Fuzzy Sets and Systems 126(3), 377–387 (2002)
13. Sakawa, M., Yano, H.: Multiobjective fuzzy linear regression analysis for fuzzy input-output data. Fuzzy Sets and Systems 47(2), 173–181 (1992)
14. Savic, D., Pedrycz, W.: Evaluation of fuzzy linear regression models. Fuzzy Sets and Systems 39(1), 51–63 (1991)
15. Tanaka, H., Hayashi, I., Watada, J.: Possibilistic linear regression analysis for fuzzy data. European Journal of Operational Research 40(3), 389–396 (1989)
16. Tanaka, H., Ishibuchi, H.: Identification of possibilistic linear systems by quadratic membership functions of fuzzy parameters. Fuzzy sets and Systems 41(2), 145–160 (1991)
17. Tanaka, H., Uejima, S., Asai, K.: Linear regression analysis with fuzzy model. IEEE Trans. Sys. Man and Cyber. 12(6), 903–907 (1982)
18. Yager, R.: Using trapezoids for representing granular objects: applications to learning and OWA aggregation. Information Sciences 178(2), 363–380 (2008)
19. Zadeh, L.: Fuzzy Sets. Information and Control 8(3), 338–353 (1965)

Fundamentals for Design and Construction of a Fuzzy Random Forest

Piero P. Bonissone, José Manuel Cadenas, María del Carmen Garrido, and R. Andrés Díaz-Valladares

Abstract. Following Breiman's methodology, we propose the fundamentals to design and construct a "forest" of randomly generated fuzzy decision trees, i.e., a *Fuzzy Random Forest*. This approach combines the robustness of multi-classifiers, the construction efficiency of decision trees, the power of the randomness to increase the diversity of the trees in the forest, and the flexibility of fuzzy logic and the fuzzy sets for data managing. A prototype for the method has been constructed and we have implemented some specific strategies for inference in the Fuzzy Random Forest. Some experimental results are given.

Keywords: Approximate Reasoning, Fuzzy Decision Trees, Random Forest, Combination Methods.

1 Introduction

Classification has always been a challenging problem, [1]. The explosion of information that is available to companies and individuals today further compounds this problem. We have witnessed a variety of methods and algorithms addressing the classification issue. In the last few years, we have also seen an increase of multi-classifiers

Piero P. Bonissone
GE Global Research
One Research Circle. Niskayuna, NY 12309, USA
e-mail: bonissone@crd.ge.com

José Manuel Cadenas and María del Carmen Garrido
Dept. Ingeniería de la Información y las Comunicaciones
University of Murcia, Murcia, Spain
e-mail: jcadenas@um.es, carmengarrido@um.es

R. Andrés Díaz-Valladares
Dept. Ciencias Computacionales
Universidad de Montemorelos, Mexico
e-mail: rdiaz@um.edu.mx

B. Bouchon-Meunier et al. (Eds.) Found. of Reas. under Uncert., STUDFUZZ 249, pp. 23–42.
springerlink.com

based approaches, which have been shown to deliver better results than individual classifiers, [22].

In this paper, we will not address the issue of how to obtain the best multi-classifier system. Rather, our focus will be on how to start from a multi-classifier system with a performance comparable to the best classifiers and extend it to handle and manipulate imperfect information (linguistic labels, missing values, etc.)

To build the multi-classifier, we follow the random forest methodology. To incorporate the processing of imperfect data, we construct the random forest using fuzzy trees as base classifiers. Therefore, we try to use the robustness of a tree ensemble, the power of the randomness to increase the diversity of the trees in the forest, and the flexibility of fuzzy logic and fuzzy sets for data managing.

In section 2, we review the major elements that constitute a multi-classifier and, given that our proposal is based on an ensemble of fuzzy trees, we will also include some comments on aspect of fuzzy decision trees. In section 3, we explain the classic algorithm to create a random forest according to Breiman [4]. In the same section we also describe the adjustments, changes and considerations needed for the construction and inference of a fuzzy random forest. We present some initial results in section 4, followed by our conclusions in section 5.

2 Multi-classifiers

When individual classifiers are combined appropriately, we usually obtain a better performance in terms of classification precision and/or speed and so find a better solution. Multi-classifiers are the result of combining several individual classifiers. Multi-classifiers differ in their diverse characteristics: (1) the number and (2) the type of the individual classifiers; (3) the characteristics of the subsets used by every classifier of the set; (4) the consideration of the decisions; and (5) the size and the nature of the training sets for the classifiers [16].

Segrera [22] divides the methods for building multi-classifiers in two groups: ensemble and hybrid methods. The first type, such as Bagging [3], Boosting [21] and Random Subspace [11], induces models that merge classifiers with the same learning algorithm, while introducing modifications in the training data set. The second type, such as Stacking [24], creates new hybrid learning techniques from different base learning algorithms.

The Bagging based ensemble uses the same base classifier a set number of times, and the training set of each individual classifier is a subset of examples which are formed by random selection and through replacement of a sample of m examples taken from the original training set made up by m examples. Since each individual classifier is trained on a bootstrap sample, the data distribution seen during training is similar to the original distribution. Thus, the individual classifiers in a bagging ensemble have relatively high classification accuracy. The only factor encouraging diversity between these classifiers is the proportion of different examples in the training samples. Although the classifier models used in Bagging are sensitive to small changes in data, the bootstrap sampling appears to lead to ensembles of low diversity compared to other methods of creating ensembles. As a result, Bagging requires larger ensemble sizes to perform well. To enforce diversity, a version of Bagging called Random Forest was proposed by Breiman [4]. In a Boosting base ensemble the individual classifiers are trained sequentially in a series, and the training set of the k-th classifier is chosen on the basis of the performance of the previous $k-1$ classifiers in the series. The probability of classifier k choosing an example for training depends on the frequency with which this was badly classified by the preceding k-1 classifiers. In other words, classifiers home in on the most complex examples for learning. Finally, the ensemble based on Random Subspaces makes a random selection of input attributes for each individual classifier.

Bagging, Boosting and Random Subspaces were evaluated in [8], where Boosting was found to be the most exact method for problems with no noise, and Bagging performed best with noise.

An ensemble uses the predictions of multiple base-classifiers, typically through majority vote or averaged prediction, to produce a final ensemble-based decision. The ensemble-based predictions typically have lower generalization error rates than those obtained by a single model. The difference depends on the type of base-classifiers used, ensemble size, and diversity or correlation between classifiers [1].

Among the hybrid multi-classifiers, the Stacking based ones stand out. These are proposed to combine heterogeneous classifiers derived from different learning algorithms and using different representation models on the same training set. In this type of multi-classifier we can talk about two stages. In the first, a set of base level classifiers is generated. In the second stage, a meta-level classifier is generated which combines the outputs of the base level classifiers.

2.1 Multi-classifiers Based on Decision Trees

Decision trees have been the basis for the most important works on multi-classifiers systems. As a result, the label "forest" has been given to an ensemble of trees working on the same classification problem.

In order to grow these ensembles, random vectors are often generated that govern the growth of each tree in the ensemble. An early example is bagging (Breiman [1996]), where to grow each tree a random selection (without replacement) is made from the examples in the training set. Another example is random split selection (Dietterich [1998]) where at each node the split is selected at random from among the K best splits. Breiman [1999] generates new training sets by randomizing the outputs in the original training set. Another approach is to select the training set from a random set of weights on the examples in the training set. Ho [1998] has written a number of papers on "the random subspace" method which makes a random selection of a subset of attributes to use to grow each tree.

The common element in all of these procedures is that for the k-th tree, a random vector θ_k is generated, independently of the past random vectors $\theta_1,\ldots,\theta_{k-1}$ but with the same distribution; and a tree is grown using the training set and θ_k. After a large number of trees has been generated, they vote for the most popular class. These procedures are called random forests [4].

As stated in the previous section, in bagging, diversity is obtained by constructing each classifier with a different set of examples, which is obtained from the original training set by re-sampling with replacement. Bagging then combines the decisions of the classifiers using uniform-weighted voting. Bagging improves the performance of single classifiers by reducing the variance error. Breiman categorizes bagging decision trees as a particular instance of random forest classification techniques. A random forest is a tree-based ensemble that uses some kind of independent randomization in the construction of every individual classifier. Many variants of bagging and random forests with excellent classification performance have been developed in [19]. Others types of random forest are: randomization [8], Forest-RI and Forest-RC [4], double-bagging [12], rotation forest [20].

Hamza [10] concludes that: 1) Random Forests are significantly better than Bagging, Boosting and a single tree; 2) their error rate is smaller than the best one obtained by other methods; and 3) they are more robust to noise than the other methods. Consequently, random forest is a very good classification method with the following characteristics: (1) it is easy to use; (2) it does not require models, nor parameters to select, except for the number of attributes to choose at random at each node; and (3) it is relatively robust to noise.

Given that decision trees play an important role in constructing multi-classifiers, below we will comment on some of their aspects and their relation to fuzzy logic.

2.1.1 Fuzzy Logic and Decision Trees

Decision tree techniques have proved to be interpretable, efficient and capable of treating with applications of great scale. However, they are highly unstable when small disturbances are introduced in data learning. Fuzzy logic offers an improvement in these aspects, due to the elasticity of the fuzzy set's formalism. In previous works [13, 14, 15] we can find approaches in which fuzzy sets and their underlying approximate reasoning capabilities have been successfully combined with decision trees. This combination has preserved the advantages of both components: uncertainty management with the comprehensibility of linguistic variables, and popularity and easy application of decision trees. The resulting trees show an increased immunity to noise, an extended applicability to uncertain or vague contexts, and support for the comprehensibility of the tree structure, which remains the principal representation of the resulting knowledge.

In the literature, we can find several proposals for building trees of fuzzy information starting with algorithms already known for building traditional trees. Fuzzy CART [13] was one of the first examples of this approach, being based on the CART algorithm. However, most authors have preferred to use the ID3 algorithm to construct trees for recursive partition of the data set of agreement to the values of the selected attribute. To use the ID3 algorithm in the construction of fuzzy trees, we need to develop attribute value space partitioning methods, branching attribute selection method, branching test method to determine the degree to which data follow the branches of a node, and leaf node labeling methods to determine classes. Fuzzy decision trees have two major components: a procedure for building fuzzy decision trees and an inference procedure for decision-making [15].

Fuzzy decision trees are constructed in a top-down manner by recursive partitioning of the training set into subsets. Some particular features of fuzzy tree learning are: the membership degree of examples, the selection of test attributes, the fuzzy tests (to determine the membership degree of the value of an attribute to a fuzzy set), and the stop criteria (besides the classic criteria when the measure of the information is under a specific threshold).

As mentioned earlier, when we use the ID3 algorithm to construct fuzzy trees, an important aspect is the partitioning or discretisation of the numerical attributes that describe the examples. This partitioning has the same potential advantages of reduced search space size, improved classifier performance and reduced risk of overfitting. However, these advantages may only be realised if the performance of the discretisation method is good; a classification method may well give inferior performance when using data which have been poorly discretised [7]. The categories into which discretisation algorithms are divided, as defined in [9], are:

- Global or local. A local discretisation method produces discretisations which are applied to localised regions of the database. A discretisation is formed separately for each of these subsets. Global methods discretise the entire database instead.
- Supervised or unsupervised. Unsupervised methods form a discretisation based purely upon the values of a given attribute. Supervised methods make use of the class values of each example when forming the discretisation for any field. The regions produced by such processes should each be relatively homogeneous with respect to class (i.e. most of the examples contained within an interval should belong to the same class). Unsupervised discretisation techniques run the risk of losing classification information by placing examples in the same interval which are very similar with respect to attribute value but belong to different classes. This is because such techniques form the discretisation without making any use of class information.
- Static or dynamic. Both methods require the definition of the maximum number of intervals k into which each attribute may be discretised. Within static discretisation, each attribute in turn is then split into k intervals. Dynamic discretisation methods search for discretisations into at most k intervals for all attributes simultaneously. Such methods should therefore be able to take into account interdependencies between attributes when performing discretisation.

3 Towards a Fuzzy Random Forest

Since multi-classifiers have been built based on decision trees, and these have been combined with fuzzy sets and their underlying approximate reasoning capabilities, it is of interest and appropriate to design multi-classifiers based on fuzzy classifiers. In this section, we present the concepts for the design of a forest of fuzzy decision trees generated randomly (Fuzzy Random Forest), following Breiman's methodology [4]. We specify the adjustments, changes and considerations needed to construct these multi-classifiers.

3.1 Random Forest (Following Breiman's Methodology)

As we observed above bagging requires larger ensemble sizes to perform well. To enforce diversity, a version of Bagging called Random Forest was proposed by Breiman [4]. The ensemble consists of decision trees again built on bootstrap samples. The difference lies in the construction of the decision tree. The attribute to split a node is selected as the best attribute from a subset of F randomly chosen attributes (from a total of M attributes), where F is a parameter of the algorithm. This small alteration appeared to be a winning heuristic, in that diversity was introduced without compromising the accuracy of the individual classifiers much. This procedure was denoted by Breiman Forest-RI.

The algorithm of learning and inference in a random forest constructed with the Forest-RI procedure, according to Breiman is the following:

Algorithm 1 - Forest-RI Algorithm following Breiman's Methodology

---

ForestRI(in:DataSet,out:Random Forest)

begin

1. Take a random sample of N examples from the data set with replacement of the complete dataset of N examples. Some examples will be selected more than once, and others will not be chosen. Approximately $2/3$ of the examples will be selected. The remaining $1/3$ of the cases is called "out of bag" (OOB).
   For each constructed tree, a new random selection of examples is performed.
2. Using the examples selected in the previous step, construct a tree (to the maximum size and without pruning). During this process, every time that it is needed to split a node, only consider a subset F of the total set of attributes M. Select the set of attributes as a random subset of the total set of available attributes. Perform a new random selection for each split. Some attributes (inclusive the best) cannot be considered for each split, but a attribute excluded in one split may be used by other splits in the same tree.
3. Repeat steps 1 and 2 to construct a forest, i.e. a collection of trees.
4. To classify an example, run it through each tree in the forest and record the predicted value. Use the predicted categories for each tree as "votes" for the best class, and use the class with the most votes as the predicted class.

end

Forest-RI have two stochastic elements: 1) the selection of data set used as input for each tree; and 2) the set of attributes considered as candidates for each node split. These randomizations, along with combining the predictions from the trees, significantly improve the overall predictive accuracy.

When we constructed a random forest using the previous algorithm, about $1/3$ of the examples are excluded from each tree in the forest. These examples are called "out of bag" (OOB); each tree will have a different set of OOB examples. The OOB examples are not used to build the tree and constitute an independent test sample for the tree. To measure the generalization error of the tree, the OOB for each tree are run through the tree and the error rate of the prediction is computed. The error rates for the trees in the forest are then averaged to give the overall generalization error rate for the decision tree model.

There are several advantages to this method of computing generalization error: (1) all cases are used to construct the model, and none have to be held back as a separate test set; (2) the testing is fast because only one forest has to be constructed (as compared to cross-validation where additional trees have to be constructed).

Some advantages reported by Breiman in [4] of Forest-RI procedure to construct a random forest include the improved speed in the construction of the multi-classifier, compared to Adaboost and Bagging. This improvement is produced because in each tree construction it is only necessary to evaluate F attributes from the total available for each node to expand. In data sets with many attributes the improvement in computation time is very significant. Moreover, the results obtained with Forest-RI compare favorably with Adaboost.

3.2 Approach and Considerations to Construct a Fuzzy Random Forest

In this work we propose to use Algorithm 1 to generate a random forest whose trees are fuzzy decision trees, proposing, therefore, a basic algorithm to generate a Fuzzy Random Forest (FRF).

Algorithm 2 - Fuzzy Decision Tree Learning

FuzzyDecisionTree(in:Examples,out:Fuzzy Tree)

begin

1. Start with examples set of entry, having the weights of the examples (in root node) equal to 1.
2. At any node N still to be expanded, compute the number of examples of each class. The examples are distributed in part or in whole by branches. The amount of each example distributed to a branch is obtained as the product of its current weight and the membership degree to the node.
3. Compute the standard information content.
4. At each node search the set of remaining attributes to split the node.
   4.1. Select with any criteria, the candidate attributes set to split the node.
   4.2. Compute the standard information content to each child node obtained from each candidate attribute.
   4.3. Select the candidate attribute such that the information gain is maximum.
5. Divide N in sub-nodes according to possible outputs of the attribute selected in the previous step.
6. Repeat steps 2-5 to stop criteria being satisfied in all nodes.

end

Each tree in the forest will be a fuzzy tree generated following the guidelines of [14], adapting it where is necessary.

Algorithm 2 shows the general steps to follow to construct a fuzzy tree. Before constructing the trees, it is necessary to carry out a partitioning of the numerical attributes which describe the examples of the databases. The partitioning algorithm we use here is based on decision trees and is global, supervised and dynamic, in accordance with the definition given above for such algorithms. A fuller description of the algorithm can be found in [6].

In algorithm 3 we show the learning process of an FRF using the fuzzy trees random generator which follows Breiman's methodology.

Algorithm 3 - FRF Learning

FRFlearning(in:Dataset,out:Fuzzy Random Forest)

begin

1. Divide the examples set of entry in subsets according to step 1 of algorithm 1.
2. For each subset of examples, apply algorithm 2 (construct a fuzzy tree). This algorithm has been adapted so that the trees can be constructed without considering all the attributes to split the nodes. This is done to be able to apply step 2 of algorithm 1.
3. Repeat steps 1 and 2 until all fuzzy trees are built. At the end, we will have constructed a fuzzy random forest.

end

With this basic algorithm, we integrate the concept of fuzzy tree within the design philosophy of Breiman's random forest. In this way, we augment the capacity of diversification of random forests with the capacity of approximate reasoning of fuzzy logic.

3.3 Proposal and Considerations For the Inference in Fuzzy Random Forest

In this section we are going to present the main aspects to be taken into account when classifying an input example with an FRF.

We will begin by briefly explaining the combination methods frequently used in multi-classifiers to obtain the final decision starting from the decision of each base classifier. Next, we present some notations that we will use throughout the section and we will describe some general aspects of the FRF inference process, adapting the combination methods from the literature. Finally, we show some specific implementation for an FRF.

3.3.1 Combination Methods

Various combination methods have been used in the literature [16, 17, 18], to take a final decision in a multi-classifier starting from individual classifiers.

In his papers, Kuncheva classifies the basic combination methods into two groups:

- Combination methods without learning, which are those with which is possible to work directly, once the classifiers have been learnt, i.e. no type of learning on the ensemble is required.
- Combination methods with learning, which try to give greater importance in the final decision to the more precise base classifiers in the case that the ensemble classifiers do not have identical precision.

In this preliminary work with the FRF multi-classifier, we will only incorporate the basic combination methods without learning to the inference with FRF. We highlight the following (Figure 1):

- majority vote, which takes as input the class vector of each element to be voted for. It generates its vote for the majority class and returns a vector which indicates the most voted class. Figure 1a) shows how this combination method works. Given the class vectors of x elements, the vote of each element is generated to its majority class and the vector $\overline{o}$ is provided with all its elements at 0, except element c, which is 1, with c being the most voted class.
- the simple minimum, which takes as input the class vector of each element. It obtains the minimum value for each class in these vectors and returns a vector which indicates the class with the highest minimum value. Figure 1b) shows this combination method. Given the class vectors of x elements, the minimum value

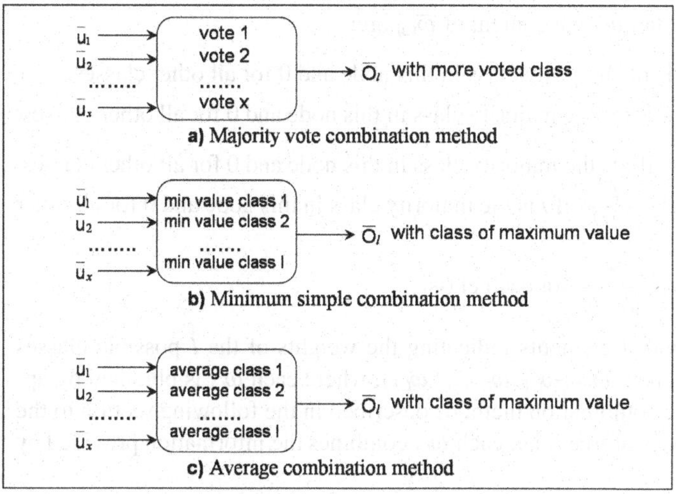

Fig. 1 Simple combination methods

for each class is calculated and the vector $\overline{o}$ is provided with all its elements at 0, except element c, which is 1, with c being the class with the highest minimum value. The simple maximum method is defined in a similar way.
- the average combination method takes as input the class vector of each element. It calculates the average value for each class and provides a vector which indicates the class of highest value. Figure 1c) shows this combination method. The product combination method is defined in a similar way. It obtains the product of the values of each class and so finally provides the class with the highest value.

3.3.2 Notations

Now, we introduce the notation that we will use:

- T is the trees' number of the forest. We will use the index t to refer to a particular tree in the forest.
- N_t is the number of leaf nodes reached by an example, in the tree t. A characteristic inherent in fuzzy trees is that the classification of an example can derive into two or more leaves due to the overlapping of the fuzzy sets. We will use the index n to refer to a particular leaf in a tree of the forest.
- I is the number of classes that we consider. We will use the index i to refer to a particular class.
- e is an example from the dataset used to build and to infer with the Fuzzy Random Forest.
- $\chi_{t,n}(e)$ is the grade which the example e activates the leaf n from tree t.
- $\overline{\omega}_{t,n}$ is a vector with I elements indicating the weights of the I possible classes in the leaf n of tree t, $\overline{\omega}_{t,n}=(\omega_{t,n,1},\omega_{t,n,2},\ldots,\omega_{t,n,I})$, where $\omega_{t,n,i}=\frac{E_i}{\sum_{j=1}^{I}E_j}$ and E_j is the sum of the weights of examples with class j in the leaf.

  Other ways of obtaining the content of $\overline{\omega}_{t,n}$ are:

  - $\omega_{t,n,i}=1$ if i is the majority class in this node and 0 for all other classes.
  - $\omega_{t,n,i}=\frac{E_i}{\sum_{j=1}^{I}E_j}$ if i is the majority class in this node and 0 for all other classes.
  - $\omega_{t,n,i}=\chi_{t,n}(e)$ if i is the majority class in this node and 0 for all other classes.
  - $\omega_{t,n,i}=\chi_{t,n}(e)\cdot\frac{E_i}{\sum_{j=1}^{I}E_j}$ if i is the majority class in this node and 0 for all other classes.
  - $\omega_{t,n,i}=\chi_{t,n}(e)\cdot\frac{E_i}{\sum_{j=1}^{I}E_j}$ for each class.

- $\overline{\omega}_t$ is a vector with I elements indicating the weights of the I possible classes in the tree t of forest, $\overline{\omega}_t=(\omega_{t,1},\omega_{t,2},\ldots,\omega_{t,I})$, where each $\omega_{t,i}$ is obtained by applying one of the combination methods described in the following section to the different leaf nodes of tree t, i.e. each $\omega_{t,i}$ combines the information provided by each leaf of t.
- $\overline{\omega}_{FRF}$ is a vector with I elements indicating the weights of the I possible classes in the forest FRF.

3.3.3 Inference in FRF

Given that the use of fuzzy trees means many leaves are reached when we try to classify an input example, the fuzzy classifier module operate on fuzzy trees of the forest with two possible strategies:

Strategy 1: Combining the information from the different leaves reached in each tree to obtain the decision of each individual tree and then applying the same one or another combination method to generate the global decision of the forest.

Strategy 2: Combining the information from all leaves reached from all trees to generate the global decision of the forest.

In both strategies, we use the function *Faggre*, which is defined as a frequently used multi-classifiers combination method [16], e.g., majority vote, minimum, maximum, average, product, etc.

In Algorithm 4, *Faggre* is used to obtain $\overline{\omega}_t$ (the weight of each tree for each class). Later, the values obtained in each tree t, will be aggregated by the function *Faggre* (again, can be any combination method mentioned previously, which is able to adapt to take into account about some considerations commented ahead) to obtain the vector $\overline{\omega}_{FRF}$ that contains the weight proposed by fuzzy random forest for the different classes.

For implementing strategy 2 (algorithm 5), the previous algorithm 4 is simplified so that it does not add the information for tree, but directly provides the information of all leaves reached by the example e in the different trees of the forest.

Algorithm 4 - FRF Inference (Strategy 1)

---

begin
  TreeClasification.
  ForestClasification.
end

TreeClasification $(in: e, in: Random\ Forest, out: \overline{\omega}_1, \overline{\omega}_2, ..., \overline{\omega}_T)$
begin
for each Tree t **do**
  $\overline{\omega}_t = Faggre(\overline{\omega}_{t,1}, \overline{\omega}_{t,2}, ..., \overline{\omega}_{t,N_t})$
end for
end

ForestClasification $(in: \overline{\omega}_1, \overline{\omega}_2, ..., \overline{\omega}_T, out: \overline{\omega}_{FRF})$
begin
  $\overline{\omega}_{FRF} = Faggre(\overline{\omega}_1, \overline{\omega}_2, ..., \overline{\omega}_T)$
end

Algorithm 5 - FRF Inference (Strategy 2)

ForestClasification $(in : e, in : Random\ Forest, out : \overline{\omega}_{FRF})$
begin
$\overline{\omega}_{FRF} = Faggre(\overline{\omega}_{1,1}, \overline{\omega}_{1,2}, ..., \overline{\omega}_{1,N_1}, \overline{\omega}_{2,1}, \overline{\omega}_{2,2}, ..., \overline{\omega}_{2,N_2}, \overline{\omega}_{T,1}, \overline{\omega}_{T,2}, ..., \overline{\omega}_{T,N_T})$
end

3.3.4 Some Specific Implementations of Inference Methods

Now, we present some specific implementations of inference in the FRF multi-classifier. Each will include some simple combination method from those described in Figure 1 in either of the general strategies for inference described in the previous section.

- Implementations based in Majority vote (MV).
  This method assigns *e* to the class label most represented among the forest's trees outputs.
  In strategy 1 the combination method based on majority vote is applied first to the leaf nodes reached by example *e* to obtain the decision of each tree. It is then applied to the decisions of the trees to obtain the forest's decision. Figure 2a) shows a schema for this implementation.
  In case of the second strategy every leaf reached of the forest is considered to vote. The forest will decide the class *c* where *c* is the only element of the vector $\overline{\omega}_{FRF}$ which takes a value 1. This implementation is shown illustratively in Figure 2b).

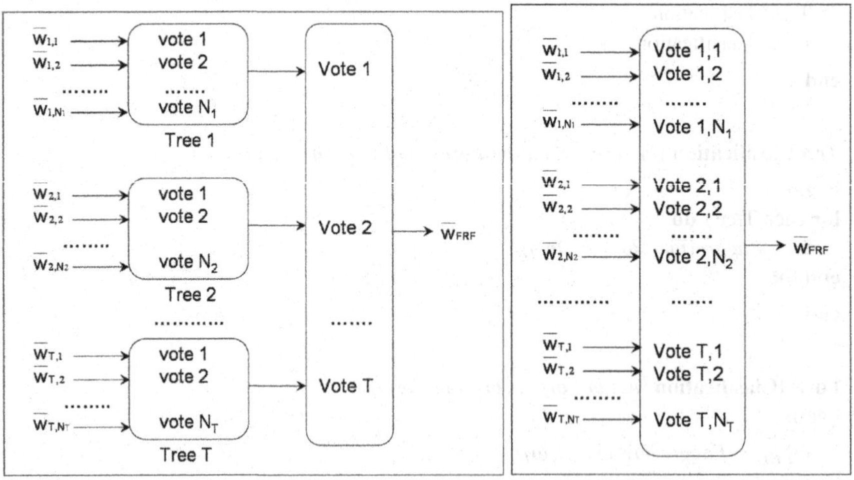

a) to strategy 1 b) to strategy 2

Fig. 2 Majority vote applied to strategy 1 and strategy 2

- Implementations based in Minimum simple (MS).
  This method uses the minimum combination method to take the minimum in each class of leaves and trees.
  In strategy 1, we implementation the inference:
  1) We apply at decision level of every tree the minimum simple as combination method.
  2) Then, we apply the majority vote combination method to the resulting vectors.

  We show this implementation illustratively in Figure 3.

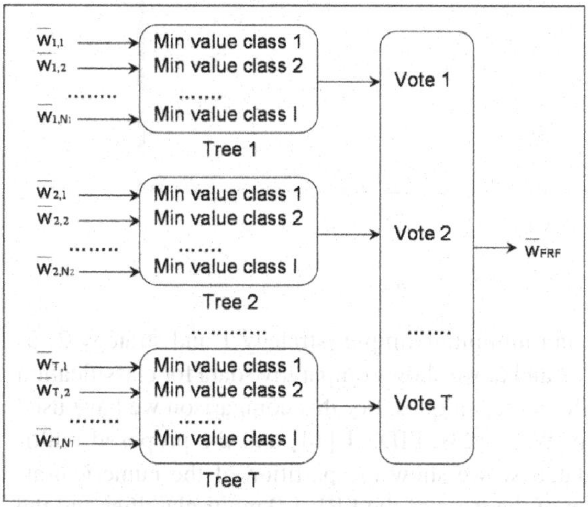

Fig. 3 Minimum-majority vote applied to strategy 1

Strategy 2 is simpler to apply since it only has one decision level. In this case the vector $\overline{\omega}_{FRF}$, is obtained by taking vectors $\overline{\omega}_{t,N_t}, \forall t$ as input in Figure 1b).

- In a similar way, we can calculate the class support from the decision profile taking Maximum, Average and Product.
- Weighted versions of the simpler combination methods will be implemented by taking into account aspects as the weight of leaves reached, standard error of each tree, the amount of imperfect information to construct each tree, etc.

4 Experiments and Preliminary Results

We are currently developing an application that is (1) capable of generating the random databases (Random Forest Generator), and (2) able to infer the classification of the forest (Fuzzy Classifiers). Figure 4 shows a screen of this application, which we have labeled FRF 1.0. We used FRF 1.0 to obtain our preliminary results.

We used the Iris database, Appendicitis database and Glass database from UCI repository [2] and realized diverse tests. We constructed the fuzzy random forest

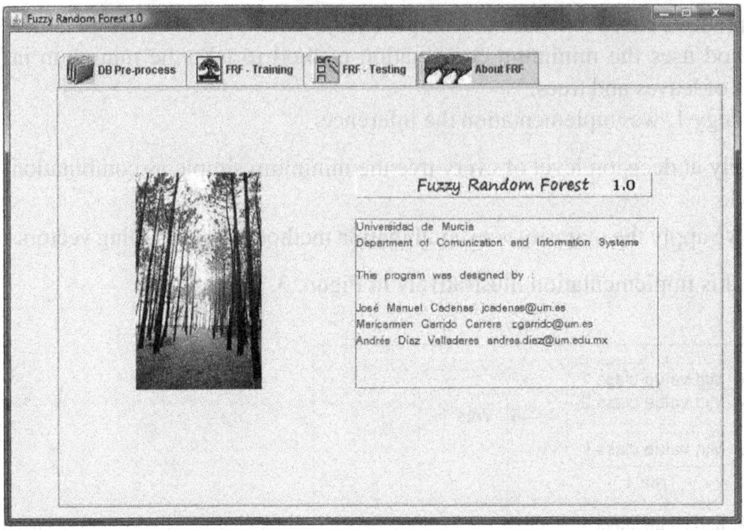

Fig. 4 Screen of FRF 1.0

considering the majority vote and minimum simple (strategy 1 and strategy 2) as inference methods. In Tables 2, 4 and 6, we show comparative data for classification of the three databases from different techniques. For this comparison we have used a set of techniques of platform Weka [23], FID3.4 [14] and the proposed multi-classifier FRF 1.0. For each database we show the partition of the numerical attributes used in the construction of the trees of the FRF 1.0 multi-classifier and the results obtained with the original database and with the following versions of the same:

- Introducing 5% of missing values.
- Introducing 10% of linguistic labels.
- Introducing 15% of linguistics labels.

In both the 10% and 15% introductions of linguistic labels, the values of the original database are substituted by the corresponding linguistic label, defined by the partition used.

Tables 2, 4 and 6 show the results of the various techniques carried out, and for FRF 1.0 (strategy 1 and strategy 2), we also indicate in brackets which inference method obtains the best value. The results are obtained by 10-fold cross validation, showing the mean value and the standard deviation obtained.

4.1 Iris Database

This set of data contains three classes with 50 examples in each. Class refers to the type of iris plant: Iris-Setosa, Iris-Versicolour and Iris-Virginica. One class is

attributes=4 || candidates = 2 $\rightarrow attrib_1, attrib_3$ || selection $\rightarrow attrib_3$
$attrib_3$ $\in$ling. label$_1$ with membership degree α_{31}: Set=38.00 Ver=0.00 Vir=0.00
$attrib_3$ $\in$ling. label$_2$ with membership degree α_{32}: Set=0.00 Ver=32.49 Vir=0.00
$attrib_3$ $\in$ling. label$_3$ with membership degree α_{33}: Set=0.00 Ver=14.51 Vir=6.00
| attributes=3 || candidates =2 $\rightarrow attrib_1, attrib_4$ || selection $\rightarrow attrib_4$
| $attrib_4$ $\in$ling. label$_1$ with membership degree α_{41}: Set=0.00 Ver=14.51 Vir=0.00
| $attrib_4$ $\in$ling. label$_2$ with membership degree α_{42}: Set=0.00 Ver=0.00 Vir=4.90
| $attrib_4$ $\in$ling. label$_3$ with membership degree α_{43}: Set=0.00 Ver=0.00 Vir=1.10
$attrib_3$ $\in$ling. label$_4$ with membership degree α_{34}: Set=0.00 Ver=0.00 Vir=4.00
$attrib_3$ $\in$ling. label$_5$ with membership degree α_{35}: Set=0.00 Ver=1.00 Vir=39.00

Fig. 5 A fuzzy tree of the Fuzzy Random Forest FRF 1.0 for IRIS

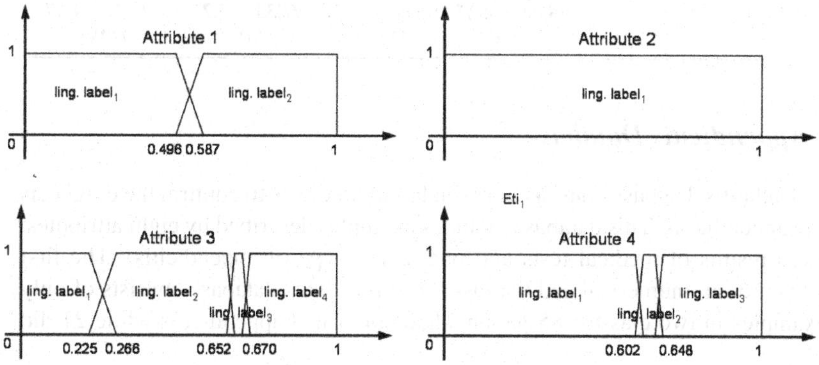

Fig. 6 Attributes partition

linearly separable from the other two, but these are not linearly separable between themselves. The database includes examples described by five attributes, of which 4 are numerical and one, class, is discrete. The numerical attributes have the following meaning: z_1 is the flower sepal length, z_2 is the flower sepal width, z_3 is the flower petal length, z_4 is the flower petal width, and Cl is the class.

Figures 5, 6 and table 1 show a fuzzy tree of the Fuzzy Random Forest FRF 1.0 for this database and the partitions of numerical attributes used by this tree, respectively. Table 2 shows the results obtained.

Table 1 Attributes partition of Iris database

	ling. label$_1$	ling. label$_2$	ling. label$_3$	ling. label$_4$
attr. 1	(0,0,0.496,0.587)	(0.496,0.587,1,1)		
attr. 2	(0,0,1,1)			
attr. 3	(0,0,0.225,0.266)	(0.225,0.266,0.652,0.653)	(0.652,0.653,0.669,0.670)	(0.669,0.67,1,1)
attr. 4	(0,0,0.602,0.606)	(0.602,0.606,0.644,0.648)	(0.644,0.648,1,1)	

Table 2 Preliminary results from FRF 1.0

IRIS				
Technique	without imperf.	5% miss.	10% ling. labels	15% ling. labels
Naives Bayes (NaiveBayes)	96.67 ± 4.47	96.00 ± 4.42	—	—
C4.5 (J48)	96.00 ± 5.33	93.33 ± 7.89	—	—
Neuronal Net (MultiplayerPerc.)	97.33 ± 3.27	92.00 ± 6.53	—	—
Ripple-Down-Rule (Ridor)	94.67 ± 4.99	86.67 ± 9.43	—	—
Random Forest (RandomForest)	95.33 ± 4.27	96.00 ± 4.47	—	—
Boosting (AdaBoostM1)	95.33 ± 5.21	92.67 ± 6.29	—	—
Bagging	96.00 ± 5.33	96.00 ± 4.42	—	—
FID 3.4	96.67 ± 3.33	92.67 ± 6.96	96.67 ± 3.33	96.00 ± 4.42
FRF 1.0 (Strategy 1)	**98.00 ± 4.27** (MV,MS)	**98.66 ± 2.81** (MS)	96.67 ± 3.33 (MV,MS)	**96.67 ± 3.33** (MV,MS)
FRF 1.0 (Strategy 2)	**98.00 ± 4.27** (MV)	96.66 ± 4.71 (MV)	**97.33 ± 3.27** (MV)	96.67 ± 4.47 (MV)

4.2 *Appendicitis Database*

The appendicitis database consists of seven laboratory tests to confirm the diagnosis of acute appendicitis. This database includes examples described by eight attributes: the seven results of medical tests and one, class (type of appendicitis). The first seven features are numerical and the last is discrete. This database consists of only 106 examples in two classes: 85 patients had confirmed appendicitis while 21 did not.

Table 3 shows the attributes partition used for the Fuzzy Random Forest FRF 1.0 for the Appendicitis database. Table 4 shows the results obtained for the various techniques applied to this database.

Table 3 Attributes partition of Appendicitis database

	ling. $label_1$	ling. $label_2$	ling. $label_3$	ling. $label_4$
attr. 1	(0,0,0.038,0.046)	(0.038,0.046,0.367,0.402)	(0.367,0.402,1,1)	
attr. 2	(0,0,1,1)			
attr. 3	(0,0,0.222,0.241)	(0.222,0.241,0.410,0.429)	(0.410,0.429,0.960,0.967)	(0.960,0.967,1,1)
attr. 4	(0,0,0.116,0.120)	(0.116,0.120,0.158,0.162)	(0.158,0.162,0.219,0.225)	(0.219,0.225,1,1)
attr. 5	(0,0,0.267,0.327)	(0.267,0.327,1,1)		
attr. 6	(0,0,0.058,0.071)	(0.058,0.071,0.646,0.657)	(0.646,0.657,0.756,0.767)	(0.756,0.767,1,1)
attr. 7	(0,0,0.065,0.080)	(0.065,0.080,1,1)		

4.3 *Glass Database*

This set of data contains six classes with 70, 76, 17, 13, 9 and 29 examples in each class. Class refers to the type of glass: building_windows_float_processed, building_windows_non_float_processed, vehicle_windows_float_processed, containers,

Table 4 Preliminary results from FRF 1.0

APPENDICITIS

Technique	without imperf.	5% miss.	10% ling. labels	15% ling. labels
Naives Bayes (NaiveBayes)	87.73 ± 6.21	88.64 ± 5.94	—	—
C4.5 (J48)	85.91 ± 9.66	85.82 ± 9.79	—	—
Neuronal Net (MultiplayerPerc.)	87.64 ± 8.83	83.00 ± 7.15	—	—
Ripple-Down-Rule (Ridor)	86.82 ± 7.76	81.09 ± 7.57	—	—
Random Forest (RandomForest)	87.82 ± 8.69	85.91 ± 7.76	—	—
Boosting (AdaBoostM1)	86.91 ± 6.11	82.27 ± 9.29	—	—
Bagging	86.91 ± 10.51	87.73 ± 5.92	—	—
FID 3.4	93.45 ± 4.30	87.82 ± 10.08	**93.64 ± 8.18**	91.64 ± 8.80
FRF 1.0 (Strategy 1)	94.36 ± 4.86 (MV)	**91.64 ± 6.75** (MV)	93.36 ± 4.60 (MV,MS)	**94.36 ± 4.86** (MV,MS)
FRF 1.0 (Strategy 2)	**95.27 ± 4.99** (MV)	**91.64 ± 6.75** (MV)	93.36 ± 4.60 (MV)	**94.36 ± 4.86** (MV)

tableware and headlamps. The study of classification of types of glass was motivated by criminological investigation. The database includes examples described by ten attributes, of which nine are numerical and one, class, is discrete. The numerical attributes have the following meaning: RI-refractive index, Na-Sodium (unit measurement: weight percent in corresponding oxide), Mg-Magnesium, Al-Aluminum, Si-Silicon, K-Potassium, Ca-Calcium, Ba-Barium, Fe-Iron, and the type of glass (class attribute).

Table 5 shows the attributes partition used for the Fuzzy Random Forest FRF 1.0 for the Glass database. Table 6 shows the results obtained for the various techniques applied to this database.

Table 5 Attributes partition of Glass database

	ling. label$_1$	ling. label$_2$	ling. label$_3$
attr. 1	(0,0,0.257,0.265)	(0.257,0.265,0.298,0.306)	(0.298,0.306,1,1)
attr. 2	(0,0,0.383,0.384)	(0.383,0.384,0.385,0.386)	(0.385,0.386,1,1)
attr. 3	(0,0,0.589,0.631)	(0.589,0.631,0.802,0.844)	(0.802,0.844,1,1)
attr. 4	(0,0,0.348,0.356)	(0.348,0.356,0.390,0.398)	(0.390,0.398,1,1)
attr. 5	(0,0,0.241,0.253)	(0.241,0.253,0.302,0.314)	(0.302,0.314,1,1)
attr. 6	(0,0,0.001,0.002)	(0.001,0.002,0.039,0.048)	(0.039,0.048,1,1)
attr. 7	(0,0,0.222,0.227)	(0.222,0.227,0.248,0.254)	(0.248,0.254,1,1)
attr. 8	(0,0,0.085,0.128)	(0.085,0.128,0.569,0.672)	(0.569,0.672,1,1)
attr. 9	(0,0,0.246,0.264)	(0.246,0.264,0.334,0.352)	(0.334,0.352,1,1)

As can be observed from tables 2, 4 and 6, FRF 1.0 has a reasonably acceptable behavior, but with the advantage of the versatile management of information (from the different results obtained with the databases with imperfection we observe that the behavior of FRF 1.0 is very stable). In these tables we show in bold the best result obtained for each database. FRF 1.0 performs well in all cases and obtains the best result in all but two, although in the case of Appendicitis database with 10%

Table 6 Preliminary results from FRF 1.0

GLASS				
Technique	without imperf.	5% miss.	10% ling. labels	15% ling. labels
Naives Bayes (NaiveBayes)	50.89 ± 4.13	52.75 ± 4.36	—	—
C4.5 (J48)	69.55 ± 7.35	67.12 ± 12.49	—	—
Neuronal Net (MultiplayerPer.	70.02 ± 6.62	60.13 ± 9.24	—	—
Ripple-Down-Rule (Ridor)	68.66 ± 6.70	64.94 ± 9.19	—	—
Random Forest (RandomForest)	80.78 ± 7.57	**78.64 ± 8.14**	—	—
Boosting (AdaBoostM1)	78.03 ± 6.00	70.00 ± 8.28	—	—
Bagging	73.74 ± 6.69	74.20 ± 8.96	—	—
FID 3.4	76.65 ± 11.12	71.54 ± 6.80	66.28 ± 10.84	66.77 ± 9.26
FRF 1.0 (Strategy 1)	**81.86 ± 6.76** (MV,MS)	76.17 ± 8.42 (MS)	**81.36 ± 5.55** (MV)	79.52 ± 7.99 (MS)
FRF 1.0 (Strategy 2)	**81.86 ± 6.76** (MV)	75.30 ± 9.96 (MV)	80.91 ± 6.45 (MV)	**79.59 ± 7.94** (MV)

of linguistic labels the standard deviation in much higher that obtained by FRF 1.0, while the averages are very similar. Both general inference strategies provided here obtain good results, and therefore seen promising.

5 Conclusions

This document presents the study of a multi-classifier system called *Fuzzy Random Forest* with a reasonably acceptable behavior. Furthermore, the system has the advantages of uncertainty management and the comprehensibility of linguistic variables. Given that there are not many techniques today that handle imperfect information, we consider our proposal to be quite promising.

We have explained the underlying methodology and principal support techniques:

- We have presented a general description of a fuzzy random forest classifier.
- For *fuzzy trees random generator*, we carried out a hybridization of the techniques of *random forest* and fuzzy trees for training
- For *fuzzy classifiers* we have presented the basic idea for the consideration of the individual classifications and how they are combined to obtain the joint classification. Also, we have distinguished some considerations that will contribute greater accuracy to the classifiers.

We are currently building a fuzzy random forest prototype, with which we plan to validate our considerations to obtain efficient multi-classifiers with imperfect information.

Finally, we must clarify that throughout this presentation, we refer to the task of inference as classification, while the task of regression is implicit in this process. This is due to the fact that the numerical attributes are divided into linguistic labels and are treated as nominal ones.

Acknowledgements. Supported by the project TIN2008-06872-C04-03 of the MICINN of Spain and European Fund for Regional Development. Thanks also to the Funding Program for Research Groups of Excellence with code 04552/GERM/06 granted by the "Fundación Séneca - Agencia regional de Ciencia y Tecnología" of Spain.

References

1. Ahn, H., Moon, H., Fazzari, J., Lim, N., Chen, J., Kodell, R.: Classification by ensembles from random partitions of high dimensional data. Computational Statistics and Data Analysis 51, 6166–6179 (2007)
2. Asuncion, A., Newman, D.J.: UCI Machine Learning Repository. University of California, School of Information and Computer Science, Irvine, CA (2007), http://www.ics.uci.edu/~mlearn/MLRepository.html
3. Breiman, L.: Bagging predictors. Machine Learning 24(2), 123–140 (1996)
4. Breiman, L.: Random forests. Machine Learning 45(1), 5–32 (2001)
5. Cadenas, J.M., Garrido, M.C., Díaz-Valladares, R.A.: Hacia el Diseño y Construcción de un Fuzzy Random Forest. In: Proceedings of the II Simposio sobre Lógica Fuzzy y Soft Computing, pp. 41–48 (2007)
6. Cadenas, J.M., Garrido, M.C., Martinez España, R.: Generando etiquetas linguísticas: Un árbol de particiones. TR - DIIC 2/08, 1–20 (2008)
7. Debuse, J.C.W., Rayward-Smith, V.J.: Discretisation of Continuous Commercial Database Features for a Simulated Annealing Data Mining Algorithm. Applied Intelligence 11, 285–295 (1999)
8. Dietterich, T.G.: An experimental comparison of three methods for constructing ensembles of decision trees: bagging, boosting, and randomization. Machine Learning 40(2), 139–157 (2000)
9. Dougherty, J., Kohavi, R., Sahami, M.: Supervised and unsupervised discretization of continuous features. In: Proceedings of the Twelfth International Conference on Machine Learning, pp. 194–202 (1995)
10. Hamza, M., Larocque, D.: An empirical comparison of ensemble methods based on classification trees. Statistical Computati. & Simulation 75(8), 629–643 (2005)
11. Ho, T.K.: The random subspace method for constructing decision forests. Transactions on Pattern Analysis and Machine Intelligence 20(8), 832–844 (1998)
12. Hothorn, T., Lausen, B.: Double-bagging: combining classifiers by bootstrap aggregation. Pattern Recognition 36(6), 1303–1309 (2003)
13. Jang, J.: Structure determination in fuzzy modeling: A Fuzzy CART approach. In: Proceedings of the IEEE Conference on Fuzzy Systems, pp. 480–485 (1994)
14. Janikow, C.Z.: Fuzzy decision trees: issues and methods. Transaction on Systems, Man and Cybernetics, Part B 28(1), 1–15 (1998)
15. Koen-Myung, L., Kyung-Mi, L., Jee-Hyong, L., Hyung, L.: A Fuzzy Decision Tree Induction Method for Fuzzy Data. In: Proceedings of the IEEE Conference on Fuzzy Systems, pp. 22–25 (1999)
16. Kuncheva, L.I.: A theorical study on six classifier fusion strategies. IEEE Transaction on PAMI 24(2), 281–286 (2002)
17. Kuncheva, L.I.: Fuzzy vs Non-fuzzy in combining classifiers designed by boosting. IEEE Transactions on Fuzzy Systems 11(6), 729–741 (2003)
18. Kuncheva, L.I.: Combining Pattern Classifiers - Methods and Algorithms. John Wiley and Sons, New Jersey (2004)

19. Martínez-Muñoz, G., Suárez, A.: Switching class labels to generate classification ensembles. Pattern Recognition 38(10), 1483–1494 (2005)
20. Rodríguez, J.J., Kuncheva, L.I., Alonso, C.J.: Rotation Forest: A new classifier ensemble method. IEEE Transactions on Pattern Analysis and Machine Intelligence 28(10), 1619–1630 (2006)
21. Schapire, R.E.: The strength of weak learnability. Machine Learning 5(2), 197–227 (1990)
22. Segrera, S., Moreno, M.: An Experimental Comparative Study of Web Mining Methods for Recommender Systems. In: Proceedings of the 6th WSEAS Intern. Conference on Distance Learning and Web Engineering, pp. 56–61 (2006)
23. Witten, I.H., Frank, E.: Data Mining - Practical machine learning tools and techniques, 2nd edn. Morgan Kaufmann, San Francisco (2005)
24. Wolpert, D.: Stacked Generalization. Neural Networks 5, 241–259 (1992)

Mining Exception Rules

Miguel Delgado, María Dolores Ruiz, and Daniel Sánchez

Abstract. Mining association rules is a well known framework for extracting useful knowledge from databases. They represent a very particular kind of relation, that of co-occurrence between two sets of items. Modifying the usual definition of such rules we may find different kinds of information in the data. Exception rules are examples of rules dealing with unusual knowledge that might be of interest for the user and there exists some approaches for extracting them which employ a set of special association rules.

The goal of this paper is manyfold. First, we provide a deep analysis of the presented previous approaches. We study their advantages, drawbacks and their semantical aspects. Second, we present a new approach using the certainty factor for measuring the strength of exception rules. We also offer a unified formulation for exception rules through the GUHA formal model first presented in the middle sixties by Hájek et al. Third, we define the so called double rules as a new type of rules which in conjunction with exception rules will describe in more detail the relationship between two sets of items. Fourth, we provide an algorithm based on the previous formulation for mining exception and double rules with reasonably good performance and some interesting results.

1 Introduction

Mining association rules is a well known framework for extracting useful knowledge from databases. The kind of knowledge they extract is the appearance of a

Miguel Delgado
Dept. Computer Science and Artificial Intelligence, University of Granada,
C/Periodista Daniel Saucedo Aranda s/n, 18071 Granada, Spain
e-mail: mdelgado@ugr.es

María Dolores Ruiz and Daniel Sánchez
Dept. Computer Science and Artificial Intelligence, University of Granada,
C/Periodista Daniel Saucedo Aranda s/n, 18071 Granada, Spain
e-mail: {mdruiz,daniel}@decsai.ugr.es

B. Bouchon-Meunier et al. (Eds.) Found. of Reas. under Uncert., STUDFUZZ 249, pp. 43–63.
springerlink.com

set of items (i.e., couples $\langle attribute, value \rangle$) together in most of the transactions in a database. An example of association rule could be "most of transactions that contain bread also contain butter", and it is noted $bread \to butter$. The intensity of the above association rule is most frequently measured by the *support* and the *confidence* measures [1], although there exists many proposals which try to extract semantically or even statistically different association rules imposing new quality measures that collect such kind of semantic or statistical aspect. In this line, the certainty factor [3] has some advantages over the confidence as it extracts more accurate rules and the number of them is substantially reduced.

There are few approaches dealing with the extraction of unusual or exceptional knowledge that might be useful. We center our attention to those proposals that allow to obtain some uncommon information, referred to as peculiarities, exception or anomalous rules. In general, these approaches are able to manage with rules that being infrequent provide a specific domain information usually delimited by a strong association rule[1].

Peculiarity rules are discovered from the data by searching the relevance among the peculiar data [25]. Roughly speaking, peculiar data is given by the attributes which contain any peculiar value. A peculiar value will be recognized when it is very different from the rest of values of the attribute in the data set. Peculiarity rules represent the associations between the peculiar data. These rules will be searched among rules with low support and high confidence but having a high change of support. An attribute-oriented method for extracting peculiarity rules is presented in [25] and some recent works can be found in [16].

Exception rules were first defined as rules that contradict the user's common belief [21], more precisely looking for an exception rule consists in finding an attribute which interacting with another may change the consequent in a strong association rule [14, 23, 24]. There exists some approaches for extracting exceptional knowledge. We can find a good survey of all of them in [9]. In particular some of them define exception rules by employing association rules. We focus our attention in those which are defined in terms of a set of "special" rules for mining the exception rules [2, 14, 23].

In general terms, considering X, Y two disjoint itemsets and E a single item, the kind of knowledge the exception rules try to capture can be interpreted as follows:

X strongly implies Y,
but, in conjunction with E,
X confidently does not imply Y.

Anomalous rules are in appearance similar to exception rules, but semantically different. An anomalous association rule is an association rule that comes to the surface when we eliminate the dominant effect produced by a strong rule. In other words, it is an association rule that is verified when a common rule fails [2]. A formal definition of anomalous rule can be found in [2]. The knowledge these rules try to capture is:

---
[1] $A \to B$ is a strong association rule if it exceeds the minimum thresholds *minsupp*, *minconf* imposed by the user.

X strongly implies Y,
but, in those cases where X does not imply Y,
then X confidently implies A,

or in other words: when X, then we have either Y (usually) or A (unusually). Here, A stands for an item that represents the anomaly.

The knowledge provided by the exception and the anomalous rules are (semantically) complementary. If we are interested in the agent of the "strange" behavior, we will look for the exceptions, and if we are interested however in what is the strange or unusual behavior, we will look for the anomalies.

The aim of this paper is to analyze in detail semantics and formulation of the existing approaches for mining exception rules. The usual definition considers a triple rule set (*csr*, *exc*, *ref*) where *csr* and *ref* stand for common sense rule and reference rule. Although association rules are used for defining exception rules, all of them do not satisfy the support and confidence conditions (only the common sense rule) as the exception and the reference rule are infrequent.

One of the tools for analyzing such kind of rules is the logic model first introduced by Hájek et al. in [10], and then developed in [8] based on the GUHA (General Unary Hypotheses Automaton) method. This model has good logical and statistic foundations that help to a major understanding of both the nature of association rules and the properties of the measures used for validating them.

The model uses the simple notions of contingency table and quantifier. The contingency table called *four fold table* collects all the information about two chosen itemsets from a database, and the quantifier is a mathematical object that unifies two types of information: (1) it measures in some sense the relation between the two attributes and (2) it also says if the measure satisfies some predefined thresholds. It has been used for modeling several types of association. In particular in [8] we show how the use of different quantifiers is useful for dealing with more kinds of associations: implicational, double implicational, equivalence, etc. and they are also helpful in order to generalize the notion of rule to many kinds of situations [13, 19].

Our objective is providing a deep analysis of the formulation and the semantics of exceptions using the model developed in [8]. In [7] we used the logic model for studying some semantical aspects of exceptions and anomalies in association rule mining. Here, we discuss about the different definitions that exception rules have received, analyzing them semantically and unifying them in a single definition. We also propose a new approach for searching exception rules using the certainty factor instead of confidence. This change will report several advantages as it will decrease the number of extracted exception rules being more accurate.

We also provide a new framework for describing in more detail the relation between two sets of items. The main idea consists in the conjoint extraction of double rules with their associated exceptions. A double rule will be a rule which represents that there is a strong association in both rule's directions $X \to Y$ and $Y \to X$. For example the double rule

if a vertebrate animal flies (X), usually it is a bird (Y)
if the vertebrate animal is a bird (Y), it usually flies (X).

has some exceptions in both directions. Mining conjunctly the double rule with its associated exception rules we can describe more precisely the relation between the double rule's antecedent and consequent.

The remainder of the paper is organized as follows. First section introduces a brief description of the formal model developed by Hájek et al. for association rules. Section 3 analyzes the proposed definitions for exception rules from a semantical point of view. Section 4 proposes a new framework for mining exception rules using the certainty factor. Section 5 offers a unified formulation of some quantifiers for extracting exception rules using the logic approach. Section 6 focuses in defining double rules. We motivate their use and then we study the extraction of exception rules when the common sense rule is a double rule. Section 7 presents a simple algorithm for extracting these kinds of rules studying the time and memory requirements, and we also show some interesting results with some real datasets. Finally, section 8 concludes with the contribution of the paper pointing out some lines for future research.

2 Modeling Association Rules

Association rules capture the semantics of conjoint appearance of sets of items. There exists different proposals for measuring the strength and quality of the association. For dealing and modeling every type of measurement we will use a model based on the GUHA (General Unary Hypotheses Automaton) method. This method was first presented in the middle sixties by Hájek et al. [11, 12]. Then it has been developed in order to have a good model for association rules in several works [17, 18] and we also have established some useful properties of this model in [7, 8].

The advantage of using this model lies in its simplicity and its logical and statistic foundations that help to a major understanding of both the nature of association rules and the basic properties of measures used for assessing their accomplishment.

We also want to emphasize that recently some authors have implemented a good and fast algorithm [19] based on a bit string approach for mining association rules. We will adapt it in section 5 for mining exception rules using the association rules representation into this model.

The starting point is a database D where the rows and columns represent transactions and items respectively (see table 1). Items could be couples of the form $\langle attribute, value \rangle$ or $\langle attribute, interval \rangle$ depending on the particular data set. The entry (i, j) of D will be equal to 1 when the item is satisfied in transaction t_j and 0 otherwise.

Table 1 Example of Database D

D	i_1	i_2	$\dots$	i_j	i_{j+1}	$\dots$	i_n
t_1	1	0	$\dots$	0	1	$\dots$	0
t_2	0	1	$\dots$	1	1	$\dots$	1
$\vdots$	$\vdots$	$\vdots$	$\ddots$	$\vdots$	$\vdots$	$\ddots$	$\vdots$
t_n	1	1	$\dots$	0	1	$\dots$	1

To the logic method we are presenting, an *itemset* is an aggregation of items by means of the logic connectives $\wedge, \vee, \neg$; and an association rule is an expression of the type $\varphi \approx \psi$ where φ and ψ are itemsets (in the sense before) derived from D. The symbol $\approx$, called *quantifier*, will represent the measure used for the assessment of the rule, and it will depend on the *four fold table* associated to the pair φ and ψ.

The so called four fold table will be denoted by $\mathcal{M} = 4ft(\varphi, \psi, D) = \langle a, b, c, d \rangle$ where a, b, c and d will be non-negative integers such that a is the number of transactions satisfying both φ and ψ, b the number of those satisfying φ and not ψ, and analogously for c and d; obviously $a+b+c+d > 0$. Graphically

$\mathcal{M}$	ψ	$\neg\psi$
φ	a	b
$\neg\varphi$	c	d

When the assessment underlying $\approx$ is made from a four fold table then $\approx$ is said to be a $4ft$-*quantifier* involved in the rule $\varphi \approx \psi$. The association rule $\varphi \approx \psi$ is said to be true in the analyzed database D (or in the matrix $\mathcal{M}$) if and only if the condition associated to the 4ft-quantifier $\approx$ is satisfied for the four fold table $4ft(\varphi, \psi, D)$.

Different kinds of association between φ and ψ can be expressed by suitable 4ft-quantifiers. We can find many examples of 4ft-quantifiers in [12], [11] and [17].

The classical framework of support-confidence for assessing association rules can be expressed by the implication quantifier [8] $\Rightarrow_I$

$$\Rightarrow_I (a,b,c,d) = \frac{a}{a+b} \tag{1}$$

This quantifier is the well known confidence of the rule $\varphi \rightarrow \psi$. We will say that $\varphi \Rightarrow_I \psi$ is satisfied if and only if

$$\Rightarrow_I (a,b,c,d) \geq minconf \quad \wedge \quad \frac{a}{n} \geq minsupp \tag{2}$$

where $0 < minconf, minsupp < 1$ denote the thresholds known as minimum confidence and minimum support respectively, and $n = a+b+c+d$ is the total number of transactions in the database D. In [8] we explain deeply the relation between quantifiers and the interest measures used for assessing the validity of rules.

3 Exception Rules

Recent approaches are based on obtaining useful knowledge by means of different tools, referred to as peculiarities, infrequent rules, exceptions or anomalous rules. The knowledge captured by these new types of rules is in many cases more useful than that obtained by simple association rules as they provide specific information about the association between two itemsets that might be of interest for the user.

Exception rules where first defined as rules that contradict the user's common belief [21]. In general the exception rules are rules that contradict in some sense a strong rule satisfied in a database. In addition, the exception rules have a very strong regularity between a set of items which are satisfied by a small number of individuals.

There are several approaches for mining exception rules in databases. A usual formulation considers a triple rule set (csr, exc, ref) where the first two rules (common sense and exception rule) coincide and the reference rule changes according to the considered approach. The following table shows the general formulation for the common sense and the exception rules.

$X \rightarrow Y$	Common sense rule (frequent and confident)
$X \wedge E \rightarrow \neg Y$	Exception rule (confident)

In fact the only requirement for the exception rule is to be a confident rule because the conditions imposed to the common sense rule and to the reference rule restrict it to have low support.

Depending on the *reference* rule, we could handle with different kinds of exception rules. In general terms, the kind of knowledge these exceptions try to capture can be interpreted as follows: X strongly implies Y (and not E); but in conjunction with E, X imply $\neg Y$ (or other item in contraposition with Y).

Now we will discuss the formalization and the semantics captured by the triples (csr, exc, ref) proposed until now. After such an analysis we will propose an alternative for mining exceptions without the imposition of the reference rule.

3.1 Analyzing Semantics of Exception Rules

There are several proposals for dealing with the task of finding exception rules. We have found three different approaches which use the triple (csr, exc, ref) for mining exception rules. We present them, analyze their weak points and we will see that the imposition of the reference rule in these approaches it is not indispensable. The lack of the reference rule can lead to spurious exception rules, but this can be solved by using stronger association measurements as we will propose in section 4.

Hussain et al. [14] take the reference rule as $E \rightarrow \neg Y$ with low support or low confidence. But in fact they impose that the rule $E \rightarrow Y$ is a strong one, i.e. it has high support and confidence, which implies the fulfilment of the reference rule.

If we take into consideration this formulation of exception:

$X \rightarrow Y$ and $E \rightarrow Y$	Strong rules
$X \wedge E \rightarrow \neg Y$	Confident rule

we see that Hussain et al. present a restrictive definition of exception. They impose a *double meaning*. The former is the original semantics of exception rule associated to the common sense rule $X \rightarrow Y$, and the second is that X is also an exception to the common sense rule $E \rightarrow Y$.

Suzuki et al. work in [22, 23] and also in other papers, with a distinct approach. They consider the rule-exception pair

$$X \rightarrow y \qquad \text{(common sense rule)}$$
$$X \wedge E \rightarrow y' \qquad \text{(exception rule)}$$

where y and y' are items with the same attribute but with different values and $X = x_1 \wedge \ldots \wedge x_p$, $E = e_1 \wedge \ldots \wedge e_q$ are conjunctions of items.

They also provide several proposals for measuring the degree of interestingness of a rule-exception pair. In [23] they use an information-based measure to determine the interestingness of the above pair of rules. But they also impose the constraint of not to be confident to the reference rule $E \rightarrow y'$.

In general terms what they propose can be summarized by

$X \rightarrow Y$	Strong rule
$X \wedge E \rightarrow \neg Y$	Confident rule
$E \rightarrow \neg Y$	Not confident

This approach is in some sense objectively novel, but in consonance to the general semantics of exception rules it should not impose a restriction outside the dominance of the common sense rule. Moreover, this restriction tells us that the major percentage of transactions which contain E and $\neg Y$ also contain X, because if it does not occur, $E \rightarrow \neg Y$ will be confident. Therefore, this approach is intended for giving importance to the fact that E appears almost always when dealing with X and $\neg Y$.

In [2] Berzal et al. collect the meaning of exceptions proposing this alternative formulation:

$X \rightarrow Y$	Strong rule
$X \wedge E \rightarrow \neg Y$	Confident rule
$X \rightarrow E$	Not strong

But imposing the reference rule will usually be redundant. Suppose that we impose the opposite of the reference rule, i.e. $X \rightarrow E$ is a strong rule. This rule says that E is frequent in those transactions containing X. The common sense rule also says that Y is frequent in those transactions containing X. So there is a high probability of having that Conf($X \wedge E \rightarrow Y$) exceeds the *minconf* threshold. If $minconf > 0.5$ since Conf$(X \wedge E \rightarrow \neg Y) = 1 - $Conf$(X \wedge E \rightarrow Y)$ the condition for the exception rule will not be satisfied. So it has been showed that imposing the negation of $X \rightarrow E$ not to be strong it is followed the negation of the exception rule condition, then by contraposition we find that the conditions associated to the common sense and the exception rules "implies with high probability the satisfiability of the previous reference rule."

3.2 *Discussion about Formulation of Exception Rules*

The main semantics of exception rules is collected by the proper exception rule *exc* in the triple (csr, exc, ref) as the common sense rule gives the relation between the normal behavior and the exceptional one. Nevertheless, the reference rule is used for restricting the action area of the exception or for dropping spurious exceptions that can be found in data. Our claim is that even some false exceptions can be extracted from data, there is no necessity of imposing the reference rule fulfilment as sometimes it maybe imposes some contradictory meaning to the exception or even some redundant condition as for instance in the last case previously seen.

Table 2 Summary of definitions of exception rules

Type of rule	Approach	Rule	Supp	Conf
csr	Suzuki, Hussain, Berzal [21], [14], [2]	$X \to Y$	High	High
exc	Suzuki, Hussain, Berzal [21], [14], [2]	$X \wedge E \to Y$	Low	High
ref	Suzuki et al. [21]	$E \to Y$	High	High
ref	Hussain et al. [14]	$E \to \neg Y$	-	Low
ref	Berzal et al. [2]	$X \to E$	Low (or)	Low

The necessity of discerning between the truly exception rules can be solved by imposing more restrictive quality measures as we treat in section 4.

We are going to analyze the reference rule semantics of previous approaches (see table 2). Translating the formulation into a practical example, the role of E in the exception rule is that of an agent which interferes in the usual behavior of the common sense rule. An example of exception rule could be:

"with the help of *antibiotics*, the patient usually tends to *recover*, unless (except when) *staphylococci* appear"

in such a case, antibiotics combined with staphylococci don't lead to recovery, even sometimes may lead to death. Following this example the possible reference rules used in every approach are:

Approach	Reference Rule		Interpretation
Hussain et al.	$E \to Y$	Strong	A patient with staphylococci tends to recover
Suzuki et al.	$E \to \neg Y$	Not confident	In general staphylococci may not lead to the recovering of the patient (low confidence)
Berzal et al.	$X \to E$	Not strong	The use of antibiotics in a patient does not imply the appearance of staphylococci (low support or confidence)

The approaches given by Hussain and Suzuki (first and second in the table) does not give any reasonably semantics to the reference rule, nevertheless in the Berzal et al's approach (last in the table) the reference rule says that antibiotics and staphylococci doesn't have a direct relation. This example shows that the reference rule in all these cases does not contribute to the semantics of exceptions even sometimes give a contradictory information. Therefore we only take the common sense and the exception rules for defining exceptions in the following, but we will study some other measures of interest instead of confidence in the following section.

4 Certainty Factor for Measuring the Strength of Exception Rules

In the ambit of searching good measures for assessing the validity of association rules many proposals had been described. One of them is known as the certainty factor framework [3, 6]. The theory of certainty factors was first introduced by Shortliffe and Buchanan in [20] and then was used by Berzal et al. [3] as an alternative to confidence. When mining exception rules there is a necessity of searching real exceptions dropping those that are spurious. For that we use the certainty factor since it is stronger than the confidence.

In particular, the support-certainty factor framework reduces the number of rules obtained being the extracted rules stronger than those obtained using the support-confidence framework. The *certainty factor*, *CF*, can be defined as follows:

$$CF(\varphi \to \psi) = \begin{cases} \dfrac{\text{Conf}(\varphi \to \psi) - \text{supp}(\psi)}{1 - \text{supp}(\psi)} & \text{if Conf}(\varphi \to \psi) > \text{supp}(\psi) \\ \dfrac{\text{Conf}(\varphi \to \psi) - \text{supp}(\psi)}{\text{supp}(\psi)} & \text{if Conf}(\varphi \to \psi) < \text{supp}(\psi) \\ 0 & \text{otherwise.} \end{cases} \tag{3}$$

The certainty factor is interpreted as a measure of variation of the probability that ψ is in a transaction when we consider only those transactions where φ is. More specifically, a positive CF measures the decrease of the probability that ψ is not in a transaction, given that φ is. A similar interpretation can be done for negative CFs.

The certainty factor measure has very interesting properties and we describe some of them which will be of interest:

1. $\text{Conf}(\varphi \to \psi) = 1$ if and only if $CF(\varphi \to \psi) = 1$.
   This property guarantees that the certainty factor of an association rule achieves its maximum possible value, 1, if and only if the rule is totally accurate [6].
2. Let $\varphi \to \psi$ be an association rule with positive certainty factor. Then the following equality holds [3]
   $$CF(\varphi \to \psi) = CF(\neg\psi \to \neg\varphi).$$
   The confidence does not fulfil this property, and it is very interesting since the rules $\varphi \to \psi$ and $\neg\psi \to \neg\varphi$ represent the same knowledge from a logic point of view.

So, using the certainty factor we can extract more accurate exception rules as we will drop those that do not exceed the *minCF* threshold. The reformulation of our approach for mining exception rules is given by

$$\frac{\begin{array}{ll} X \to Y & \text{Strong rule in } D \\ E \to \neg Y & \text{Certain rule in } D_X \end{array}}{}$$

where $D_X = \{t \in D : X \subset t\}$ and we have called *certain* to the rule whose certainty factor exceeds the imposed threshold for the certainty factor.

When restricting to D_X mining exception rules using the confidence coincides with the previous approach (without restricting to D_X) as $\text{Conf}(X \wedge E \to \neg Y) = \text{Conf}_X(E \to \neg Y)$ since

$$\text{supp}_X(E) = \frac{|X \cap E|}{|X|} = \frac{|X \cap E| \,/\, |D|}{|X| \,/\, |D|} = \frac{\text{supp}(X \cup E)}{\text{supp}(X)} \tag{4}$$

$$\begin{aligned}\text{Conf}_X(E \to \neg Y) &= \frac{\text{supp}_X(E \cup \neg Y)}{\text{supp}_X(E)} = \frac{|X \cap E \cap \neg Y| \,/\, |X|}{|X \cap E| \,/\, |X|} = \frac{|X \cap E \cap \neg Y|}{|X \cap E|} = \\ &= \text{Conf}(X \wedge E \to \neg Y)\end{aligned} \tag{5}$$

where we have noted by supp_X and Conf_X the support and the confidence restricted to the database D_X.

Another problem when mining exception rules is the high number of extracted rules that increments the complexity when mining exceptions since for each strong rule we want to discover the associated exceptions. The certainty factor can also help us reducing the number of extracted rules being stronger than those found using the confidence. So, in this case it is convenient to use the following approach:

$X \to Y$	Frequent and certain in D
$E \to \neg Y$	Certain in D_X

In section 7 we will show how the certainty factor achieves a considerable reduction in the number of extracted rules and in consequence in the number of discovered exceptions.

5 Analysis of Exception Rules through the Logic Model

The search of a logic for dealing with exceptions or default values has received a lot of attention during many years [4, 5]. Several proposals were introduced as forms of default reasoning for building a complete and consistent knowledge base. Our approach consists in modifying the existing formal model seen in section 2 for (1) representing the knowledge collected by exception rules, and (2) for taking into account the measures used in the task of mining exception rules which involves the pair (csr, exc).

In [7] we presented a first approach for explaining the true semantics of exception and anomalous rules using the previous logic model. Moreover the model provides a unified framework for working conjunctly with association rules and their exceptions.

This section is devoted to offer a new representation for the concept of exception using the introduced formal model. Following our last approach we will treat two different approaches. The former will use the confidence and the second the certainty factor. Let be X, Y and E three itemsets (or attributes in the logic model) in a database D. We regard the frequency of appearance of E and Y in D_X, where D_X contains those transactions in D satisfying X. The associated four fold table ($4ft$) is in the following table:

D_X	Y	$\neg Y$	
E	e	f	
$\neg E$	g	h	
			$a+b$

where e is the number of transactions in D_X satisfying Y and E; f the number of transactions in D_X satisfying E and not Y and so on. The sums of these frequencies correspond to the a and b frequencies seen in the previous section, i.e. $a = e+g$, $b = f+h$. We also use n for the number of transactions in D.

Using the predefined $4ft$-quantifiers for association rules we adapt them for the particular case of mining exception rules. We say that a quantifier $\approx^E$ is an E-quantifier if it involves only the four frequencies introduced in the previous table. If we use confidence, we define the implication E-quantifier, $\Rightarrow^E$ by

$$\Rightarrow^E (e,f,g,h) = \frac{f}{e+f} \tag{6}$$

which measures the strength (in particular, the confidence) of the exception rule. For completing the task of mining the exception rules the implication E-quantifier must be higher than the *minconf* threshold:

$$\Rightarrow^E (e,f,g,h) \geq minconf. \tag{7}$$

So, for mining exception rules, first we have to impose the usual restrictions to the common sense rule by means of the implication quantifier previously seen, that is,

$$\Rightarrow_I (e,f,g,h) = \frac{e+g}{e+f+g+h} \geq minconf \wedge \frac{e+g}{n} \geq minsupp \tag{8}$$

and then the restriction associated to the E-quantifier.

It should be noted that an E-quantifier does not require the imposition of the *minsupp* threshold. This is an important difference between *simple* quantifiers and E-quantifiers. Simple quantifiers are usually *based* which means that there must be enough transactions supporting the fulfilment of the quantifier.

For our approach the E-quantifier into consideration is the *certainty factor* E-equivalence based on the equivalence $\equiv_{CF}$ defined in [8]. The resulting E-quantifier, $\equiv^E_{CF}$ is defined as follows

$$\equiv^E_{CF} (e,f,g,h) = \begin{cases} \dfrac{\Rightarrow^E (e,f) - supp_X(\neg Y)}{1 - supp_X(\neg Y)} & \text{if } \Rightarrow^E (e,f) > supp_X(\neg Y) \\ 0 & \text{if } \Rightarrow^E (e,f) = supp_X(\neg Y) \\ \dfrac{\Rightarrow^E (e,f) - supp_X(\neg Y)}{supp_X(\neg Y)} & \text{if } \Rightarrow^E (e,f) < supp_X(\neg Y) \end{cases} \tag{9}$$

which is equivalent to

$$\equiv^E_{CF}(e,f,g,h) = \begin{cases} \dfrac{fg-eh}{(e+f)(e+g)} & \text{if } fg > eh \\ 0 & \text{if } fg = eh \\ \dfrac{fg-eh}{(e+f)(f+h)} & \text{if } fg < eh. \end{cases} \quad (10)$$

Using the logic model, the proposed procedure is well summarized saying that first we use the implication quantifier $\Rightarrow_I$ as usual for assessing the validity of the strong rule $X \rightarrow Y$ in D (as we see at the end of section 2). Then we impose the implication $\Rightarrow^E$ if we want to use the confidence or the equivalence $\equiv^E_{CF}$ for our new approach. As we proved in [8] the certainty factor belongs to the class of equivalence quantifiers [17] and it satisfies a set of desirable properties for a good interestingness measure.

6 Double Rules

When mining rules, sometimes we may find that two itemsets are very related and there is no difference about the direction of that relationship. We can take advantage of this situation for extracting a new kind of knowledge from the database. We propose a toy example for elucidating a prototypical ambient where this kind of "bidirectional" rules are useful.

Example 1. Imagine we have a database collecting information about the vertebrate animals and their characteristics in a national park. One of the strong rules we can extract is:

if the animal flies, it is a bird.

But this rule can also be extracted in the other direction, i.e. the rule

if the animal is a bird, it flies

is also a strong rule.

In this context, we need a new kind of rules which collects this new type of knowledge.

Definition 1. *An association rule $X \rightarrow Y$ is* double strong *if both $X \rightarrow Y$ and $Y \rightarrow X$ are strong.*

Here and subsequently the double directional arrow $X \leftrightarrow Y$ will denote double strong rules. We also consider that X is the double rule's antecedent if $Conf(X \rightarrow Y) \geq Conf(Y \rightarrow X)$. If we have an equality, we can choose X or Y indistinctly. According to this definition, we propose the analogous definition for a 4ft-quantifier.

Definition 2. *A quantifier $\approx$ is called* double strong *if it is defined by the conditions:*

$$\approx(a,b,c,d),\ \approx(a,c,b,d) \geq p \wedge \frac{a}{n} \geq minsupp$$

where $0 < p < 1$ and $0 < minsupp < 1$.

Note that the support of both rules $X \to Y$ and $Y \to X$ is the same, and p represents a threshold for the value of the quantifier. In our case, $p = minconf$ or $minCF$.

We remember that a 4ft-quantifier $\approx$ is *symmetric* [11] if $\approx(a,b,c,d) = \approx(a,c,b,d)$. Then, it is easy to see that a symmetric strong quantifier is always a doble strong quantifier. This is resumed in the following corollary.

Corollary 1. *A symmetric quantifier* $\approx$ *will be used in a* double strong *way if it satisfies the following conditions:*

$$\approx(a,b,c,d) \geq p \wedge a \geq Base$$

where $0 < p < 1$ *and* $0 < Base < a+b+c+d$.

This property reduces the number of threshold impositions when mining double rules for every symmetric quantifier.

6.1 Exceptions for Double Rules

As double rules are constituted by two different strong rules they are able to have several exceptions in both directions as we show in the following example.

Example 2. In the ambient of example 1 two kinds of exceptions can be discovered. The first type contains the exceptions for the consequent of the double strong rule, for example:

if the animal flies, it is a bird, *except* bats

and the second type will be the exceptions for the antecedent of the double rule:

if the animal it is a bird, it flies,
except penguin, ostrich, cock and hen.

This section is devoted to show the different situations when mining exception rules in the special case of double rules analyzing some interesting cases.

In general, we are going to deal with some different alternatives.

1. It could happen that there are no exceptions for the double strong rule.
2. We find an exception in only one direction of the double rule.

either

$X \leftrightarrow Y$	Double strong rule
$X \wedge E \to \neg Y$	Confident rule

or

$X \leftrightarrow Y$	Double strong rule
$Y \wedge E \to \neg X$	Confident rule

3. We have exceptions in both double rule's directions

$X \leftrightarrow Y$	Double strong rule
$X \wedge E \to \neg Y$	Confident rule
$Y \wedge E' \to \neg X$	Confident rule

In this case, we have two possibilities:

a. The double strong rule has different exceptions in each rule's direction, i.e. $E \neq E'$.
b. The double strong rule has the same exceptions in both rule's directions, i.e. we have that $E = E'$:

$X \leftrightarrow Y$	Double strong rule
$X \wedge E \rightarrow \neg Y$	Confident rule
$Y \wedge E \rightarrow \neg X$	Confident rule

In order to illustrate this last case, let us consider the relation shown in table 3 containing twelve transactions. From this dataset, we obtain: $\text{supp}(X \leftrightarrow Y) \simeq 0.583$, $\text{Conf}(X \rightarrow Y) \simeq 0.78$, and $\text{Conf}(Y \rightarrow X) \simeq 0.78$, which show that $X \leftrightarrow Y$ is a double strong rule if we impose the *minsupp* threshold to 0.5 and the *minconf* threshold to 0.6. We also obtain: $\text{Conf}(X \wedge E \rightarrow \neg Y) \simeq 0.67$, and $\text{Conf}(Y \wedge E \rightarrow \neg X) \simeq 0.67$, thus we obtain that E is the same exception for both sides of the double rule.

Table 3 Database with a double exception

X	Y	F	$\cdots$
X	Y	F	$\cdots$
X	Y	F	$\cdots$
X	Y	F	$\cdots$
X	Y	F	$\cdots$
X	Y	F	$\cdots$
X	Y	E	$\cdots$
X	Y'	E	$\cdots$
X	Y'	E	$\cdots$
X'	Y	E	$\cdots$
X'	Y	E	$\cdots$
X'	Y'	E	$\cdots$

Semantically we could say that we have found an "agent" which affects both sides of the common sense rule, i.e. its presence disturbs the double strong rule's $X \leftrightarrow Y$ usual behavior.

One interesting case is the last one where the interaction of a "strange" (in the sense of unfrequent) factor changes the double strong rule's behavior. When this happen, we say that E is a *double exception*. For cases 2 and 3, depending on the grade of satisfiability given by the measure (confidence or certainty factor) we can extract a better description about the relation of two variables. This is the case of our example, where the relationship between bird and flying is totally described when we take into account its exceptions. In other fields as medicine, we can find a strict relation between a disease and its symptoms when we also extract every posible exception.

7 Experimental Evaluation

We have proposed some new approaches for mining some new and useful knowledge from a database. A new approach using the certainty factor has been proposed for mining exception rules. We also have introduced a novel kind of rules that will be useful when for describing in more detail the relation between a set of items. Mining double rules in conjunction with the associated exceptions offers a clarification about the agents that perturbs the strong rule's usual behavior.

In the remainder of the section we will present a simple algorithm for mining exception rules optimized by using the fast-mining-bit string approach [15] and using the defined quantifiers. We have performed experiments to assess some claims: (1) the proposed algorithm is feasible in practice, and (2) the time complexity and memory requirements are also analyzed by means of a wide range of experiments with some real datasets. We conclude presenting some interesting exception and double rules extracted from real databases.

7.1 Algorithm and Implementation Issues

The algorithm we propose is able to mine conjunctly the set of strong rules in a database with their associated exceptions. We only consider exceptions given by a single item. For mining the association rules we use a simple implementation of the Apriori algorithm modifying it for dealing with a set of items given by their representation using BitSets.

Previous works [15, 19] have implemented the Apriori algorithm using a bit-string representation of items. In both works the results obtained are quite good with respect to time. Other advantage is the speediness when performing the logical operations such as conjunction or cardinality.

In particular in our implementation instead of strings of bits we use the java class `java.util.BitSet` which contains the implementation of the object `BitSet` and some useful operations. The BitSet object stores a set of bits (zero or one) in each position. The main idea of our implementation consists in storing the whole database into a vector of BitSets with size equal to the number of transactions and dimension equal to the number of items. Each BitSet will contain if an item (or itemset when dealing with conjunction of items) is satisfied (1) or not (0) in each transaction.

The general framework we have implemented for mining exceptions is described in algorithm 1.

First step can be done in different ways for creating the suitable set of items depending of the particular database.

Steps 1.1 and 1.2 are conjunctly processed in order to read the database only once. The created BitSet vector will have dimension equal to the number of items in the database and each element of the vector will contain a BitSet with the value one or zero in position i if the item appears or not in the ith transaction.

The last part describes the process for mining exceptions. For that we will use the antecedent and consequent BitSet representations for the common sense rule

Algorithm 1: Mining Exceptions

Input: Transactional database, minsupp, minconf or minCF
Output: Set of association rules with their associated exception rules.

1. Database Preprocessing
   1.1 Transformation of the transactional database into a boolean database.
   1.2 Database storage into a vector of BitSets.
2. Mining Process (similar to Apriori methodology).
   2.1 Mining Common Sense Rules
      Searching the set of candidates (frequent itemsets) for extracting the common sense rule.
      Storage of BitSet vector indexes associated to the frequent itemsets and their supports.
      Common sense rules extraction exceeding the minsupp and minconf thresholds
   2.2.1 Mining Exception Rules
      For every common sense rule $X \rightarrow Y$ we compute the possible exceptions:
      For each item $E \subset I$ (except those in the common sense rule)
         Compute $X \wedge E \wedge \neg Y$ and its support
         Compute $X \wedge E$ and its support
         Using confidence:
         Compute $Conf(X \wedge E \rightarrow \neg Y)$ **if** it is $\geq$ *minconf* **then** we have an exception
         Using certainty factor:
         Compute $supp_X(\neg Y)$
         Compute $CF_X(E \rightarrow \neg Y)$ **if** it is $\geq$ *minCF* **then** we have an exception

and for the whole set of items. The crucial point in the step 2.2.1 is that of computing the exception rule's confidence/certainty factor that is, Conf($X \wedge E \rightarrow \neg Y$) or $CF(X \wedge E \rightarrow \neg Y)$ where E can be any single item (even not frequent) which is high time-consuming, but using bitsets the conjunction and cardinality operations are fast. The problem is the computation consequent's negation $\neg Y$ which is slower than conjunction, but we solve this problem using what we know from the logic model. Instead of doing the negation of the consequent we compute two frequencies: supp($X \cup E \cup \neg Y$) and supp($X \cup E$) which are the f and $e+f$ frequencies of $4ft(E,Y,D_X)$ involved in the E-quantifier $\Rightarrow^E$ as follows:

$$f+e = \text{supp}(X \cup E), \quad f = \text{supp}(X \cup E \cup \neg Y) = \text{supp}(X \cup E) - \text{supp}(X \cup E \cup Y).$$

The proposed algorithm is also useful for mining double rules as it extracts all the strong rules without taking care of the direction of the association. For a better results exploration we search the common sense rule and just then its "opposite" for obtaining double rules together.

7.2 Time and Space Complexity

The proposed algorithm integrates the extraction of exception rules in the principal mining process. The Apriori implementation has been used for mining the common

Table 4 Databases description used in the experiments

Database	Size	Items
Barbora	6181	33
Nursery	12960	27
Car	1728	21

sense rules but it has been modified for containing the exception rules mining. For each mined rule we call to the exception rule mining process. The complexity of our approach will depend on the complexity of the principal rule mining algorithm, in the Apriori case is $O(n2^i)$ where n is the number of transactions and i the number of items. It also depends linearly on the number of extracted rules says r and linearly on the number of items. So in our case, we have a theoretic complexity of $O(nri2^i)$. For improving the efficiency the principal rule mining algorithm can be changed by a faster one, but our purpose here is showing that mining exception rules is feasible even for large databases as we will see in some real experimentation.

Concerning memory, the size of memory requirements for standard databases is acceptable. For example for the Barbora bank database (`http://lispminer.vse.cz/download`) that is a part of Discovery challenge of the conference PKDD99 held in Prague, the required memory for the database consisting in 6181 transactions and 12 attributes (33 items) is 107 kb [19], and for 61810 transactions is 1.04 MB.

To illustrate the time complexity we have discussed, and to show the performance of our proposals we have performed some experiments with the Barbora bank database and with two databases (Car and Nursery) from the UCI repository (see table 4). We used a 1.73GHz Intel Core 2Duo notebook with 1024MB of main memory, running Windows XP. The extraction of exceptions is in general computationally expensive as we have to test for every strong rule if there exists any exception in the set of items. Our approach plays an essential role in this task. Imposing the certainty factor for the common sense rule mining, the number of mined rules decreases (obtaining more accurate association rules [3]) and in consequence, the time consumed for mining exception rules is also decreased. In table 5 and figure 1

Table 5 Time (in seconds) and rule comparison between Confidence and Certainty Factor in 20 times the Barbora Bank database (123620 transactions) with $minsupp = 5\%$

minconf	Rules	Exceptions	Time	*mincf*	Rules	Exceptions	Time
0.95	9	8	141	0.95	4	3	127
0.90	46	56	236	0.90	4	3	127
0.85	139	175	445	0.85	6	5	132
0.80	191	286	575	0.80	9	9	140
0.75	258	435	751	0.75	10	10	143
0.70	313	583	818	0.60	13	13	150

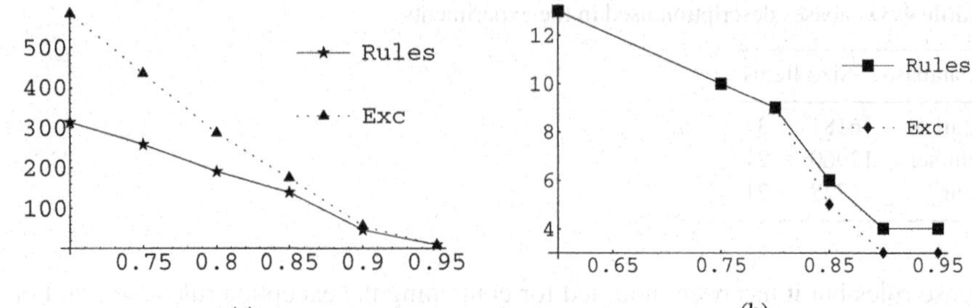

Fig. 1 (a) Number of rules and exceptions (y-axis) as a function of *minconf* (x-axis) in 20 times the Barbora Bank data set (123620 transactions). (b) Number of rules and exceptions (y-axis) as a function of *minCF* (x-axis) in 20 times the Barbora Bank data set (123620 transactions).

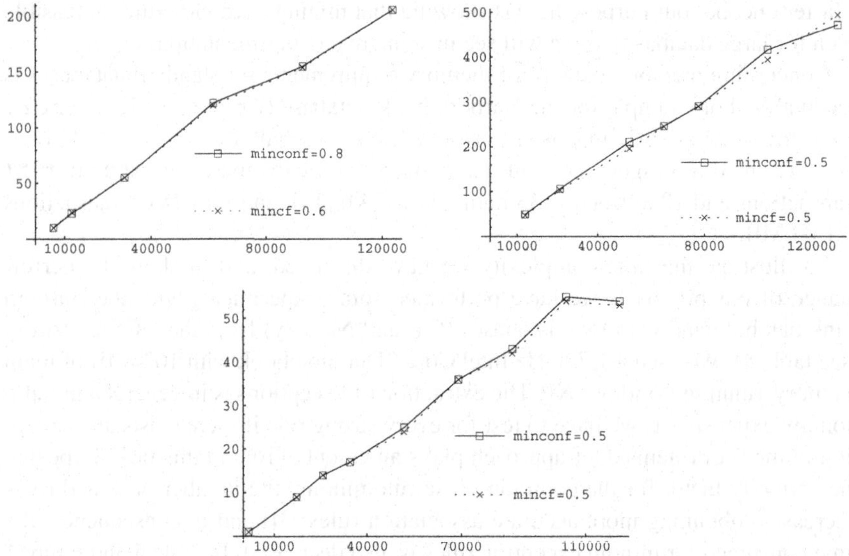

Fig. 2 Time in seconds (y-axis) as a function of the number of transactions (x-axis) for three data sets (left Barbora, right Nursery, bottom Car).

we can see the differences between using confidence and certainty factor for mining the common sense rules as well as the exception rules.

We want to remark that computing certainty factor does not increase the time execution as we can see in figure 2. In these experiments we only increased the number of transactions (repeating several times the databases) and we considered the same *minconf* threshold for the common sense rule mining (for obtaining the same number of common sense rules) repeating them changing the imposition of confidence by certainty factor for mining the exceptions.

7.3 Some Interesting Results

Now we are interested in how using our approach we can find interesting exception and double rules. In the previous section we show how imposing the certainty factor reduces substantially the number of obtained exception rules as we also reduce the number of common sense rules. We will show some interesting rules obtained using our approach (*csr*, *exc*) by imposing confidence. We will use *OR* for enumerating the different extracted exceptions.

In the Barbora bank data set we have found several exception rules as for instance:

"*IF* the sex = woman & age $\in [55-65)$ years old
THEN the status of the mortgage = good ($supp = 0.106$ and $conf = 0.866$),
EXCEPT when the amount of money > 500000 czech crowns ($conf = 0.889$)
OR the payment is between 8000 and 9000 czech crowns ($conf = 0.8$)".

In this case we have found two different exceptions for the same strong rule, although we can see that both exceptions are related since when the amount of money is high it is usual that the payment is also high.

Using the Car database we have a considerably big amount of exception rules. This phenomenon is sometimes caused by the presence of a frequent itemset which is complementary to one that appears in the consequent of the rule, as for instance:

"*IF* the price of buying the car is high
THEN its maintenance is low ($supp = 0.125$ and $conf = 0.5$),
EXCEPT if its maintenance is high ($conf = 1$) *OR* medium ($conf = 1$)".

This rule confirms that our method works as it also extracts the obvious exception rules.

Concerning to double rules the probability of finding double rules decreases when we increase the minimum confidence threshold. As we have already mention, the relation between two sets of items is better described by the strength of association in both directions and it is complemented by the exceptions extracted in each direction.

In the Barbora bank database we have found a double strong rule which also has exceptions in one rule direction, in particular in that with the lesser confidence. The double rule which has support equal to 0.68 is:

"*IF* the salary $\in [8000, 10000)$ czech crowns
THEN the mortgage status is good ($conf = 0.879$) and vice versa ($conf = 0.771$),
EXCEPT when the salary is bigger than 10000 czech crowns ($conf \simeq 0.778$)".

This rule shows the high grade of relationship between a "small" salary and the good status of the mortgage.

8 Conclusions and Future Research

Mining exception rules can be useful in several domains. We tried to do a wide analysis about semantics and formulation of some previous approaches obtaining that the reference rule could be removed under some reasonably conditions. We also offer a unified formulation behind the GUHA model which uses the concept of quantifier. The proposed algorithm takes advantage of the formulation as we can mine different exception rules by only changing the quantifier. In our approach we only deal with the implication quantifiers $\Rightarrow_I$ and $\Rightarrow^E$ and the certainty factor equivalences $\equiv_{CF}, \equiv_{CF}^E$.

We present a new framework for describing in more detail the relation between a set of items by means of double rules and their associated exceptions. This can be useful in some area domains such as medicine where we know there is a straight relationship between some sets of items (illness, symptoms and medicaments for example).

We also have carried out several experiments for showing the reliability of our approach which can be improve by using the certainty factor and a faster rule extraction algorithm.

Note that exceptions only offers the "agent" that causes the unusual common sense rule behavior, but there exists other approaches that focus in the complementary semantics like the anomalous rules.

Our future plans include an analogous analysis about anomalous rules and their interaction with double rules. We also plan to offer a unified framework for mining conjunctly exception and anomalous rules by means of the formulation of suitable quantifiers.

Acknowledgements. We would like to acknowledge support for this work from the Ministerio de Educación y Ciencia of Spain by the project grants TIN2006-15041-C04-01 and TIN2006-07262.

References

1. Agrawal, R., Manilla, H., Sukent, R., Toivonen, A., Verkamo, A.: Fast discovery of Association rules, pp. 307–328. AAA Press (1996)
2. Berzal, F., Cubero, J., Marín, N., Gámez, M.: Anomalous association rules. In: IEEE Int. Conf. on Data Mining (2004)
3. Berzal, F., Delgado, M., Sánchez, D., Vila, M.: Measuring accuracy and interest of association rules: A new framework. Intelligent Data Analysis 6(3), 221–235 (2002)
4. Brachman, R., Levesque, H.: Knowledge Representation and Reasoning. Elsevier, Morgan Kaufmann publishers (2004)
5. Brewka, G.: Nonmonotonic Reasoning: Logical Foundations of Commonsense. University Press, Cambridge (1991)
6. Delgado, M., Marín, N., Sánchez, D., Vila, M.: Fuzzy association rules: General model and applications. IEEE Transactions on Fuzzy Systems 11(2), 214–225 (2003)
7. Delgado, M., Ruiz, M., Sánchez, D.: A logic approach for exceptions and anomalies in association rules. Mathware & Soft Computing 15(3), 285–295 (2008)
8. Delgado, M., Ruiz, M., Sánchez, D.: Studying interest measures for association rules through a logical model. Int. Journal of Uncertainty, Fuzziness and Knowledge-Based Systems (2007) (submitted)

9. Duval, B., Salleb, A., Vrain, C.: On the discovery of exception rules: A survey. Studies in Computational Intelligence 43, 77–98 (2007)
10. Hájek, P., Havel, I., Chytil, M.: The guha method of automatic hypotheses determination. Computing 1, 293–308 (1966)
11. Hájek, P., Havránek, T.: Mechanising Hypothesis Formation-Mathematical Foundations for a General Theory. Springer, Heidelberg (1978)
12. Havránek, T.: Statistical quantifiers in observational calculi: An application in guha-methods. Theory and Decision 6, 213–230 (1974)
13. Holeňa, M.: Fuzzy hypotheses for guha implications. Fuzzy Sets and Systems 98, 101–125 (1998)
14. Hussain, F., Liu, H., Suzuki, E., Lu, H.: Exception rule mining with a relative interestingness measure. In: Terano, T., Chen, A.L.P. (eds.) PAKDD 2000. LNCS, vol. 1805, pp. 86–97. Springer, Heidelberg (2000)
15. Louie, E., Lin, T.: Finding association rules using fast bit computation: Machine-oriented modeling. In: Ohsuga, S., Raś, Z.W. (eds.) ISMIS 2000. LNCS (LNAI), vol. 1932, pp. 486–494. Springer, Heidelberg (2000)
16. Ohshima, M., Zhong, N., Yao, Y., Liu, C.: Relational peculiarity-oriented mining. Data Mining and Knowledge Discovery 15(2), 249–273 (2007)
17. Rauch, J.: Logic of association rules. Applied Intelligence 22, 9–28 (2005)
18. Rauch, J., Šimunek, M.: Mining for 4ft association rules. In: Morishita, S., Arikawa, S. (eds.) DS 2000. LNCS (LNAI), vol. 1967, pp. 268–272. Springer, Heidelberg (2000)
19. Rauch, J., Šimunek, M.: An alternative approach to mining association rules. Studies in Computational Intelligence 6, 211–231 (2005)
20. Shortliffe, E., Buchanan, B.: A model of inexact reasoning in medicine. Mathematical Biosciences 23, 351–379 (1975)
21. Suzuki, E.: Discovering unexpected exceptions: A stochastic approach. In: Proceedings of the fourth international workshop on RSFD, pp. 225–232 (1996)
22. Suzuki, E.: Undirected discovery of interesting exception rules. Int. Jour. of Pattern Recognition and Artificial Intelligence 16(8), 1065–1086 (2002)
23. Suzuki, E., Shimura, M.: Exceptional knowledge discovery in databases based on information theory. In: Proc. of the 2nd Int. Conf. on Knowledge Discovery and Data Mining, pp. 275–278 (1996)
24. Suzuki, E., Zytkow, J.: Unified algorithm for undirected discovery of exception rules. Int. Jour. of Intelligent Systems 20, 673–691 (2005)
25. Zhong, N., Ohshima, M., Ohsuga, S.: Peculiarity oriented mining and its application for knowledge discovery in amino-acid data. In: Cheung, D., Williams, G.J., Li, Q. (eds.) PAKDD 2001. LNCS (LNAI), vol. 2035, pp. 260–269. Springer, Heidelberg (2001)

Degrees of Truth, Ill-Known Sets and Contradiction

Didier Dubois

Abstract. In many works dealing with knowledge representation, there is a temptation to extend the truth-set underlying a given logic with values expressing ignorance and contradiction. This is the case with partial logic and Belnap bilattice logic with respect to classical logic. This is also true in three-valued logics of rough sets. It is found again in interval-valued, and type two extensions of fuzzy sets. This paper shows that ignorance and contradiction cannot be viewed as additional truth-values nor processed in a truth-functional manner, and that doing it leads to weak or debatable uncertainty handling approaches.

1 Introduction

From the inception of many-valued logics, there have been attempts to attach an epistemic flavor to truth degrees. Intermediary truth-values between true and false were often interpreted as expressing a form of ignorance or partial belief (less often, the idea of contradiction). However, multiple-valued logics are generally truth-functional. The trouble here is that, when trying to capture the status of any unknown proposition by a truth-value, the very assumption of truth-functionality (building truth-tables for all connectives) is debatable. Combining two propositions whose truth-value is unknown sometimes results in tautological or contradictory statements, whose truth-value can be asserted from the start, even without any prior knowledge. As long as p can only be either true or false, even if this truth-value cannot be computed or prescribed as of to-day, the proposition $p \wedge \neg p$ can be unmistakably at any time predicted as being false and $p \vee \neg p$ as being true while $p \wedge p$ and $p \vee p$ remain contingent. So there is no way of defining a sensible truth-table that accounts for the idea of *possible* : belief is never truth-functional [21]. Mixing up truth and belief has led to a very confusing situation in traditional many-valued logics, and has probably hampered the development of applications of these logics. The

Didier Dubois
IRIT-CNRS, Université de Toulouse, France
e-mail: dubois@irit.fr

B. Bouchon-Meunier et al. (Eds.) Found. of Reas. under Uncert., STUDFUZZ 249, pp. 65–83.
springerlink.com

epistemic understanding of truth-functional many-valued logics has been criticized by some scholars quite early, for instance by Urquhart [41]. Fuzzy logic is likewise often attacked because it is truth-functional. A well-known example is by Elkan [25] criticising the usual fuzzy connectives max, min, $1-$, as leading to an inconsistent approach. Looking at these critiques more closely, it can be seen that the root of the controversy also lies in a confusion between degrees of truth and degrees of belief. Fuzzy logic is not specifically concerned with belief representation, only with gradual (not black or white) concepts [33]. However this misunderstanding seems to come a long way. For instance, a truth-value strictly between true and false was named "possible" [38], a word which refers to uncertainty modelling and modalities. We claimed in [18] that we cannot consistently reason under incomplete or conflicting information about propositions by augmenting the set of Boolean truth-values *true* and *false* with epistemic notions like "unknown" or "contradictory", modeling them as additional genuine truth-values of their own, as done in partial logic and Belnap's allegedly useful four-valued logic.

After reminding how uncertainty due to incompleteness is handled within propositional logic, the paper summarizes the critical discussion on partial logic previously proposed in [18], showing the corresponding extension of sets to ill-known sets, whose connectives are closely related to Kleene 3-valued logic. Two examples of ill-known sets are exhibited, especially rough sets. The debatable assumption behind some three-valued logics of rough sets is laid bare. Next, a critical discussion on Belnap logic is given, borrowing from [18]. Finally, we consider the case of truth-functional extensions of fuzzy set algebras, such as interval-valued fuzzy sets and membership/nonmembership pairs of Atanassov, as well as type two fuzzy sets where truth-functionality is also taken for granted, and that suffer from the same kind of limitations.

2 Truth vs. Belief in Classical Logic

In the following, 1 stands for *true* and 0 stands for *false*. In a previous paper [22] we pointed out that while classical (propositional) logic is always presented as the logic of the true and the false, this description neglects the epistemic aspects of this logic. Namely, if a set $\mathscr{B}$ of well-formed Boolean formulae is understood as a set of propositions believed by an intelligent agent (a *belief base*) then the underlying uncertainty theory is ternary and not binary. The three situations are:

1. p is believed (or known), which is the case if $\mathscr{B}$ implies p;
2. its negation is believed (or known), which is the case if $\mathscr{B}$ implies $\neg p$;
3. neither p nor $\neg p$ is believed, which is the case if $\mathscr{B}$ implies neither $\neg p$ nor p.

In this setting belief is Boolean, in the sense that a proposition is believed or not. We can define a belief assessment procedure to propositions, by means of a certainty function N assigning value 1 to p whenever $\mathscr{B}$ implies p ($N(p) = 1$) and 0 otherwise. The third situation above indicates a proposition that is neither believed nor is disbelieved by a particular agent. N is not a truth-assignment: one may have $N(p) = N(\neg p) = 0$, when p is unknown. The N function encodes a necessity-like modality. Indeed it is not fully compositional; while $N(p \wedge q) = \min(N(p), N(q))$, $N(p \vee q) \neq \max(N(p), N(q))$, generally, and $N(p)$ is not $1 - N(\neg p)$. The latter is the possibility function, in agreement with the duality between possibility and necessity in modal logic. So even Boolean belief is not compositional.

It is clear that belief refers to the notion of validity of p in the face of $\mathscr{B}$ and is a matter of consequencehood, not truth-values. The property $N(p \wedge q) = \min(N(p), N(q))$ just expresses that the intersection of deductively closed knowledge bases is closed, while $N(p \vee q) \neq \max(N(p), N(q))$ reminds us that the union of deductively closed propositional bases is not closed.

Through inference, we can check what are the possible truth-values left for propositions when constraints expressed in the belief base are taken into account. In fact, belief is represented by means of *subsets of possible truth-values* enabled for p when taking propositions in $\mathscr{B}$ for granted. Full belief in p corresponds to the singleton $\{1\}$ (only the truth-value "true" is possible); full disbelief in p corresponds to the singleton $\{0\}$; the situation of total uncertainty relative to p for the agent corresponds to the set $\{0, 1\}$. This set is to be understood disjunctively (both truth-values for p remain possible due to incompleteness, but only one is correct). Under such conventions, the characteristic function of $\{0, 1\}$ is viewed as a possibility distribution π (Zadeh [48]). Namely, $\pi(0) = \pi(1) = 1$ means that both 0 and 1 are possible. It contrasts with other uses of subsets of truth-values, interpreted conjunctively, whereby $\{0, 1\}$ is understood as the *simultaneous* attachment of "true" and "false" to p (expressing a contradiction, see Dunn [24]). This convention is based on necessity degrees $N(0) = 1 - \pi(1); N(1) = 1 - \pi(0)$. Then clearly, $N(0) = 1 = N(1)$ indicates a strong contradiction. But this convention cannot be easily extended beyond two-valued truth sets, so we shall not use it.

It must be emphasized that $\{0\}, \{1\}$, and $\{0, 1\}$ are not truth-values of propositions in $\mathscr{B}$. They express what can be called *epistemic valuations* whereby the agent believes p, believes $\neg p$, or is ignorant about p respectively. It makes it clear at the mathematical level that confusing truth-values and epistemic valuations comes down to confusing elements of a set and singletons contained it it, let alone subsets.

Clearly, the negation of the statement *p is believed* (inferred from $\mathscr{B}$) is not the statement $\neg p$ *is believed* , it is *p is not believed*. However, the statement *p is not believed* cannot be written in propositional logic because its syntax does not allow for expressing ignorance in the object language. The latter requires a modal logic, since in classical logic, if one interprets $p \in \mathscr{B}$ as a belief, $\neg p \in \mathscr{B}$ means that $\neg p$, is believed, not that p is just not believed. Likewise, $p \vee q \in \mathscr{B}$ is believed does not mean that either p is believed or q is believed. Assigning epistemic valuation $\{1\}$ to $p \vee q$ is actually weaker than assigning $\{1\}$ to one of p or q. In the case of ignorance about p, $\{0, 1\}$ should be assigned to p and to $\neg p$. However only $\{1\}$ can be attached to their disjunction (since it is a tautology). This fact only reminds us that the union of deductively closed belief sets need not be deductively closed.

In order to capture the lack of belief or ignorance at the object level, formulas of propositional logic can be embedded within a modal-like system (Dubois, Hájek, and Prade [17], Banerjee and Dubois [6]). This embedding of classical logic into a modal logic is not the usual one: usually, propositional logic is a fragment (without modalities) of a modal logic. In the system MEL developed in [6], all wffs are made of classical propositions p prefixed by $\Box$, or their combination by means of classical connectives (formulas α of the form $\Box p$ are in MEL, and so are $\neg\alpha$, $\alpha \wedge \beta$). Boxed formulas $\Box p$ are new atoms of a (higher order) propositional logic MEL satisfies modal axoms K and D, but does not allow for nested modalities. The boxed fragment of MEL is isomorphic to propositional calculus. Any modal logic where the K axiom $\Box(p \to q) \to (\Box p \to \Box q)$ holds verifies this embedding property. So MEL is not at all a standard modal logic, in the sense that it encapsulates propositional calculus but it does not extend it. Philosophically, MEL modalities are understood *de dicto*, and not *de re*, contrary to the tradition of XXth century logic. Namely, $\Box p$ concerns the certainty of being able to assert p, not the certainty that an event referred to as p has occurred, is really true: MEL forbids direct access to the "real world", and this is consistent with the fact that propositional formulas like p (stating that p is true) cannot be expressed in MEL.

In fact, the truth-value of $\Box p$ tells whether p is believed or not: $\Box p$ *is true* precisely means that the agent's beliefs enforce $\{1\}$ as the subset of truth-values left to p, i.e. *it is true that p is believed (to be true)*. So, what belief internally means may be captured by a kind of external truth-set, say $\{\mathbf{0}, \mathbf{1}\}$. Mind that the value $\mathbf{1}$ in $t(\Box p) = \mathbf{1}$ and the value 1 in $t(p) = 1$ refer to different truth-sets (and different propositions). This trick can be used for probability theory and other non-compositional uncertainty theories (see Godo, Hájek et al. [30, 29]) and leads to a better way of legitimating the use of many-valued logics for uncertainty management: the lack of compositionality of belief is captured in the object language. For instance, the degree of probability *Prob*(p) can be modeled as the truth-value of the proposition "Probable(p)" (which expresses the statement that p is probably true), where *Probable* is a many-valued predicate, but *Prob*(p) is not the (allegedly) multivalued truth-value of the (Boolean) proposition p.

3 From Partial Logic to Ill-Known Sets

Partial logic starts from the claim that not all propositional variables need to be assigned a truth-value, thus defining partial interpretations and that such undefinedness may stem from a lack of information. This program is clearly in the scope of theories of uncertainty and partial belief, introduced to cope with limited knowledge. Other interpretations of partiality exist, that are not considered here. From a historical perspective, the formalism of partial logic is not so old, but has its root in Kleene [35]'s three-valued logic, where the third truth-value expresses the impossibility to decide if a proposition is true or false. The reader is referred to the dissertation of Thijsse [40] and a survey paper by Blamey [11].

3.1 Connectives of Partial Logic

At the semantic level, the main idea of partial logic is to change interpretations $s \in S$ into partial interpretations, also called *coherent situations* (or *situations,* for short) obtained by assigning a Boolean truth-value to some (but not all) of the propositional variables forming a set $Prop = \{a, b, c, \dots\}$. A coherent situation can be represented as any conjunction of literals pertaining to distinct propositional variables. Denote by σ a situation, $\mathscr{S}$ the set of such situations, and $V(a, \sigma)$ the partial function from $Prop \times \mathscr{S}$ to $\{0, 1\}$ such that $V(a, \sigma) = 1$ if a is true in σ, 0 if a is false in σ, and is undefined otherwise. Then, two relations are defined for the semantics of connectives, namely *satisfies* ($\models_T$) and *falsifies* ($\models_F$):

- $\sigma \models_T a$ if and only if $V(\sigma, a) = 1$; $\sigma \models_F a$ if and only if $V(\sigma, a) = 0$;
- $\sigma \models_T \neg p$ if and only if $\sigma \models_F p$; $\sigma \models_F \neg p$ if and only if $\sigma \models_T p$;
- $\sigma \models_T p \wedge q$ if and only if $\sigma \models_T p$ and $\sigma \models_T q$;
- $\sigma \models_F p \wedge q$ if and only if $\sigma \models_F p$ or $\sigma \models_F q$
- $\sigma \models_T p \vee q$ if and only if $\sigma \models_T p$ or $\sigma \models_T q$;
- $\sigma \models_F p \vee q$ if and only if s $\models_F p$ and $\sigma \models_F q$.

In partial logic a coherent situation can be encoded as a truth-assignment t_σ mapping each propositional variable to the set $\{0, \frac{1}{2}, 1\}$, understood as a partial Boolean truth-assignment in $\{0, 1\}$. Let $t_\sigma(a) = 1$ if atom a appears in σ, 0 if $\neg a$ appears in σ, and $t_\sigma(a) = \frac{1}{2}$ if a is absent from σ. The basic partial logic can thus be described by means of a three-valued logic, where $\frac{1}{2}$ (again) means *unknown*. The connectives can be expressed as follows: $1 - x$ for the negation, max for disjunction, min for the conjunction, and $\max(1 - x, y)$ for the implication. Note that if $t_\sigma(p) = t_\sigma(q) = \frac{1}{2}$, then also $t_\sigma(p \vee q) = t_\sigma(p \wedge q) = t_\sigma(p \to q) = \frac{1}{2}$ in this approach.

3.2 *Supervaluations*

Since these definitions express truth-functionality in a three-valued logic, this logic fails to satisfy all classical tautologies. But this anomaly stems from the same difficulty again, that is, no three-element set can be endowed with Boolean algebra structure! (nor is the set **3** of non-empty intervals on $\{0, 1\}$). A coherent situation σ can be interpreted as a special set $A(\sigma) \subseteq S$ of standard Boolean interpretations, and can be viewed as a disjunction thereof. A coherent situation can be encoded as a formula whose set of models $A(\sigma)$ can be built just completing σ by all possible assignments of 0 or 1 to variables not assigned yet. It represents an epistemic state reflecting a lack of information. If this view is correct, the equivalence $\sigma \models_T p \vee q$ if and only if $\sigma \models_T p$ or $\sigma \models_T q$ cannot hold under classical model semantics. Indeed $\sigma \models_T p$ supposedly means $A(\sigma) \subseteq [p]$ and $\sigma \models_F p$ supposedly means $A(\sigma) \subseteq [\neg p]$, where $[p]$ is the set of interpretations where p is true. But while $A(\sigma) \subseteq [p \vee q]$ holds whenever $A(\sigma) \subseteq [p]$ or $A(\sigma) \subseteq [q]$ holds, the converse is invalid!

This is the point made by Van Fraassen [42] who first introduced the notion of supervaluation to account for this situation. A supervaluation *SV* over a coherent situation σ is (in our terminology) a function that assigns, to each proposition in the language and each coherent situation σ, the *super-truth-value* $SV(p, \sigma) = 1$ (0) to propositions that are true (false) for all Boolean completions of σ. It is clear that p is "super-true" ($SV(p, \sigma) = 1$) if and only if $A(\sigma) \subseteq [p]$, so that supervaluation theory recovers missing classical tautologies by again giving up truth-functionality: $p \vee \neg p$ is always super-true, but $SV(p \vee q, \sigma)$ cannot be computed from $SV(p, \sigma)$ and $SV(q, \sigma)$. The term "super-true" in the sense of Van Fraassen stands for "certainly true" in the terminology of possibilistic belief management in classical logic. The belief calculus at work in propositional logic covers the semantics of partial logic as a special case. It exactly coincides with the semantics of the supervaluation approach. Assuming compositionality of epistemic annotations by means of Kleene three-valued logic provides only an imprecise approximation of the actual Boolean truth-values of complex formulas [14].

3.3 *Ill-Known Sets*

Besides, the algebra underlying this (Kleene-like) three-valued logic is isomorphic to the set **3** of non-empty intervals on $\{0,1\}$, equipped with the interval extension of classical connectives. Consider $\frac{1}{2}$ as the set $\{0,1\}$ (understood as an interval such that $0 < 1$), the other intervals being the singletons $\{0\}$ and $\{1\}$. Indeed this comes down to computing the following cases:

- For conjunction : $\{0\} \wedge \{0,1\} = \{0 \wedge 0, 0 \wedge 1\} = \{0\}$; $\{1\} \wedge \{0,1\} = \{1 \wedge 0, 1 \wedge 1\} = \{0,1\}$, etc.
- For disjunction : $\{0\} \vee \{0,1\} = \{0 \vee 0, 0 \vee 1\} = \{0,1\}$; $\{1\} \vee \{0,1\} = \{1 \vee 0, 1 \vee 1\} = \{1\}$, etc.
- For negation: $\neg\{0,1\} = \{\neg 0, \neg 1\} = \{0,1\}$.

It yields the following tables for connectives $\vee$ and $\wedge$:

Table 1 Kleene disjunction for interval-valued sets

$\vee$	$\{0\}$	$\{0,1\}$	$\{1\}$
$\{0\}$	$\{0\}$	$\{0,1\}$	$\{1\}$
$\{0,1\}$	$\{0,1\}$	$\{0,1\}$	$\{1\}$
$\{1\}$	$\{1\}$	$\{1\}$	$\{1\}$

Table 2 Kleene conjunction for interval-valued sets

$\wedge$	$\{0\}$	$\{0,1\}$	$\{1\}$
$\{0\}$	$\{0\}$	$\{0\}$	$\{0\}$
$\{0,1\}$	$\{0\}$	$\{0,1\}$	$\{0,1\}$
$\{1\}$	$\{0\}$	$\{0,1\}$	$\{1\}$

This remark suggests that sets could be extended to *ill-known subsets* of a set S, assigning to elements $s \in S$ one of the three non-empty subsets of $\{0,1\}$. It is tempting to model them by three-valued sets denoted $\hat{A}$ whose characteristic function ranges on $\mathbf{3} = 2^{\{0,1\}} - \emptyset$ with the following conventions

$$\mu_{\hat{A}}(s) = \begin{cases} \{1\} & \text{if } s \text{ belongs for sure to the set } A \\ \{0\} & \text{if } s \text{ for sure does not belongs to the set } A \\ \{0,1\} & \text{if it is unknown whether } s \text{ belongs or not to the set } A \end{cases}$$

It encodes a pair of nested sets (A_*, A^*), A_* containing the sure elements, $A^* \setminus A_*$ being the elements with unknown membership. This is called an interval-set by Yao [45]. It is possible to extend the standard set theoretic operations to such three-valued sets using Kleene three-valued logic, equivalent to the interval operations to connectives defined in Tables 1 and 2 ($\hat{A}$ looks like a kind of fuzzy set). Equivalently, one may, as done by Yao [45], consider the interval extension of Boolean connectives to interval sets $A_* \subset A \subset A^*$. Note that while the subsets of $\{0,1\}$ form a Boolean algebra (under set inclusion), the set of "intervals" $\mathbf{3} = \{\{0\},\{1\},\{0,1\}\}$ of $\{0,1\}$ form a 3-element chain, a different structure, hence the loss of tautologies, if it is used as a new truth set, so that ill-known sets have properties different from sets. However it should be clear that this algebraic structure does not address (but in a very approximate way) the issue of reasoning about the ill-known set A. For instance, the complement of $\hat{A}$ is $\hat{A}^c$ obtained by switching $\{0\}$ and $\{1\}$ in the above definition, i.e. yields the pair $((A^*)^c, (A_*)^c)$. Hence $\hat{A} \cap \hat{A}^c$ corresponds to the pair $(A_* \cap (A^*)^c, A^* \cap (A_*)^c)$, where $A_* \cap (A^*)^c = \emptyset$ while $A^* \cup (A_*)^c \neq \emptyset$, generally [45]. However, the fuzzy set $\hat{A}$ is not an object in itself, it is a representation of the incomplete knowledge of an agent about a set A, of which all that is known is that $A_* \subset A \subset A^*$. But despite the fact that A is ill-known, $A \cap A^c = \emptyset$ regardless of what is known or not, and this information is lost by the Kleene setting, considering subsets of truth-values as truth-values and acting compositionally. Kleene's three valued logic is more naturally truth-functional when viewed as a simplified variant of *fuzzy logic*, where the third truth-value means half-true. The loss of classical tautologies then looks more acceptable.

For instance, let a one-to-many mapping $\Phi : S \to 2^V$ represent an imprecise observation of some attribute $f : S \to V$. Namely, for each object $s \in S$, all that is known about the attribute value $f(s)$ is that it belongs to the set $\Phi(s) \subseteq V$. Suppose we want to describe the set $f^{-1}(T)$ of objects that satisfy a property T, namely $\{s \in S : f(s) \in S \subset V\}$. Because of the incompleteness of the information, the subset $f^{-1}(T) \subseteq S$ is an "ill-known set" [20]. In other words, $f^{-1}(T)$ can be approximated from above and from below, respectively by upper and lower inverses of A via Φ:

- $\Phi^*(T) = \{s \in S \text{ s.t. } \Phi(s) \cap A \neq \emptyset\}$ is the set of objects that possibly belong to $f^{-1}(T)$.
- $\Phi_*(T) = \{s \in S \text{ s.t. } \Phi(s) \subseteq A\}$ is the set of objects that surely belong to $f^{-1}(T)$.

The pair $(\Phi_*(T), \Phi^*(T))$ is such that $\Phi_*(T) \subseteq f^{-1}(T) \subseteq \Phi^*(T)$ and defines an ill-known set. The multi-valued mappings Φ^* and Φ_* are respectively upper and lower inverses of Φ. Clearly, connectives will not be not truth-functional, since in general, inclusions $\Phi^*(T \cap U) \subset \Phi^*(T) \cap \Phi^*(U)$ and $\Phi_*(T) \cup \Phi_*(T) \subset \Phi_*(T) \cup \Phi_*(U)$ will be strict.

4 Rough Sets and 3-Valued Logic

Another typical example of ill-known set is a rough set. Here, uncertainty takes the form of a partition of the universe S of objects, say $S_1, \dots S_k$. For instance objects are described by an insufficient number of attributes so that some objects have the same description. All that is known about any object in S is which subset of the partition it belongs to. So each subset A of S is only known in terms of its upper and lower approximations, a pair (A_*, A^*) such that

$$A^* = \cup\{S_i, S_i \cap A \neq \emptyset\}$$

and

$$A_* = \cup\{S_i, S_i \subseteq A\}.$$

It is clear that truth-functionality fails again as $(A \cap B)^* \subset A^* \cap B^*$ and $A_* \cup B_* \subset (A \cup B)^*$, in general (e.g. Yao [46]).

4.1 Three-Valued Settings for Rough Sets

However, various authors have tried to capture the essential features of rough sets by means of a three-valued compositional calculus (for instance Banerjee [7, 5], Itturioz [34], etc.). This is due to the existence of several points of view on rough sets, some of which are compatible with a less stringent interpretation. The most standard view is to call *rough relatively to an equivalence relation* R a subset A of S such that $A_* \neq A^*$; on the contrary, a set A such that $A_* = A^*$ is said to be *exact*. The next definition considers rough sets as equivalence classes *of subsets* of S that have the same upper and lower approximations. In this view, two sets A and B such that $A_* = B_*$ and $A^* = B^*$ are considered indistinguishable, and one is led to study nested pairs of exact sets (E, F) with $E \subseteq S$ as primitive objects representing equivalence classes of indistinguishable sets. Note that (E, F) is indeed a pair of upper and lower approximation only if $F \setminus E$ does not contain any singleton of S (since such a singleton can never overlap a subset of S without being included in it). So defining a rough set as any nested pair of exact sets (E, F) is not really faithful to the basic framework.

It is nevertheless tempting to see approximation pairs of subsets as naturally 3-valued entities. The basic justification for this move is the existence of some underlying sets C, D such that (Bonikowski [12]):

$$C^* = A^* \cap B^* \text{ and } D_* = A_* \cup B_* \tag{1}$$

that depend on the original sets A, B. Moreover $C_* = A_* \cap B_*$ and $D^* = A^* \cup B^*$ as well. Banerjee and Chakraborty [7] use

$$C = A \sqcap B = (A \cap B) \cup (A \cap B^* \cap (A \cap B)^{*c})$$

and

$$D = A \sqcup B = (A \cup B) \cap (A \cup B_* \cup (A \cup B)^c_*).$$

Likewise, noticing that $(A^c)_* = (A_*)^c$, an implication $\Rightarrow$ can be defined such that $(E,F) \Rightarrow (E',F')$ holds if and only if $E \subseteq E'$ and $F \subseteq F'$, namely, if $E = A_*$, $F = A^*$, $E' = B_*$, $F' = B^*$, the pair of nested exact sets $(E,F) \Rightarrow (E',F')$ is made of the upper and lower approximations of $((A_*)^c \cup B_*) \cap ((A^*)^c \cup B^*)$. This framework for rough equivalence classes is the one of what Banerjee and Chakraborty [7] call prerough algebra. It is shown to be equivalent to a 3-valued Łukasiewicz algebra by Banerjee [5].

4.2 On the Language-Dependent Definition of Sets

However, it must be noticed that the sets C and D as defined by Banerjee and Chakraborty [7] (also Iturrioz [34]) so as to ensure the validity of equation (1) do not depend on operands A and B only: since C is defined using an upper approximation and D involves a lower approximation, C and D depend on the partition used to define exact sets. In fact Bonikowski [12] shows that the set C is always of the form $(A_* \cap B_*) \cup Y$ where Y is obtained as follows: Let the exact set $(A^* \cap B^*) \setminus (A_* \cap B_*)$ be made of union $S_1 \cup \cdots \cup S_k$ of equivalence classes of objects in S, and consider proper non-empty subsets T_i of $S_i, i = 1 \ldots k$. Then take $Y = T_1 \cup \cdots \cup T_k$. Note that by construction $Y_* = \emptyset$, while $Y^* = S_1 \cup \cdots \cup S_k$. These properties ensure that $C^* = A^* \cap B^*$ while $C_* = A_* \cap B_*$. Besides, this construct makes it clear that no such equivalence class S_i should be a singleton of the original set S (otherwise S_i has no proper non-empty subset, and $Y_* = \emptyset$ may be impossible.)

Rough sets are induced by the existence of several objects that cannot be told apart because of having the same description in a certain language used by an observer. However, subsets of S defined in extension exist independently of whether they can be described exactly or not in this language. The set C, laid bare above, whose upper and lower approximations are $A^* \cap B^*$ and $A_* \cap B_*$ depends on the number of attributes used to describe objects. Moreover, this set is not even uniquely defined. Here lies the questionable assumption: sets A and B are intrinsically independent of the higher level language: they are given subsets of objects that can be defined in extension (perhaps using a lower level more precise language). On the contrary, their upper and lower approximations depend on the higher level language used to describe these sets: the more attributes the finer the descriptions. In other words, pairs of exact sets (A_*,A^*) and (B_*,B^*) are not existing entities, they are mental constructs representing A and B using attributes. They are observer-dependent, while A and B can be viewed as actual subsets. On the contrary the above discussion shows that C and D are not actual entities, as these subsets are observer-dependent as well, and can be chosen arbitrarily to some extent. Adding one attribute will not affect A nor B but it will change the equivalence relation, hence the partition, hence C and D as well. So the algebraic construct leading to a three-valued logic does away with the idea that approximation pairs stem from well-defined intrinsic subsets of the original space, and that logical combinations of such approximation pairs should reflect the corresponding combination of lower level ("objective") entities, that should not be affected by the discrimination power of the observer or the higher level language used to describe the objects : to be objective entities, C and D should be well-defined and depend only on A and B in S.

But then as recalled earlier, truth-functionality is lost, i.e. we cannot exactly represent the combination of subsets of S by the combination of their approximations.

The pre-rough algebras and the corresponding 3-valued logic studied by Banerjee and Chakraborty [7] are tailored for manipulating equivalence classes of subsets of S, all consisting of *all* sets having the same upper and lower approximations, without singling out any of them as being the "real" one in each such equivalence class. More recently, Avron and Konikowska [3] have tried to suggest a more relaxed three-valued setting for rough sets using non-deterministic truth-tables, that accommodate the inclusions $(A \cap B)^* \subset A^* \cap B^*$ and $A_* \cup B_* \subset (A \cup B)^*$, admitting the idea that if an element belongs to both boundaries of two upper approximations, it may or not belong to the boundary of the upper approximation of their intersection.

5 Belnap Four-Valued Logic

Two seminal papers of Belnap [9, 10] propose an approach to reasoning both with incomplete and with inconsistent information. It relies on a set of truth-values forming a bilattice, further studied by scholars like Ginsberg [28] and Fitting [26] (see Konieczny et al. [36] for a recent survey). Belnap logic, considered as a system for reasoning under imperfect information, suffers from the same difficulties as partial logic, and for the same reason. Indeed one may consider this logic as using the three epistemic valuations already considered in the previous sections (*certainly true, certainly false and unknown*), along with an additional one that accounts for epistemic conflicts.

5.1 The Contradiction-Tolerant Setting

Belnap considers an artificial information processor, fed from a variety of sources, and capable of anwering queries on propositions of interest. In this context, inconsistency threatens, all the more so as the information processor is supposed never to subtract information. The basic assumption is that the computer receives information about atomic propositions in a cumulative way from outside sources, each asserting for each atomic proposition whether it is true, false, or being silent about it. The notion of *epistemic set-up* is defined as an assignment, of one of four values denoted $\mathbf{T}, \mathbf{F}, \mathbf{BOTH}, \mathbf{NONE}$, to each atomic proposition $a, b, \ldots$:

1. Assigning **T** to a means the computer has only been told that a is true.
2. Assigning **F** to a means the computer has only been told that a is false.
3. Assigning **BOTH** to a means the computer has been told at least that a is true by one source and false by another.
4. Assigning **NONE** to a means the computer has been told nothing about a.

In view of the previous discussion, the set $\mathbf{4} = \{\mathbf{T}, \mathbf{F}, \mathbf{BOTH}, \mathbf{NONE}\}$ coincides with the power set of $\{0, 1\}$, namely $\mathbf{T} = \{1\}$, $\mathbf{F} = \{0\}$, the encoding of the other values depending on the adopted convention: under Dunn Convention, $\mathbf{NONE} = \emptyset$;

BOTH $= \{0,1\}$. It expresses accumulation of information by sources. This convention uses Boolean necessity degrees, i.e. **BOTH** means $N(0) = N(1) = 1$, **NONE** means $N(0) = N(1) = 0$. According to the terminology of possibility theory, **NONE** $= \{0,1\}$; **BOTH** $= \emptyset$. These subsets represent constraints, i.e., mutually exclusive truth-values, one of which is the right one. **NONE** means $\pi(0) = \pi(1) = 1$, **BOTH** means $\pi(0) = \pi(1) = 0$. Then $\emptyset$ corresponds to no solution.

The approach relies on two orderings in **4**:

- *The information ordering*, $\sqsubset$, such that **NONE** $\sqsubset$ **T** $\sqsubset$ **BOTH**;**NONE** $\sqsubset$ **F** $\sqsubset$ **BOTH**. This ordering reflects the inclusion relation of the sets $\emptyset$, $\{0\}, \{1\}$, and $\{0,1\}$, using Dunn convention. It intends to reflect the amount of (possibly conflicting) data provided by the sources. **NONE** is at the bottom because (to quote) "it gives no information at all". **BOTH** is at the top because (following Belnap) it gives too much information.
- *The logical ordering*, $\prec$, according to which **F** $\prec$ **BOTH** $\prec$ **T** and **F** $\prec$ **NONE** $\prec$ **T** each reflecting the truth-set of Kleene's logic. It corresponds to the idea of "less true than", even if this may sound misleadingly suggesting a confusion with the idea of graded truth. In fact **F** $\prec$ **BOTH** $\prec$ **T** canonically extends the ordering $0 < 1$ to the set **3** of non-empty intervals on $\{0,1\}$, under Dunn convention and **F** $\prec$ **NONE** $\prec$ **T** does the same under possibility degree convention.

Then, connectives of negation, conjunction and disjunction are defined truth-functionally on the bilattice. The set **4** is isomorphic to $2^{\{0,1\}}$ equipped with two lattice structures:

- *the information lattice*, a Scott approximation lattice based on union and intersection of sets of truth-values using Dunn convention. For instance, in this lattice the maximum of **T** and **F** is **BOTH**;
- *the logical lattice*, based on the interval extension of min, max and $1-$ from $\{0,1\}$ to $2^{\{0,1\}} \setminus \{\emptyset\}$ respectively under Dunn Convention (for **BOTH**) and possibility degree convention (for **NONE**).

These logical connectives respect the following constraints:

1. They reduce to classical negation, conjunction and disjunction on $\{\mathbf{T},\mathbf{F}\}$;
2. They are monotonic w.r.t. the information ordering $\sqsubset$;
3. $p \wedge q = q$ if and only if $p \vee q = p$;
4. They satisfy commutativity, associativity of $\vee, \wedge$, De Morgan laws.

For instance, the first property enforces $\neg\mathbf{T} = \mathbf{F}$ and $\neg\mathbf{F} = \mathbf{T}$ and then, the monotonicity requirement forces the negation $\neg$ to be such that $\neg$**BOTH** $=$ **BOTH** and $\neg$**NONE** $=$ **NONE**. It can be shown that the restrictions of all connectives to the subsets $\{\mathbf{T},\mathbf{F},\mathbf{NONE}\}$ and $\{\mathbf{T},\mathbf{F},\mathbf{BOTH}\}$ coincide with Kleene's three-valued truth-tables, encoding **BOTH** and **NONE** as $\frac{1}{2}$. The conjunction and disjunction operations $\vee$ and $\wedge$ exactly correspond to the lattice meet and joint for the logical lattice ordering. In fact, **BOTH** and **NONE** cannot be distinguished by the logical ordering $\prec$ and play symmetric roles in the truth-tables. The major new point is the result of combining conjunctively and disjunctively **BOTH** and **NONE**.

The only possibility left for such combinations is that **BOTH** $\wedge$ **NONE** $=$ **F** and **BOTH** $\vee$ **NONE** $=$ **T**. This looks intuitively surprising but there is no other choice and this is in agreement with the information lattice.

5.2 *Is It How a Computer Should Think ?*

Belnap's calculus is an extension of partial logic to the truth-functional handling of inconsistency. In his paper, Belnap does warn the reader on the fact that the four values are not ontological truth-values but epistemic ones. They are qualifications referring to the state of knowledge of the agent (here the computer). The set-representation of Belnap truth-values after Dunn [24] rather comforts the idea that these are not truth-values. Again, $\{1\}$ is a subset of $\{0,1\}$ while 1 is an element thereof.

Belnaps explicitly claims that the systematic use of the truth-tables of **4** "tells us how the computer should answer questions about complex formulas, based on a set-up representing its 'epistemic state"([9], p. 41). However, since the truth-tables of conjunction and disjunction extend the ones of partial logic so as to include the value **BOTH**, Belnap's logic inherits all difficulties of partial logic regarding the truth-value **NONE**. Moreover, equalities **BOTH** $\wedge$ **BOTH** $=$ **BOTH**, **BOTH** $\vee$ **BOTH** $=$ **BOTH** are hardly acceptable when applied to propositions of the form p and $\neg p$, if it is agreed that these are classical propositional formulas.

Another issue is how to interpret the results **BOTH** $\wedge$ **NONE** $=$ **F** and **BOTH** $\vee$ **NONE** $=$ **T**. One may rely on bipolar reasoning and argumentation to defend that when p is **BOTH** and q is **NONE**, $p \wedge q$ should be **BOTH** $\wedge$ **NONE** $=$ **F**. Suppose there are two sources providing information, say S_1 and S_2. Assume S_1 says p is true and S_2 says it is false. This is why p is **BOTH**. Both sources say nothing about q, so q is **NONE**. So one may consider that S_1 would have nothing to say about $p \wedge q$, but one may legitimately assert that S_2 would say $p \wedge q$ is false. In other words, $p \wedge q$ is **F**: one may say that there is one reason to have $p \wedge q$ false, and no reason to have it true. However, suppose two atomic propositions a and b with $E(a) =$ **BOTH** and $E(b) =$ **NONE**. Then $E(a \wedge b) =$ **F**. But since Belnap negation is such that $E(\neg a) =$ **BOTH** and $E(\neg b) =$ **NONE**, we also get $E(\neg a \wedge b) = E(a \wedge \neg b) = E(\neg a \wedge \neg b) =$ **F**. Hence $E((a \wedge b) \vee (\neg a \wedge b) \vee (a \wedge \neg b) \vee (\neg a \wedge \neg b)) =$ **F** that is, $E(\top) =$ **F** which is hardly acceptable again. See Fox [27] for a related critique.

More recently Avron et al. [2] have reconsidered the problem of a computer collecting and combining information from various sources in a wider framework, where sources may provide information about complex formulas too. The combination of epistemic valuations attached to atoms or formulas is dictated by rules that govern the properties of connectives and their interaction with valuation assignments in a more transparent way than Belnap truth-tables. Various assumptions on the combination strategy and the nature of propositions to be inferred (the possibly true ones or the certainly true ones) lead to recover various more or less strong logics, including Belnap formalism. The proposed setting thus avoids making the confusion between truth-values (that can be Boolean or not in the proposed approach, according to the properties chosen) and epistemic valuations.

6 Interval-Valued Fuzzy Sets

IVFs were introduced by Zadeh [50], along with some other scholars, in the seventies (see [23] for a bibliography), as a natural truth-functional extension of fuzzy sets. Variants of these mathematical objects exist, under various names (vague sets [13] for instance). The IVF calculus has become popular in the fuzzy engineering community of the USA because of many recent publications by Jerry Mendel and his colleagues [39]. This section points out the fact that if intervals of membership grades are interpreted as partial ignorance about precise degrees, the calculus of IVFs suffers from the same flaw as partial logic, and the truth-functional calculus of ill-known sets, of which it is a many-valued extension.

6.1 Definitions

An interval-valued fuzzy set is defined by an interval-valued membership function. Independently, Atanassov [1] introduced the idea of defining a fuzzy set by ascribing a membership function and a non-membership function separately, in such a way that an element cannot have degrees of membership and non-membership that sum up to more than 1. Such a pair was given the misleading name of "Intuitionistic Fuzzy Sets" as it seems to be foreign to intuitionism [23]. It also corresponds to an intuition that differs from the one behind IVFs, although both turned out to be mathematically equivalent notions (e.g. G. Deschrijver, E. Kerre [15]).

An IVF is defined by a mapping F from the universe S to the set of closed intervals in $[0,1]$. Let $F(s) = [F_*(s), F^*(s)]$. The union, intersection and complementation of IVF's are obtained by canonically extending fuzzy set-theoretic operations to interval-valued operands in the sense of interval arithmetic. As such operations are monotonic, this step is mathematically obvious. For instance, the most elementary fuzzy set operations are extended as follows, for conjunction $F \cap G$, disjunction $F \cup G$ and negation F^c, respectively:

$$[F \cap G](s) = [\min(F_*(s), G_*(s)), \min(F^*(s), G^*(s))];$$

$$[F \cup G](s) = [\max(F_*(s), G_*(s)), \max(F^*(s), G^*(s))];$$

$$F^c(s) = [1 - F^*(s), 1 - F_*(s)].$$

Considering IVFs as a calculus of intervals on $[0,1]$ equipped with such operations, they are a special case of L-fuzzy sets in the sense of Goguen [31], so as mathematical objects, they are not of special interest. An IVF is also a special case of type two fuzzy set (also introduced by Zadeh [49]). Of course all connectives of fuzzy set theory were extended to interval-valued fuzzy sets and their clones. IFVs are being studied as specific abstract algebraic structures [16], and a multiple-valued logic was recently proposed for them, called the triangle logic [43].

6.2 The Paradox of Truth-Functional Interval-Valued Connectives

Paradoxes of IVFs are less blatant than those of Kleene and Łukasiewicz three-valued logics (when the third truth-value refers to ideas of incomplete knowledge) because in the latter case, the lack of excluded-middle law on Boolean propositions is a striking anomalous feature. In the case of fuzzy logic, some laws of classical logic are violated anyway. However, the fact that interval-valued fuzzy sets have a weaker structure than the fuzzy set algebra they extend should act as a warning. Indeed, since fuzzy sets equipped with fixed connectives have a given well-defined structure, this structure should be valid whether the membership grades are known or not.

For instance, the fact that $\min(F(s), F^c(s)) \leq 0.5$ should hold whether $F(s)$ is known or not. This is a weak form of the contradiction law. However, applying the truth-tables of interval-valued fuzzy sets to the case when $F(s) = [0,1]$ (total ignorance) leads to $\min(F(s), 1 - F(s)) = [0,1]$, which means a considerable loss of information. The same feature appears with the weak excluded middle law, where again $\max(F(s), F^c(s)) = [0,1]$ is found, while $\max(F(s), F^c(s)) \geq 0.5$ should hold in any case. More generally, if the truth-value $t(p) = F(s)$ is only known to belong to some subinterval $[a,b]$ of the unit interval, the truth-functional calculus yields $t(p \wedge \neg p) = \min(F(s), 1 - F(s)) \in [\min(a, 1-b), \min(b, 1-a)]$, sometimes not included in $[0, \frac{1}{2}]$.

In fact, treating fuzzy sets with ill-known membership functions as a truth-functional calculus of IFVs is similar to the paradoxical calculus of ill-known sets based on Kleene's three-valued logics, where the third truth value is interpreted as total ignorance. Indeed, as shown above, operations on ill-known sets as well as partial logic are debatably construed as an interval-valued truth-functional extension of Boolean logic that is isomorphic to Kleene logic. Ill-known sets are to classical sets what IVFs are to fuzzy sets.

The basic point is that IVFs lead to a multiple-valued logic where the truth set [0, 1] is turned into the set of intervals on [0, 1], i.e. *intervals are seen as genuine truth-values*. This approach does not address the issue of ill-known membership grades, where the latter are nevertheless supposed to be precise, even if out of reach. Choosing intervals for truth-values is a matter of adopting a new convention for truth, while reasoning about ill-known membership grades does not require a change of the truth set. When reasoning about ill-known membership grades, the truth set remains [0, 1] and truth-values obey the laws of some multiple-valued calculus, while intervals model epistemic states about truth-values, just like elements in Belnap **4**. A logic that reasons about ill-known membership grades cannot be truth-functional. It should handle weighted formulas where the weight is an interval representing our knowledge about the truth-value of the formula, similar to Pavelka's logic [33], Lehmke's weighted fuzzy logic [37]. Then, the algebraic properties of the underlying logic should be exploited as constraints. Interval-weighted formulas are also signed formulas in many-valued logic. Reasoning about ill-known membership grades is then a matter of constraint propagation, especially interval analysis, and not only simple interval arithmetics on connectives. Automated reasoning methods based on signed formulae in multiple-valued logics follow this line and turn inference into optimization problems [32].

6.3 *Reasoning about Ill-Known Truth-Values*

The generic reasoning problem in interval-valued fuzzy logic is of the following form: Given a set of weighted many-valued propositional formulas $\{p_i, [a_i, b_i]), i = 1, \ldots, n\}$, the problem of inferring another proposition p comes down to finding the most narrow interval $[a, b]$ such that $(p, [a, b])$ can be deduced from $\{p_i, [a_i, b_i]), i = 1, \ldots, n\}$. It corresponds to the following optimization problem:

maximize (resp. minimize) $t(p)$ under the constraints $t(p_i) \in [a_i, b_i], i = 1, \ldots, n$.

This problem cannot be solved by a truth-functional interval-valued fuzzy logic. A simpler instance of this problem is the one of finding the membership function of a complex combination of IVFs. It comes down to finding the interval containing the truth-value of a many-valued formula, given intervals containing the truth-values of its atoms. For instance, using the most basic connectives, finding the membership function of $F \cap F^c$ when F is an IVF comes down to solving for each element of the universe of discourse the following problem:

maximize (resp. minimize) $f(x) = \min(x, 1-x)$ under the constraint $x \in [a, b]$.

Since the function f is not monotonic, the solution is obviously not (always) the interval $[\min(a, 1-b), \min(b, 1-a)]$ suggested by IVF connectives, it is as follows:

$$f(x) \in [a, b] \text{ if } b \leq 0.5;$$
$$f(x) \in [\min(a, 1-b), 0.5] \text{ if } a \leq 0.5 \leq b;$$
$$f(x) \in [1-a, 1-b] \text{ if } a \geq 0.5.$$

Only the first and the third case match the IVF connectives solution.

In Łukasiewicz logic, using the bounded sum and linear product connectives, inferring in the interval-valued setting comes down to solving linear programming problems [32]. Especially the condition $F \cap F^c = \emptyset$ is always trivially valid using linear product, even if F is an IFV, since $\max(0, x + (1-x) - 1) = 0$.

6.4 *Type 2 Fuzzy Sets vs. Fuzzy Truth-Values*

The next step beyond interval-valued fuzzy sets is the case of type two fuzzy sets. It is then assumed that the truth value $F(s)$ of element $s \in S$ is changed into a fuzzy set of the unit interval. Generally, it is supposed to be a fuzzy interval on the unit interval, that for clarity we can denote by $\tilde{F}(s)$, with membership function $\mu_{\tilde{F}(s)}$ for each $s \in S$. The rationale for such a notion is again the idea that membership grades to linguistic concepts are generally ill-known, or that several different persons will provide different membership grades. On such a basis connectives for fuzzy sets are extended to type two fuzzy sets using the extension principle [50, 19], for instance using extended versions of min, max and $1 - \cdot$:

$$\mu_{\tilde{F}(s)\cap\tilde{G}(s)}(t) = \sup_{t=\min(t',t'')} \min(\mu_{\tilde{F}(s)}(t'), \mu_{\tilde{G}(s)}(t''))$$

$$\mu_{\tilde{F}(s)\cup\tilde{G}(s)}(t) = \sup_{t=\max(t',t'')} \min(\mu_{\tilde{F}(s)}(t'), \mu_{\tilde{G}(s)}(t''))$$

$$\mu_{\tilde{F}^c(s)}(t) = \mu_{\tilde{F}(s)}(1-t)$$

See [44] for a careful study of connectives for type two fuzzy sets; their results apply as well to the special case of IVFs. An operational setting where this truth-functional calculus makes sense is yet to come.

In fact, this calculus is partially at odds with the most usual interpretation of type two membership grades, namely fuzzy truth-values proposed by Zadeh [49]. It corresponds to a fuzzification of the ill-known attribute situation of section 3.3. Bellman and Zadeh [8] defined the *fuzzy truth-value* of a fuzzy statement "x is F" given that another one, "x is B", is taken for granted. When $B = \{s_0\}$, i.e. "$x = s_0$", the degree of truth of "x is F" is simply $F(s_0)$, the degree of membership of s_0 to the fuzzy set F. More generally, the information on the degree of truth of "x is F" given "x is B" will be described by a fuzzy set $\tau(F;B)$ of the unit interval with membership function:

$$\mu_{\tau(F;B)}(t) = \begin{cases} \sup\{B(s) \mid F(s) = t\}, \text{ if } F^{-1}(t) \neq \emptyset \\ 0, \quad \text{otherwise} \end{cases} \tag{2}$$

for all $t \in [0,1]$. As can be checked, $\tau(S;D)$ is a fuzzy subset of truth-values and $\mu_{\tau(F;B)}(t)$ is the degree of possibility, according to the available information B, that there exists an interpretation that makes "x is F" true at degree t.

We can apply this approach to interpret type two fuzzy sets as stemming from an ill-known attribute f described by means of a *fuzzy* mapping $\Phi : S \to V$ such that $\Phi(s)$ is a fuzzy subset of possible values of the actual attribute $f(s)$. The degree $F(s)$ to which an element $s \in S$ satisfies a prescribed fuzzy property P_F defined on V is ill-known and can be represented by a fuzzy membership grade $\tilde{F}(s) = \tau(P_F, \Phi(s))$. Again, it will not be possible to apply the truth-functional calculus of type two fuzzy sets to this case where membership grades are ill-known. Generally, considering the case of two fuzzy properties P_F, P_G on V, the fuzzy truth-value $\tau(P_F \cap P_G, \Phi(s))$ is not a function of $\tau(P_F, \Phi(s))$ and $\tau(P_G, \Phi(s))$; $\tau(P_F \cup P_G, \Phi(s))$ is not a function of $\tau(P_F, \Phi(s))$ and $\tau(P_G, \Phi(s))$. This lack of compositionality is one more proof that fuzzy truth-values are not full-fledged truth-values in the sense of a compositional many-valued logic.

7 Conclusion

In conclusion, there is a pervasive confusion between truth-values and the epistemic valuations an agent may use to describe a state of knowledge: the former are compositional by assumption, the latter cannot be consistently so. This paper suggests that such difficulties appear in partial logic, three-valued logics of rough sets, Belnap logic, interval-valued and type two fuzzy logic. In logical approaches to incompleteness and contradiction, the goal of preserving tautologies of the underlying logic (classical or multivalued) should supersede the goal of maintaining a truth-functional setting. Considering subsets or fuzzy subsets of a truth-set as genuine truth-values leads to new many-valued logics that do not address the issue of uncertain reasoning on the underlying original logic. Such "powerset logics" are special cases of lattice-valued logic that need another motivation than reasoning under uncertainty. Our critique encompasses the truth-functional calculus of type two fuzzy sets [39] as well, since it again considers fuzzy sets of truth-values as truth-values. In that respect, the meaning of "fuzzy truth-values" proposed in [49] is sometimes misunderstood, as they cannot be at the same time genuine truth-values and ill-known ones.

References

1. Atanassov, K.: Intuitionistic fuzzy sets. Fuzzy Sets & Syst. 20, 87–96 (1986)
2. Avron, A., Ben-Naim, J., Konikowska, B.: Processing Information from a Set of Sources. In: Makinson, D., Malinowski, J., Wansing, H. (eds.) Towards Mathematical Philosophy, vol. 28, pp. 165–185 (2009)
3. Avron, A., Konikowska, B.: Rough Sets and 3-Valued Logics. Studia Logica 90(1), 69–92 (2008)
4. Avron, A., Lev, I.: Non-Deterministic Multiple-valued Structures. Journal of Logic and Computation 15, 241–261 (2005)
5. Banerjee, M.: Rough Sets and 3-Valued Łukasiewicz Logic. Fundamenta Informaticae 31(3/4), 213–220 (1997)
6. Banerjee, M., Dubois, D.: A simple modal logic for reasoning about revealed beliefs. In: ECSQARU 2009. LNCS (LNAI), vol. 5590, pp. 805–816. Springer, Heidelberg (2009)
7. Banerjee, M., Chakraborty, M.K.: Rough Sets Through Algebraic Logic. Fundamenta Informaticae 28(3-4), 211–221 (1996)
8. Bellman, R.E., Zadeh, L.A.: Local and fuzzy logics. In: Dunn, J.M., Epstein, G. (eds.) Modern Uses of Multiple-Valued Logic, pp. 103–165. D. Reidel, Dordrecht (1977)
9. Belnap, N.D.: How a computer should think. In: Ryle, G. (ed.) Contemporary Aspects of Philosophy, pp. 30–56. Oriel Press (1977)
10. Belnap, N.D.: A useful four-valued logic. In: Dunn, J.M., Epstein, G. (eds.) Modern Uses of Multiple-Valued Logic. D.Reidel, Dordrecht (1977)
11. Blamey, S.: Partial Logic. In: Handbook of Philosophical Logic, vol. 3, pp. 1–70. D. Reidel, Dordrecht (1985)
12. Bonikowski, Z.: A Certain Conception of the Calculus of Rough Sets. Notre Dame Journal of Formal Logic 33(3), 412–421 (1992)
13. Bustince, H., Burillo, P.: Vague sets are intuitionistic fuzzy sets. Fuzzy Sets & Syst. 79, 403–405 (1996)
14. De Cooman, G.: From possibilistic information to Kleene's strong multi-valued logics. In: Dubois, D., et al. (eds.) Fuzzy Sets, Logics and Reasoning about Knowledge. Kluwer Academic Publishers, Dordrecht (1999)
15. Deschrijver, G., Kerre, E.: On the relationship between some extensions of fuzzy set theory. Fuzzy Sets and Syst. 133, 227–235 (2004)

16. Deschrijver, G., Kerre, E.: Advances and challenges in interval-valued fuzzy logic. Fuzzy Sets & Syst. 157, 622–627 (2006)
17. Dubois, D., Hájek, P., Prade, H.: Knowledge-Driven versus data-driven logics. Journal of Logic, Language, and Information 9, 65–89 (2000)
18. Dubois, D.: On ignorance and contradiction considered as truth-values. Logic Journal of IGPL 16(2), 195–216 (2008)
19. Dubois, D., Prade, H.: Operations in a fuzzy-valued logic. Information & Control 43(2), 224–240 (1979)
20. Dubois, D., Prade, H.: Twofold fuzzy sets and rough sets - Some issues in knowledge representation. Fuzzy Sets & Syst. 23, 3–18 (1987)
21. Dubois, D., Prade, H.: Can we enforce full compositionality in uncertainty calculi? In: Proc. Nat. Conference on Artificial Intelligence (AAAI 1994), Seattle, WA, pp. 149–154 (1994)
22. Dubois, D., Prade, H.: Possibility theory, probability theory and multiple-valued logics: A clarification. Ann. Math. and AI 32, 35–66 (2001)
23. Dubois, D., Gottwald, S., Hajek, P., Kacprzyk, J., Prade, H.: Terminological difficulties in fuzzy set theory - The case of Intuitionistic Fuzzy Sets. Fuzzy Sets & Syst. 156, 485–491 (2005)
24. Dunn, J.M.: Intuitive semantics for first-degree entailment and coupled trees. Philosophical Studies 29, 149–168 (1976)
25. Elkan, C.: The paradoxical success of fuzzy logic. In: Proc. Nat. Conference on Artificial Intelligence (AAAI 1993), Washington, DC, pp. 698–703 (1993); Extended version (with discussions): IEEE Expert 9(4), 2–49 (1994)
26. Fitting, M.: Bilattices and the Semantics of Logic Programming. J. Logic Programming 11(1-2), 91–116 (1991)
27. Fox, J.: Motivation and Demotivation of a Four-Valued Logic. Notre Dame Journal of Formal Logic 31(1), 76–80 (1990)
28. Ginsberg, M.L.: Multivalued logics: A uniform approach to inference in artificial intelligence. Computational Intelligence 4(3), 256–316 (1992)
29. Godo, L., Hájek, P., Esteva, F.: A Fuzzy Modal Logic for Belief Functions. Fundamenta Informaticae 57(2-4), 127–146 (2003)
30. Hájek, P., Godo, L., Esteva, F.: Fuzzy logic and probability. In: Proc. 11th Annual Conference on Uncertainty in Artificial Intelligence, Montreal, pp. 237–244. Morgan Kaufmann, San Francisco (1995)
31. Goguen, J.A.: L-fuzzy sets. J. Math. Anal. Appl. 8, 145–174 (1967)
32. Haehnle, R.: Proof Theory of Many-Valued Logic - Linear Optimization - Logic Design: Connections and Interactions. Soft Computing 1, 107–119 (1997)
33. Hájek, P.: The Metamathematics of Fuzzy Logics. Kluwer Academic, Dordrecht (1998)
34. Iturrioz, L.: Rough sets and three-valued structures. In: Orlowska, E. (ed.) Logic at Work, pp. 596–603. Physica-Verlag, Heidelberg (1998)
35. Kleene, S.C.: Introduction to Metamathematics. North Holland, Amsterdam (1952)
36. Konieczny, S., Marquis, P., Besnard, P.: Bipolarity in bilattice logics. Int. J. Intelligent Syst. 23(10), 1046–1061 (2008)
37. Lehmke, S.: Logics which Allow Degrees of Truth and Degrees of Validity. PhD, Universität Dortmund, Germany (2001)
38. Łukasiewicz, J.: Philosophical remarks on many-valued systems of propositional logic (1930). In: Borkowski, J. (ed.) Selected Works, pp. 153–179. North-Holland, Amsterdam (1970)
39. Mendel, J.M.: Advances in type-2 fuzzy sets and systems. Information Sci. 177(1), 84–110 (2007)

40. Thijsse, E.G.C.: Partial Logic and Knowledge Representation. PhD thesis, University of Tilburg, The Netherlands (1992)
41. Urquhart, A.: Many-Valued Logic. In: Gabbay, D.M., Guenthner, F. (eds.) Handbook of Philosophical Logic. Alternatives to Classical Logic, vol. III, pp. 71–116. D.Reidel, Dordrecht (1986)
42. van Fraassen, B.C.: Singular Terms, Truth -value Gaps, and Free Logic. Journal of Philosophy 63, 481–495 (1966)
43. Van Gasse, B., Cornelis, C., Deschrijver, G., Kerre, E.E.: Triangle algebras: A formal logic approach to interval-valued residuated lattices. Fuzzy Sets & Syst. 159, 1042–1060 (2008)
44. Walker, C.L., Walker, E.A.: The algebra of fuzzy truth values. Fuzzy Sets & Syst. 149, 309–347 (2005)
45. Yao, Y.Y.: Interval-Set Algebra for Qualitative Knowledge Representation. In: Proc. Int. Conf. Computing and Information, pp. 370–374 (1993)
46. Yao, Y.Y.: A Comparative Study of Fuzzy Sets and Rough Sets. Information Sci. 109(1-4), 227–242 (1998)
47. Zadeh, L.A.: Fuzzy sets. Information and Control 8, 338–353 (1965)
48. Zadeh, L.A.: Fuzzy sets as a basis for a theory of possibility. Fuzzy Sets & Syst. 1, 3–28 (1978)
49. Zadeh, L.A.: Fuzzy Logic and approximate reasoning. Synthese 30, 407–428 (1975)
50. Zadeh, L.: The concept of a linguistic variable and its application to approximate reasoning-I. Information Sci. 8, 199–249 (1975)

Algorithms for Trapezoidal Approximations of Fuzzy Numbers Preserving the Expected Interval

Przemysław Grzegorzewski

Abstract. Fuzzy number approximation by trapezoidal fuzzy numbers which preserves the expected interval is discussed. New algorithms for calculating the proper approximations are proposed. It is shown that the adequate approximation operator is chosen with respect both to the global spread of a fuzzy number and the size of possible asymmetry between the spread of the left-hand and right-hand part of a fuzzy number.

Keywords: fuzzy numbers, approximation of fuzzy numbers, expected interval, expected value, width, ambiguity, value of a fuzzy number.

1 Introduction

Trapezoidal approximation of fuzzy numbers was considered by many authors (see, e.g. [1]–[4], [7], [14]–[18], [22]–[25]). In [16] a list of criteria which trapezoidal approximation operators should possess was formulated and a new approach to trapezoidal approximation that lead to, so-called, the nearest trapezoidal approximation operator preserving the expected interval was suggested. Then in [17] a corrected version of that operator was given but the ultimate shape of that approximation operator was presented in [4] and [23]. It appears that the form of the nearest trapezoidal approximation operator preserving the expected interval depends on the particular shape of a fuzzy number to be approximated. Actually, a given fuzzy number might be approximated by one of the four admissible approximation operators. Which one

Przemysław Grzegorzewski
Systems Research Institute, Polish Academy of Sciences, Newelska 6, 01-447 Warsaw, Poland
e-mail: pgrzeg@ibspan.waw.pl
and
Faculty of Mathematics and Information Science, Warsaw University of Technology, Plac Politechniki 1, 00-661 Warsaw, Poland
e-mail: pgrzeg@mini.pw.edu.pl

B. Bouchon-Meunier et al. (Eds.) Found. of Reas. under Uncert., STUDFUZZ 249, pp. 85–98.
springerlink.com

should be used depends on parameters that characterize the location and spread of a fuzzy number, i.e. on its value, weighted expected value, ambiguity and width, respectively. These conditions together with natural algorithms for computing the nearest trapezoidal approximation preserving the expected interval were given in [14].

One may ask why the preservation of the expected interval is so exposed. There are many reasons to do so starting from the important properties of the expected interval itself (see, e.g. [10,20]). But there are also some other interesting properties of fuzzy numbers which remain invariant under approximation provided the expected interval remains unchanged (see [16]).

In the present paper we suggest modified conditions for choosing the proper approximation operator. They are both simpler than discussed previously and they have more natural interpretation. One of the suggested algorithms indicates that the adequate approximation operator is chosen with respect both to the global spread of a fuzzy number and the size of possible asymmetry between the spread of the left-hand and right-hand part of a fuzzy number. Moreover, another mathematical formulae for the operators under discussion are proposed.

2 Concepts and Notations

Let A denote a *fuzzy number*, i.e. such fuzzy subset A of the real line $\mathbb{R}$ with membership function $\mu_A : \mathbb{R} \to [0,1]$ which is (see [9]):

- normal (i.e. there exist an element x_0 such that $\mu_A(x_0) = 1$),
- fuzzy convex (i.e. $\mu_A(\lambda x_1 + (1-\lambda)x_2) \geq \mu_A(x_1) \wedge \mu_A(x_2)$, $\forall x_1, x_2 \in \mathbb{R}$, $\forall \lambda \in [0,1]$),
- μ_A is upper semicontinuous,
- supp$A$ is bounded, where supp$A = cl(\{x \in \mathbb{R} : \mu_A(x) > 0\})$, and cl is the closure operator.

A space of all fuzzy numbers will be denoted by $\mathbb{F}(\mathbb{R})$.

Moreover, let $A_\alpha = \{x \in R : \mu_A(x) \geq \alpha\}$, $\alpha \in (0,1]$, denote an α-cut of a fuzzy number A. As it is known, every α-cut of a fuzzy number is a closed interval, i.e. $A_\alpha = [A_L(\alpha), A_U(\alpha)]$, where

$$A_L(\alpha) = \inf\{x \in \mathbb{R} : \mu_A(x) \geq \alpha\} \tag{1}$$

$$A_U(\alpha) = \sup\{x \in \mathbb{R} : \mu_A(x) \geq \alpha\}. \tag{2}$$

The *expected interval* $EI(A)$ of a fuzzy number A is given by (see [10], [20])

$$EI(A) = \left[\int_0^1 A_L(\alpha)d\alpha, \int_0^1 A_U(\alpha)d\alpha\right]. \tag{3}$$

The middle point of the expected interval given by

$$EV(A) = \frac{1}{2}\left(\int_0^1 A_L(\alpha)d\alpha + \int_0^1 A_U(\alpha)d\alpha\right) \tag{4}$$

is called the *expected value* of a fuzzy number and it represents the typical value of the fuzzy number A (see [10], [20]). Sometimes its generalization, called *weighted expected value*, might be interesting. It is defined by

$$EV_q(A) = (1-q)\int_0^1 A_L(\alpha)d\alpha + q\int_0^1 A_U(\alpha)d\alpha, \tag{5}$$

where $q \in [0,1]$ (see [11]).

Another useful parameter characterizing the nonspecifity of a fuzzy number is called the *width* of a fuzzy number (see [6]) and is defined by

$$\begin{aligned} w(A) &= \int_{-\infty}^{\infty} \mu_A(x)dx \qquad (6) \\ &= \int_0^1 (A_U(\alpha) - A_L(\alpha))d\alpha. \end{aligned}$$

To simplify the representation of fuzzy numbers Delgado et al. [7] suggested two parameters – value and ambiguity – which represent some basic features of fuzzy numbers and hence they were called a canonical representation of fuzzy numbers. The first notion

$$Val(A) = \int_0^1 \alpha(A_L(\alpha) + A_U(\alpha))d\alpha \tag{7}$$

is called the *value* of fuzzy number A and might be seen as a point that corresponds to the typical value of the magnitude that the fuzzy number A represents. The next index, called the *ambiguity* is given by

$$Amb(A) = \int_0^1 \alpha(A_U(\alpha) - A_L(\alpha))d\alpha, \tag{8}$$

and it characterizes the global spread of the membership function and hence is a measure of vagueness of fuzzy number A.

For two arbitrary fuzzy numbers A and B with α-cuts $[A_L(\alpha), A_U(\alpha)]$ and $[B_L(\alpha), B_U(\alpha)]$, respectively, the quantity

$$d(A,B) = \sqrt{\int_0^1 (A_L(\alpha) - B_L(\alpha))^2 d\alpha + \int_0^1 (A_U(\alpha) - B_U(\alpha))^2 d\alpha} \tag{9}$$

is the distance between A and B (for more details we refer the reader to [11]). There are, of course, some other measures of the distance between fuzzy numbers (e.g. [5]), however (9) is not only very popular but it seems to be especially useful in relation with the expected interval (see [12]).

3 Trapezoidal Approximation

Suppose we want to substitute a fuzzy number A by a "suitable" trapezoidal fuzzy number $T(A)$, i.e. by a fuzzy number with linear sides and the membership function having the following form

$$\mu_{T(A)}(x) = \begin{cases} 0 & \text{if} \quad x < t_1, \\ \frac{x-t_1}{t_2-t_1} & \text{if} \quad t_1 \le x < t_2, \\ 1 & \text{if} \quad t_2 \le x \le t_3, \\ \frac{t_4-x}{t_4-t_3} & \text{if} \quad t_3 < x \le t_4, \\ 0 & \text{if} \quad t_4 < x. \end{cases} \tag{10}$$

A family of all trapezoidal fuzzy number will be denoted by $\mathbb{F}^T(\mathbb{R})$.

Here one may ask a natural question: *Why are we interested in trapezoidal approximation of fuzzy numbers?*

Besides the immediate answer that it is a quite interesting mathematical problem there are, of course, many practical reasons for such approximations. The most important motivation is to simplify the representation of fuzzy numbers which

- makes calculations easier
- simplifies computer applications
- gives more intuitive and more natural interpretation
- enables the first step of the defuzzification process, i.e.

$$A \in \mathbb{F}(\mathbb{R}) \Rightarrow T(A) \in \mathbb{F}^T(\mathbb{R}) \Rightarrow C(T(A)) \in \mathbb{P}(\mathbb{R}) \Rightarrow W(C(T(A))) \in \mathbb{R} \tag{11}$$

where $\mathbb{P}(\mathbb{R})$ is a family of all intervals, while C and W denote an operator that produces an interval approximation and defuzzification operator, respectively.

If we agree that such trapezoidal approximation is worth of trouble, then the following important question arises: *How to construct an optimal trapezoidal approximation of a fuzzy number?* And here one can suggest many possible solutions. However, as it was motivated in [16], a suitable operator should possess some desired properties and should fulfill some necessary and minimal requirement. For the broad list of such postulated characteristics we refer the reader to [16]. Moreover, in this paper and in some further ones (see, e.g., [17, 4, 23]) is was shown that the approximation operator which guarantees many desired properties can be obtained as the operator T which produces a trapezoidal fuzzy number $T(A)$ that is the closest with respect to distance (9) to given original fuzzy number A among all trapezoidal fuzzy numbers having identical expected interval as the original one. More precisely, we get the following problem.

Problem:

Find such operator $T : \mathbb{F}(\mathbb{R}) \to \mathbb{F}^T(\mathbb{R})$, which minimizes

$$d(A,T(A)) = \sqrt{\int_0^1 (A_\alpha^L - T_\alpha^L(A))^2 d\alpha + \int_0^1 (A_\alpha^U - T_\alpha^U(A))^2 d\alpha} \tag{12}$$

and preserves the expected interval of a fuzzy number, i.e. fulfills the following condition:

$$EI(T(A)) = EI(A). \tag{13}$$

However, since a trapezoidal fuzzy number is completely described by four real numbers that are borders of its support and core, our goal reduces to finding such real numbers $t_1 \le t_2 \le t_3 \le t_4$ that characterize $T(A) = T(t_1,t_2,t_3,t_4)$. Such operator is called *the nearest trapezoidal approximation operator preserving the expected interval* (actually, T is d-nearest trapezoidal approximation operator, where D is given by (9), however further on we call it, in brief, just the nearest one).

The solution of the above mentioned problem was suggested firstly in [16] and later it was improved in [17]. Although operators given in these papers generally produce proper approximations, one can construct such fuzzy number that they do not work correctly and the output is not a trapezoidal fuzzy number. Therefore, Ban [4] and Yeh [23] removed this gap and proposed a final solution containing four possible operators $T_i(A) = T_i(t_1,t_2,t_3,t_4)$, $i = 1,\ldots,4$:

(a) If

$$-\int_0^1 A_L(\alpha)d\alpha + \int_0^1 A_U(\alpha)d\alpha + 3\int_0^1 \alpha A_L(\alpha)\,d\alpha - 3\int_0^1 \alpha A_U(\alpha)\,d\alpha \le 0 \tag{14}$$

then the solution $T_1(A) = T_1(t_1,t_2,t_3,t_4)$ is given by

$$t_1 = 4\int_0^1 A_L(\alpha)\,d\alpha - 6\int_0^1 \alpha A_L(\alpha)\,d\alpha \tag{15}$$

$$t_2 = -2\int_0^1 A_L(\alpha)\,d\alpha + 6\int_0^1 \alpha A_L(\alpha)\,d\alpha \tag{16}$$

$$t_3 = -2\int_0^1 A_U(\alpha)\,d\alpha + 6\int_0^1 \alpha A_U(\alpha)\,d\alpha \tag{17}$$

$$t_4 = 4\int_0^1 A_U(\alpha)\,d\alpha - 6\int_0^1 \alpha A_U(\alpha)\,d\alpha \tag{18}$$

(b) If

$$-\int_0^1 A_L(\alpha)d\alpha + \int_0^1 A_U(\alpha)d\alpha + 3\int_0^1 \alpha A_L(\alpha)\,d\alpha - 3\int_0^1 \alpha A_U(\alpha)\,d\alpha > 0 \tag{19}$$

and

$$2\int_0^1 A_L(\alpha)d\alpha + \int_0^1 A_U(\alpha)d\alpha - 3\int_0^1 \alpha A_L(\alpha)\,d\alpha - 3\int_0^1 \alpha A_U(\alpha)\,d\alpha \le 0 \tag{20}$$

$$-\int_0^1 A_L(\alpha)d\alpha - 2\int_0^1 A_U(\alpha)d\alpha + 3\int_0^1 \alpha A_L(\alpha)\,d\alpha + 3\int_0^1 \alpha A_U(\alpha)\,d\alpha \le 0 \tag{21}$$

then we get $T_2(A) = T_2(t_1, t_2, t_3, t_4)$, where

$$t_1 = 3\int_0^1 A_L(\alpha)d\alpha + \int_0^1 A_U(\alpha)d\alpha - 3\int_0^1 \alpha A_L(\alpha)\, d\alpha - 3\int_0^1 \alpha A_U(\alpha)\, d\alpha \tag{22}$$

$$t_2 = -\int_0^1 A_L(\alpha)d\alpha - \int_0^1 A_U(\alpha)d\alpha + 3\int_0^1 \alpha A_L(\alpha)\, d\alpha + 3\int_0^1 \alpha A_U(\alpha)\, d\alpha \tag{23}$$

$$t_3 = t_2 \tag{24}$$

$$t_4 = \int_0^1 A_L(\alpha)d\alpha + 3\int_0^1 A_U(\alpha)d\alpha - 3\int_0^1 \alpha A_L(\alpha)\, d\alpha - 3\int_0^1 \alpha A_U(\alpha)\, d\alpha \tag{25}$$

(c) If

$$2\int_0^1 A_L(\alpha)d\alpha + \int_0^1 A_U(\alpha)d\alpha - 3\int_0^1 \alpha A_L(\alpha)\, d\alpha - 3\int_0^1 \alpha A_U(\alpha)\, d\alpha > 0 \tag{26}$$

then we get $T_3(A) = T_3(t_1, t_2, t_3, t_4)$ given by

$$t_1 = t_2 = t_3 = \int_0^1 A_L(\alpha)d\alpha \tag{27}$$

$$t_4 = 2\int_0^1 A_U(\alpha)d\alpha - \int_0^1 A_L(\alpha)d\alpha \tag{28}$$

(d) If

$$-\int_0^1 A_L(\alpha)d\alpha - 2\int_0^1 A_U(\alpha)d\alpha + 3\int_0^1 \alpha A_L(\alpha)\, d\alpha + 3\int_0^1 \alpha A_U(\alpha)\, d\alpha > 0 \tag{29}$$

then we obtain $T_4(A) = T_4(t_1, t_2, t_3, t_4)$ such that

$$t_1 = 2\int_0^1 A_L(\alpha)d\alpha - \int_0^1 A_U(\alpha)d\alpha \tag{30}$$

$$t_2 = t_3 = t_4 = \int_0^1 A_U(\alpha)d\alpha. \tag{31}$$

Therefore, we have received four different operators providing the nearest trapezoidal fuzzy number that preserves the expected value of the original fuzzy number, where T_1 leads to trapezoidal (but not triangular) fuzzy number, T_2 stands for the operator that leads to triangular fuzzy number with two sides, while T_3 and T_4 produce triangular fuzzy numbers with the right side only or with the left side only, respectively (note, that in [17] operators T_1 and T_2 were given only).

Which operator should be used in a particular situation depends on a given fuzzy number, i.e. it depends on conditions (14), (19)–(21), (26) or (29) that seem to be very artificial and technical. Hence there was a great need for some further considerations to make these conditions more clear and to find their better interpretation. It was done by Grzegorzewski [14] who simplified the requirements for choosing the proper approximation operators. According to [14] we get the following algorithm for computing the nearest trapezoidal approximation preserving the expected interval.

Algorithm 1

Step 1. If $Amb(A) \geq \frac{1}{3}w(A)$ then apply operator T_1 given by (15)-(18), else
Step 2. if $EV_{\frac{1}{3}}(A) \leq Val(A) \leq EV_{\frac{2}{3}}(A)$ then apply operator T_2 given by (22)-(25), else
Step 3. if $Val(A) < EV_{\frac{1}{3}}(A)$ then apply operator T_3 given by (27)-(28), else
Step 4. apply operator T_4 given by (30)-(31).

As it is seen we approximate a fuzzy number A by the trapezoidal approximation operator T_1 provided ambiguity of this fuzzy number is greater than one third of its width. Otherwise, we approximate A by a triangular number. It means that for less vague fuzzy numbers the solution is always a triangular fuzzy number.

Thus, to sum up, the distinction between possible solutions - either trapezoidal $T_1(A)$ or triangular $T_2(A)$ - depends on the relationship between two parameters of the original fuzzy number that describe its dispersion. In other words, to approximate a fuzzy number A we apply operator T_2 provided A has only slight ambiguity and its typical value is located neither close to the left nor to the right border of its support. However, a fuzzy number with its value *Val* located close to the left border of its support would be approximated by a triangular fuzzy number with the right side only, produced by operator T_3, while a fuzzy number with its value *Val* located close to the right border of its support would be approximated by a triangular fuzzy number with the left side only, produced by operator T_4.

We can also obtain an equivalent algorithm for choosing a proper approximation operator using parameter $\overline{y}(A)$ called the y-coordinate of the centroid point of a fuzzy number A. In [21] the authors showed that

$$\overline{y}(A) = \frac{\int_0^1 \alpha\,(A_U(\alpha) - A_L(\alpha))\,d\alpha}{\int_0^1 (A_U(\alpha) - A_L(\alpha))\,d\alpha}. \tag{32}$$

It is easily seen that

$$\overline{y}(A) = \frac{Amb(A)}{w(A)}. \tag{33}$$

Therefore, we get immediately that our condition (14) is equivalent to the following one

$$\overline{y}(A) \geq \frac{1}{3}. \tag{34}$$

It means that we approximate a fuzzy number A by the trapezoidal approximation operator T_1 if the y-coordinate of the centroid point of A is not smaller than one third. Otherwise, we apply operator T_2 or T_3 or T_4. The consecutive steps for choosing a suitable operator remains as before. Thus we get another algorithms (which are in fact a conjunction of our Algorithm 1 and 2 and the algorithm given in [23]).

Algorithm 2

Step 1. If $\overline{y}(A) \geq \frac{1}{3}$ then apply operator T_1 given by (15)-(18), else
Step 2. if $EV_{\frac{1}{3}}(A) \leq Val(A) \leq EV_{\frac{2}{3}}(A)$ then apply operator T_2 given by (22)-(25), else
Step 3. if $Val(A) < EV_{\frac{1}{3}}(A)$ then apply operator T_3 given by (27)-(28), else
Step 4. apply operator T_4 given by (30)-(31).

4 Discussion and New Algorithms

Although the given above explanation of the conditions that delimits situations corresponding to different approximation operators is correct, yet it sounds slightly insufficient. Especially conditions related to the location parameters do not have clear interpretation. However, it appears that we can propose equivalent conditions which seem to be more satisfactory and more natural.

Since by (5)

$$EV_{\frac{2}{3}}(A) = \frac{1}{3}\int_0^1 A_L(\alpha)d\alpha + \frac{2}{3}\int_0^1 A_U(\alpha)d\alpha$$
$$EV_{\frac{1}{3}}(A) = \frac{2}{3}\int_0^1 A_L(\alpha)d\alpha + \frac{1}{3}\int_0^1 A_U(\alpha)d\alpha$$

then according to (4) and (6) we get

$$EV_{\frac{2}{3}}(A) + EV_{\frac{1}{3}}(A) = \int_0^1 A_L(\alpha)d\alpha + \int_0^1 A_U(\alpha)d\alpha = 2EV(A)$$

and

$$EV_{\frac{2}{3}}(A) - EV_{\frac{1}{3}}(A) = \frac{1}{3}\int_0^1 A_U(\alpha)d\alpha - \frac{1}{3}\int_0^1 A_L(\alpha)d\alpha = \frac{1}{3}w(A).$$

Hence

$$EV_{\frac{2}{3}}(A) = EV(A) + \frac{1}{6}w(A)$$

and

$$EV_{\frac{1}{3}}(A) = EV(A) - \frac{1}{6}w(A).$$

Thus conditions (20)-(21) is equivalent to

$$|EV(A) - Val(A)| \le \frac{1}{6}w(A). \tag{35}$$

Similarly, requirement (26) might be replaced by

$$Val(A) < EV(A) - \frac{1}{6}w(A), \tag{36}$$

while (29) is equivalent to

$$Val(A) > EV(A) + \frac{1}{6}w(A). \tag{37}$$

Thus we get another algorithm.

Algorithm 3

Step 1. If $Amb(A) \ge \frac{1}{3}w(A)$ then apply operator T_1 given by (15)-(18), else
Step 2. if $|EV(A) - Val(A)| \le \frac{1}{6}w(A)$ then apply operator T_2 given by (22)-(25), else
Step 3. if $Val(A) > EV(A) + \frac{1}{6}w(A)$ then apply operator T_4 given by (30)-(31), else
Step 4. apply operator T_3 given by (27)-(28).

As it is seen now to find a proper approximation using Algorithm 2 we have to calculate at most 4 parameters, while in previous Algorithm 1 we had 5 parameters.

To emphasize much more that the distinction among operators T_2, T_3 and T_4 is based on the asymmetry of spread of the membership function let us introduce the following notions.

Definition 1. *The left-hand ambiguity of a fuzzy number A with α-cuts* $A_\alpha = [A_L(\alpha), A_U(\alpha)]$ *is defined by*

$$Amb_L(A) = \int_0^1 \alpha[EV(A) - A_L(\alpha)]d\alpha, \tag{38}$$

while the right-hand ambiguity of a fuzzy number A is given by

$$Amb_U(A) = \int_0^1 \alpha[A_U(\alpha) - EV(A)]d\alpha. \tag{39}$$

One may notice that our definition of the left-hand and right-hand ambiguity differs from the definitions proposed in [8] where the center point of the core of a fuzzy number is placed instead of $EV(A)$. Assuming that the expected value of a fuzzy number $EV(A)$ characterizes its typical value the left-hand and right-hand ambiguity describe the spread of the the left-hand and right-hand part of a fuzzy number, respectively. Moreover, as it is easily seen, both characteristics give the total spread of a fuzzy number, i.e.

$$Amb_L(A) + Amb_U(A) = Amb(A). \tag{40}$$

For our further considerations the following notion would be useful.

Definition 2. *The difference between the left-hand and right-hand ambiguity of a fuzzy number A is defined by*

$$\Delta Amb(A) = Amb_U(A) - Amb_L(A) \tag{41}$$

By (7) and (4) we get immediately that

$$\Delta Amb(A) = Val(A) - EV(A). \tag{42}$$

Therefore we get another condition equivalent both to (20)-(21) and (35), i.e.

$$|\Delta Amb(A)| \leq \frac{1}{6}w(A). \tag{43}$$

Similarly, condition

$$\Delta Amb(A) > \frac{1}{6}w(A) \tag{44}$$

is equivalent to (29) and (37), while condition

$$\Delta Amb(A) < -\frac{1}{6}w(A), \tag{45}$$

i.e.

$$Amb_L(A) - Amb_U(A) > \frac{1}{6}w(A) \tag{46}$$

is equivalent to (26) and (36).

Using these requirements we obtain another algorithm for computing the nearest trapezoidal approximation preserving the expected interval.

Algorithm 4

Step 1. If $Amb(A) \geq \frac{1}{3}w(A)$ then apply operator T_1 given by (15)-(18), else

Step 2. if $|\Delta Amb(A)| \leq \frac{1}{6}w(A)$ then apply operator T_2 given by (22)-(25), else

Step 3. if $\Delta Amb(A) > \frac{1}{6}w(A)$ then apply operator T_4 given by (30)-(31), else

Step 4. apply operator T_3 given by (27)-(28).

It is worth noticing that Algorithm 4 utilizes 3 parameters only and a final decision for the proper choice of the approximation operator depends both on the global spread of a fuzzy number and the size of possible asymmetry between the spread of the left-hand and right-hand part of a fuzzy number.

Last of all let us notice that we may express formulae describing operators $T_1 - T_4$ using suitable parameters describing fuzzy numbers instead of relevant integrals given in Section 3. In particular we obtain a very natural formulae especially for operators T_3 and T_4, namely:

- $T_1(A) = T_1(t_1, t_2, t_3, t_4)$, where

$$t_1 = EV(A) - 2w(A) + 6Amb_L(A) \tag{47}$$
$$t_2 = EV(A) + w(A) - 6Amb_L(A) \tag{48}$$
$$t_3 = EV(A) - w(A) + 6Amb_U(A) \tag{49}$$
$$t_4 = EV(A) + 2w(A) - 6Amb_U(A). \tag{50}$$

- $T_2(A) = T_2(t_1, t_2, t_3, t_4)$, where

$$t_1 = EV(A) - w(A) - 3\Delta Amb(A) \tag{51}$$
$$t_2 = t_3 = EV(A) - \Delta Amb(A) \tag{52}$$
$$t_4 = EV(A) + w(A) - 3\Delta Amb(A) \tag{53}$$

- $T_3(A) = T_3(t_1, t_2, t_3, t_4)$ is given by

$$t_1 = t_2 = t_3 = EV(A) - \frac{1}{2}w(A) \tag{54}$$
$$t_4 = EV(A) + \frac{3}{2}w(A) \tag{55}$$

- $T_4(A) = T_4(t_1, t_2, t_3, t_4)$ is given by

$$t_1 = EV(A) - \frac{3}{2}w(A) \tag{56}$$
$$t_2 = t_3 = t_4 = EV(A) + \frac{1}{2}w(A). \tag{57}$$

As it is seen, in all cases the crucial point of the trapezoidal fuzzy number obtained as the approximation is the expected value the original fuzzy number which is invariant under approximation (see [14]). Then all points $t_1 - t_4$ that describe the trapezoidal fuzzy number are obtained by adding or subtracting some multiplicities of different measures of spread of the original fuzzy number.

5 Properties

Before we discuss the properties of our trapezoidal approximation operators let us notice that we can consider the family $\mathbb{F}(\mathbb{R})$ of all fuzzy numbers as a union of four subfamilies $\mathbb{F}_i(\mathbb{R})$ corresponding to different approximation operators to be used. Namely, we may say that a fuzzy number A belongs to subfamily $\mathbb{F}_i(\mathbb{R})$ if and only if T_i $(i = 1, \ldots, 4)$ is an appropriate operator that should be used for getting a proper trapezoidal approximation. Thus, according to the previous sections we can prove the following lemmas (which are extended versions of those given in [14]).

Lemma 1. *The following conditions are equivalent:*

(a) $A \in \mathbb{F}_1(\mathbb{R})$,
(b) *condition (14) holds,*
(c) $Amb(A) \geq \frac{1}{3}w(A)$,
(d) $\overline{y}(A) \geq \frac{1}{3}$.

Lemma 2. *The following conditions are equivalent:*

(a) $A \in \mathbb{F}_2(\mathbb{R})$,
(b) *conditions (19), (20) and (21) hold,*
(c) $Amb(A) < \frac{1}{3}w(A)$ *and* $EV_{\frac{1}{3}}(A) \leq Val(A) \leq EV_{\frac{2}{3}}(A)$,
(d) $\overline{y}(A) < \frac{1}{3}$ *and* $EV_{\frac{1}{3}}(A) \leq Val(A) \leq EV_{\frac{2}{3}}(A)$,
(e) $Amb(A) < \frac{1}{3}w(A)$ *and* $|EV(A) - Val(A)| \leq \frac{1}{6}w(A)$,
(f) $Amb(A) < \frac{1}{3}w(A)$ *and* $|\Delta Amb(A)| \leq \frac{1}{6}w(A)$.

Lemma 3. *The following conditions are equivalent:*

(a) $A \in \mathbb{F}_3(\mathbb{R})$,
(b) *condition (26) holds,*
(c) $Val(A) < EV_{\frac{1}{3}}(A)$,
(d) $Val(A) < EV(A) + \frac{1}{6}w(A)$,
(e) $\Delta Amb(A) < \frac{1}{6}w(A)$.

Lemma 4. *The following conditions are equivalent:*

(a) $A \in \mathbb{F}_4(\mathbb{R})$,
(b) *condition (29) holds,*
(c) $Val(A) > EV_{\frac{2}{3}}(A)$,
(d) $Val(A) > EV(A) + \frac{1}{6}w(A)$,
(e) $\Delta Amb(A) > \frac{1}{6}w(A)$.

One may notice that subfamilies $\mathbb{F}_1(\mathbb{R}), \ldots, \mathbb{F}_4(\mathbb{R})$ form a partition of a family of all fuzzy numbers $\mathbb{F}(\mathbb{R})$. Actually, by lemmas given above, we may conclude immediately that

$$\mathbb{F}_1(\mathbb{R}) \cup \ldots \cup \mathbb{F}_4(\mathbb{R}) = \mathbb{F}(\mathbb{R}) \tag{58}$$

and

$$\mathbb{F}_i(\mathbb{R}) \cap \mathbb{F}_j(\mathbb{R}) = \emptyset \quad \text{for} \quad i \neq j. \tag{59}$$

Introducing this useful notation we can now turn back to properties of the trapezoidal approximation operators. It can be shown that for $A \in \mathbb{F}_i(\mathbb{R})$ the nearest trapezoidal approximation operator T_i, $i = 1, \ldots, 4$, preserving expected interval is invariant to translations and scale invariant, is monotonic and fulfills identity criterion, preserves the expected value and the weighted expected value and fulfills the nearness criterion with respect to metric (9) in the subfamily of all trapezoidal fuzzy numbers with fixed expected interval. Moreover, it is continuous and compatible with the extension principle, is order invariant with respect to some preference fuzzy relations, is correlation invariant and it preserves the width. For more details we refer the reader to [4], [16] and [17]. It has been also shown ([14]) that T_1 and T_2 preserve the value of a fuzzy number, while $Val(T_3(A)) < Val(A)$ and $Val(T_4(A)) > Val(A)$. However the ambiguity is preserved only by T_1 and $Amb(T_i(A)) > Amb(A)$ for $i = 2,3,4$.

6 Conclusions

In the present contribution we have continued the discussion on the problem of trapezoidal approximation of fuzzy numbers showing another algorithms for computing the proper nearest trapezoidal approximation preserving the expected interval. It seems that these new algorithms are simpler and have more clear interpretation than the algorithms proposed before. Especially Algorithm 4 is the most concise and shows that the choice of the adequate approximation operator depends both on the global spread of a fuzzy number and the size of possible asymmetry between the spread of the left-hand and right-hand part of the original fuzzy number.

Trapezoidal approximation, of course, is not the only possible way for simplifying the shape of the membership function of the fuzzy numbers under study. Even greater simplification can be obtained through the interval approximation. The readers interested in this approach are referred to [6, 12, 13]. On the other hand one may need a nonlinear approximation. Such attempt was proposed in [19].

References

1. Abbasbandy, S., Asady, B.: The nearest approximation of a fuzzy quantity in parametric form. Applied Mathematics and Computation 172, 624–632 (2006)
2. Abbasbandy, S., Amirfakhrian, M.: The nearest trapezoidal form of a generalized LR fuzzy number. International Journal of Approximate Reasoning 43, 166–178 (2006)
3. Allahviranloo, T., Firozja, M.A.: Note on trapezoidal approximations of fuzzy numbers. Fuzzy Sets and Systems 158, 755–756 (2007)
4. Ban, A.: Approximation of fuzzy numbers by trapezoidal fuzzy numbers preserving the expected interval. Fuzzy Sets and Systems 159, 1327–1344 (2008)
5. Bertoluzza, C., Corral, N., Salas, A.: On a new class of distances between fuzzy numbers. Mathware and Soft Computing 2, 71–84 (1995)
6. Chanas, S.: On the interval approximation of a fuzzy number. Fuzzy Sets and Systems 122, 353–356 (2001)
7. Delgado, M., Vila, M.A., Voxman, W.: On a canonical representation of a fuzzy number. Fuzzy Sets and Systems 93, 125–135 (1998)
8. Delgado, M., Vila, M.A., Voxman, W.: A fuzziness measure for fuzzy number: applications. Fuzzy Sets and Systems 94, 205–216 (1998)
9. Dubois, D., Prade, H.: Operations on fuzzy numbers. Int. J. Syst. Sci. 9, 613–626 (1978)
10. Dubois, D., Prade, H.: The mean value of a fuzzy number. Fuzzy Sets and Systems 24, 279–300 (1987)

11. Grzegorzewski, P.: Metrics and orders in space of fuzzy numbers. Fuzzy Sets and Systems 97, 83–94 (1998)
12. Grzegorzewski, P.: Nearest interval approximation of a fuzzy number. Fuzzy Sets and Systems 130, 321–330 (2002)
13. Grzegorzewski, P.: Approximation of a Fuzzy Number Preserving Entropy-Like Nonspecifity. Operations Research and Decisions 4, 49–59 (2003)
14. Grzegorzewski, P.: Trapezoidal approximations of fuzzy numbers preserving the expected interval - algorithms and properties. Fuzzy Sets and Systems 159, 1354–1364 (2008)
15. Grzegorzewski, P.: New algorithms for trapezoidal approximation of fuzzy numbers preserving the expected interval. In: Magdalena, L., Ojeda-Aciego, M., Verdegay, J.L. (eds.) Proceedings of the Twelfth International Conference on Information Processing and Management of Uncertainty in Knowledge-Based Systems, IPMU 2008, Spain, Torremolinos, Málaga, June 22-27, pp. 117–123 (2008)
16. Grzegorzewski, P., Mrówka, E.: Trapezoidal approximations of fuzzy numbers. Fuzzy Sets and Systems 153, 115–135 (2005)
17. Grzegorzewski, P., Mrówka, E.: Trapezoidal approximations of fuzzy numbers - revisited. Fuzzy Sets and Systems 158, 757–768 (2007)
18. Grzegorzewski, P., Pasternak-Winiarska, K.: Weighted trapezoidal approximations of fuzzy numbers. In: Proceedings of IFSA World congress and Eusflat Conference IFSA/Eusflat 2009, Lisbon, Portugal, July 20-24, pp. 1531–1534 (2009)
19. Grzegorzewski, P., Stefanini, L.: Non-linear shaped approximation of fuzzy numbers. In: Proceedings of IFSA World congress and Eusflat Conference IFSA/Eusflat 2009, Lisbon, Portugal, July 20-24, pp. 1535–1540 (2009)
20. Heilpern, S.: The expected value of a fuzzy number. Fuzzy Sets and Systems 47, 81–86 (1992)
21. Wang, Y.M., Yang, J.B., Xu, D.L., Chin, K.S.: On the centroids of fuzzy numbers. Fuzzy Sets and Systems 157, 919–926 (2006)
22. Yeh, C.T.: A note on trapezoidal approximations of fuzzy numbers. Fuzzy Sets and Systems 158, 747–754 (2007)
23. Yeh, C.T.: Trapezoidal and triangular approximations preserving the expected interval. Fuzzy Sets and Systems 159, 1345–1353 (2008)
24. Yeh, C.T.: On improving trapezoidal and triangular approximations of fuzzy numbers. International Journal of Approximate Reasoning 48, 297–313 (2008)
25. Zeng, W., Li, H.: Weighted triangular approximation of fuzzy numbers. International Journal of Approximate Reasoning 46, 137–150 (2007)

Possibilistic Similarity Measures

Ilyes Jenhani, Salem Benferhat, and Zied Elouedi

Abstract. This paper investigates the problem of measuring the similarity degree between two normalized possibility distributions encoding preferences or uncertain knowledge. In a first part, basic natural properties of such similarity measures are proposed. Then a survey of the existing possibilistic similarity indexes is presented and in particular, we analyze which existing similarity measure satisfies the set of basic properties. The second part of this paper goes one step further and provides a set of extended properties that any similarity relation should satisfy. Finally, some definitions of possibilistic similarity measures that involve inconsistency degrees between possibility distributions are discussed.

1 Introduction

The concept of similarity is a very hot topic for many research fields and the comparison of objects represents a fundamental task in many real-world application areas such as, decision making, medicine, meteorology, psychology, molecular biology, data mining, case-based reasoning, etc.

Uncertainty and imprecision are often inherent in modeling knowledge for most real-world problems including the above mentioned areas. Uncertainty about values of given variables (e.g. the *source* of a car breakdown, the *temperature* of a patient,

Ilyes Jenhani
LARODEC, Institut Supérieur de Gestion de Tunis, Tunisia,
CRIL, Université d'Artois, Lens, France
e-mail: ilyes.j@lycos.com
Salem Benferhat
CRIL, Université d'Artois, Lens, France
e-mail: benferhat@cril.univ-artois.fr
Zied Elouedi
LARODEC, Institut Supérieur de Gestion de Tunis, Tunisia
e-mail: zied.elouedi@gmx.fr

B. Bouchon-Meunier et al. (Eds.) Found. of Reas. under Uncert., STUDFUZZ 249, pp. 99–123.
springerlink.com

the *property_value* of a client asking for a loan, etc.) can result from some errors and hence from non-reliability (in the case of experimental measures) or from different background knowledge. As a consequence, it is possible to obtain different uncertain pieces of information about a given value from different sources. Obviously, comparing these pieces of information could be of a great interest and should be accomplished effectively in order to solve many problems in the presence of imperfect data, like in fusion, clustering, classification, etc.

Comparing pieces of imperfect information has attracted a lot of attention in probability theory [4,21], in belief function theory [13,19,27,32], in fuzzy set theory (e.g. [2,3,14,29]), in credal set theory (e.g. [1]), etc. In particular, to compare probability distributions, one can use the well-known Minkowski (and its derivatives, i.e., Euclidean, Manhattan, and Maximum distances) and KL-divergence [21]. Another distance has been proposed by Chan et al. [4] for bounding probabilistic belief change. In belief function theory [24], several distance measures between bodies of evidence deserve to be mentioned. Some distances have been proposed as measures of performance (MOP) of identification algorithms [13,19]. A distance was used for the optimization of the parameters of a belief *k*-nearest neighbor classifier [32]. In [27], authors proposed a distance for the quantification of errors resulting from basic probability assignment approximations. Many similarity/distance measures used in belief function theory go through the pignistic transformation [26] of a basic belief assignment into a probability distribution, then apply probabilistic distances.

Many contributions on measures of similarity between fuzzy sets have already been made [2,3,14,29]. For instance, in the work by Bouchon-Meunier et al. [3], the authors proposed a similarity measure between fuzzy sets as an extension of Tversky's model on crisp sets [28]. The measure was then used to develop an image search engine. In [29], the authors have made a comparison between existing classical similarity measures for fuzzy sets and proposed the sameness degree which is based on fuzzy subsethood and fuzzy implication operators. Moreover, in [2] and [14], the authors have proposed many fuzzy distance measures which are fuzzy versions of classical cardinality-based distances.

Despite the popularity of possibility theory, only few works are dedicated to similarity measures within this framework [16,20,22]. Existing works have provided definitions of possibilistic similarity functions without natural properties of such measures. Moreover, none of the proposed measures have taken into account the concept of inconsistency when measuring possibilistic similarity. In fact, this latter concept plays an important role in evaluating the similarity degree between two possibilistic pieces of information as it will be shown later.

In this paper, we will mainly focus on measures for the comparison of uncertain information represented by normalized possibility distributions. We propose natural properties that these measures should satisfy and provide an inconsistency-based possibilistic similarity function satisfying the proposed axioms. This paper is an extended and a revised version of the conference papers [17] and [18].

The rest of the paper is organized as follows: Section 2 gives necessary backgrounds concerning possibility theory. Section 3 presents a set of natural properties that a possibilistic similarity measure (PSM for short) should satisfy. Section 4 gives a brief overview of existing similarity and distance measures for possibilistic imperfect information. After showing that existing functions do not entirely satisfy the proposed basic properties, we propose, in Section 5, additional properties that take into account the inconsistency degree between possibilistic pieces of information. Section 6 proposes a new PSM that satisfies all the new properties. Finally, Section 7 concludes the paper.

2 Possibility Theory: A Refresher

Possibility theory represents a non-classical uncertainty theory, first introduced by Zadeh [31] and then developed by several authors (e.g. Dubois and Prade [7]). In this section, we will give a brief refresher on possibility theory.

Possibility distribution

Given a universe of discourse $\Omega = \{\omega_1, \omega_2, ..., \omega_n\}$, one of the fundamental concept of possibility theory is the notion of *possibility distribution* denoted by π. π is a function which associates to each element ω_i from the universe of discourse Ω a value from a bounded and linearly ordered valuation set (L,<). This value is called a *possibility degree*: it encodes our knowledge on the real world. Note that, in possibility theory, the scale can be numerical (e.g. L=[0,1]): in this case we have numerical possibility degrees from the interval [0,1] and hence we are dealing with the quantitative setting of the theory. In the qualitative setting, it is the ordering between different possible values that is important. For more details on discussions between quantitative and qualitative possibility theory; see [8]. In the following, the uncertainty scale will be represented by the unit interval [0,1].

By convention, $\pi(\omega_i) = 1$ means that it is fully possible that ω_i is the real world, $\pi(\omega_i) = 0$ means that ω_i cannot be the real world (is impossible). Flexibility is modeled by allowing to give a possibility degree from]0,1[. In possibility theory, extreme cases of knowledge are given by:

- *Complete knowledge*: $\exists \omega_i,\ \pi(\omega_i) = 1$ *and* $\forall\ \omega_j \neq \omega_i,\ \pi(\omega_j) = 0$.
- *Total ignorance*: $\forall\ \omega_i \in \Omega,\ \pi(\omega_i) = 1$ (all values in Ω are possible).

For the sake of simplicity, for the rest of the paper, a possibility distribution π on a finite set $\Omega = \{\omega_1, \omega_2, ..., \omega_n\}$ will be denoted by $\pi[\pi(\omega_1), \pi(\omega_2), ..., \pi(\omega_n)]$.

Possibility and Necessity measures

From a possibility distribution, two dual measures can be derived [7] : *Possibility* and *Necessity* measures. Given a possibility distribution π on the universe of discourse Ω, the corresponding possibility and necessity measures of any event $A \subseteq 2^{\Omega}$ are, respectively, determined by the formulas: $\Pi(A) = \max_{\omega \in A}\ \pi(\omega)$ and $N(A) = \min_{\omega \notin A}\ (1 - \pi(\omega)) = 1 - \Pi(\overline{A})$. $\Pi(A)$ evaluates at which level A is *consistent* with our knowledge represented by π while $N(A)$ evaluates at which level A is *certainly* implied by our knowledge represented by π.

Normalized possibility distributions and inconsistency

A possibility distribution π is said to be *normalized* if there exists at least one state $\omega_i \in \Omega$ which is totally possible (i.e. $\max_{\omega \in \Omega}\{\pi(\omega)\} = \pi(\omega_i)$=1). Otherwise, π is considered as sub-normalized and in this case

$$Inc(\pi) = 1 - \max_{\omega \in \Omega}\{\pi(\omega)\} \tag{1}$$

is called the *inconsistency degree* of π. It is clear that, for normalized π, $\max_{\omega \in \Omega}$ $\{\pi(\omega)\} = 1$, hence Inc(π)=0. The measure *Inc* is very useful in assessing the degree of conflict between two distributions π_1 and π_2 which is given by $Inc(\pi_1, \pi_2) = Inc(\pi_1 \wedge \pi_2)$, where $\wedge$ is a conjunctive t-norm operator (satisfying the axioms of : boundary conditions, monotonicity, commutativity and associativity) [9]. For sake of simplicity, we take the *minimum* and *product* conjunctive ($\wedge$) operators, which have been largely used in a possibility theory framework. Obviously, when $\pi_1 \wedge \pi_2$ gives a sub-normalized possibility distribution, it indicates that there is a conflict between π_1 and π_2 ($Inc(\pi_1, \pi_2) \in]0,1]$). On the other hand, when $\pi_1 \wedge \pi_2$ is normalized, there is no conflict and hence $Inc(\pi_1, \pi_2) = 0$.

Let π_1[0.5,0,1,0.8] and π_2[1,0.7,0.3,1] be two normalized possibility distributions. If we take the minimum as the conjunctive ($\wedge$) operator, we obtain: Inc(π_1,π_2)=Inc([0.5,0,0.3,0.8])=1-0.8=0.2. We can conclude that the two sources are slightly inconsistent with each others.

Non-specificity

Possibility theory is driven by the principle of *minimum specificity*: A possibility distribution π_1 is said to be *more specific than* π_2 if and only if for each state of affairs $\omega_i \in \Omega$, $\pi_1(\omega_i) \leq \pi_2(\omega_i)$ [30]. Clearly, the more specific π, the more informative it is.

Another definition of non-specificity has been introduced in [15]. Given a permutation σ of the degrees of a possibility distribution $\pi[\pi_{\sigma(1)}, \pi_{\sigma(2)}, ..., \pi_{\sigma(n)}]$ such that $\pi_{\sigma(1)} \geq \pi_{\sigma(2)} \geq ... \geq \pi_{\sigma(n)}$, the non-specificity of a possibility distribution π, so-called *U-uncertainty* is given by:

$$U(\pi) = \sum_{i=2}^{n} (\pi_{\sigma(i)} - \pi_{\sigma(i+1)}) \, log_2 \, i + (1 - \pi_{\sigma(1)}) \, log_2 \, n \tag{2}$$

where $\pi_{\sigma(n+1)} = 0$ by convention. Note that if $\forall \omega_i \in \Omega$, $\pi_1(\omega_i) \leq \pi_2(\omega_i)$, then we necessarily have $U(\pi_1) \leq U(\pi_2)$. However, the converse is false. Namely, $U(\pi_1) \leq U(\pi_2)$ does not necessarily imply that $\forall \omega_i \in \Omega$, $\pi_1(\omega_i) \leq \pi_2(\omega_i)$. Indeed, let us consider the following possibility distributions: π_1[1,1,0] and π_2[1,0.5,0.5]. We have U(π_2)=0.792<U(π_1)=1 but it is not true that $\pi_2 < \pi_1$.

3 Basic Natural Properties of a Possibilistic Similarity Measure

The issue of comparing imperfect pieces of information depends on the way these pieces of information are represented. In fact, each kind of imperfection has to be modeled by the appropriate uncertainty theory otherwise, resulting models and results of all operations on that models (fusion, inference, etc.) will not be faithful to the reality. Hence, in the case of possibility theory, comparing uncertain pieces of information comes down to comparing possibility distributions representing these pieces of information. Hence, we need a measure to quantify the amount of similarity between two possibility distributions. A fundamental question that arises is: what are natural properties that such measures should satisfy?

Let π_1 and π_2 be two normalized possibility distributions on the same universe of discourse Ω. Possibility distributions considered in this paper are assumed to be normalized. A possibilistic similarity measure (PSM), denoted by $s(\pi_1, \pi_2)$, should satisfy the following basic properties [17]:

Property 1. Non-negativity
$s(\pi_1, \pi_2) \geq 0$.
Namely, similarity between two possibility distributions should never be negative.

Property 2. Symmetry
$s(\pi_1, \pi_2) = s(\pi_2, \pi_1)$.
This property means that the similarity of π_2 compared to π_1 or π_1 compared to π_2 is strictly the same.

Property 3. Upper bound and Non-degeneracy
$\forall\ \pi_i,\ s(\pi_i, \pi_i) = 1$ and $\forall \pi_i,\ \pi_j,\ s(\pi_i, \pi_j) \leq 1$.
Namely, identity implies full similarity, which is represented by the degree 1.

Property 4. Lower bound
If $\forall \omega_i \in \Omega$,
i) $\pi_1(\omega_i) \in \{0,1\}$, $\pi_2(\omega_i) \in \{0,1\}$ and *ii)* $\pi_2(\omega_i) = 1 - \pi_1(\omega_i)$ then $s(\pi_1, \pi_2)$=0.
Maximally contradictory possibility distributions have the lowest similarity degree. Item *i)* means that π_1 and π_2 should be binary. Since we deal with normalized possibility distributions, items *i)* and *ii)* imply:

iii) $\exists\ \omega_q \in \Omega\ s.t.\ \pi_1(\omega_q) = 1$
iv) $\exists\ \omega_p \in \Omega\ s.t.\ \pi_1(\omega_p) = 0$

The following example illustrates *properties 3* and *4* :

Example 1. Let X be a variable with an unknown value and let $\Omega = \{\omega_1, \omega_2, \omega_3, \omega_4\}$ be the set representing the possible values of X. Let us take the possibility distribution representing the total knowledge on X given by an agent, i.e., $\pi_1[1,0,0,0]$. Thus, a most similar possibility distribution to π_1 according to *Property 3* is $\pi_2[1,0,0,0]$ (the best case) and according to *Property 4*, the least similar possibility distribution to π_1 is $\pi_3[0,1,1,1]$ (the worst case).

Next property says that if we have three information sources: the first one is giving an information π_1 which is more specific than the one given by the second source π_2 which is in turn more specific than the one given by the third source π_3. Moreover, we assume that all sources agree for at least one state. Then, π_1 is closer to π_2 than to π_3.

Property 5. Large inclusion (specificity)
If $Inc(\pi_1,\pi_2) = Inc(\pi_2,\pi_3) = Inc(\pi_1,\pi_3) = 0$ and $\forall \omega_i \in \Omega$, $\pi_1(\omega_i) \leq \pi_2(\omega_i)$ and $\pi_2(\omega_i) \leq \pi_3(\omega_i)$, then $s(\pi_1,\pi_2) \geq s(\pi_1,\pi_3)$.

Property 6 says that the similarity between two possibility distributions π_1 and π_2 remains the same if we change the order of elements having the same indexes in both possibility distributions π_1 and π_2.

Property 6. Permutation
Let π_1 and π_2 be two possibility distributions and let ρ be a permutation of indexes in π_1 and π_2 (leading to $\pi_{\rho(1)}$ and $\pi_{\rho(2)}$).
Then, $\forall\ \pi_1, \pi_2$, $s(\pi_1,\pi_2)=s(\pi_{\rho(1)},\pi_{\rho(2)})$.

Next corollary follows from the permutation property for the case of four possibility distributions. The corollary says that if the similarity between two pieces of uncertain information π_1 and π_2 is greater than the similarity between π_3 and π_4, switching or permuting the elements having the same indexes in π_1 and π_2 and doing the same thing for π_3 and π_4 (but not necessarily the same indexes used with π_1 and π_2), then similarities will remain unchanged. Hence, similarity between π_1 and π_2 is still greater than the similarity between π_3 and π_4. This expresses the fact that elements of each possibility distribution have the same degree of importance.

Corollary 1. Let s be a similarity measure satisfying the permutation property and π_1, π_2, π_3 and π_4 be four possibility distributions and let ρ (resp. ρ') be a permutation of indexes in π_1 and π_2 (resp. in π_3 and π_4).
Then, $\forall\ \pi_1, \pi_2, \pi_3, \pi_4$, $s(\pi_1,\pi_2) \geq s(\pi_3,\pi_4)$ iff $s(\pi_{\rho(1)},\pi_{\rho(2)}) \geq s(\pi_{\rho'(3)},\pi_{\rho'(4)})$.

The proof is immediate since following Property 6, $s(\pi_1,\pi_2) = s(\pi_{\rho(1)},\pi_{\rho(2)})$ and $s(\pi_3,\pi_4) = s(\pi_{\rho'(3)},\pi_{\rho'(4)})$.
The following example illustrates *properties 5-6*:

Example 2. Let X be a variable with an unknown value and let $\Omega = \{\omega_1,\omega_2,\omega_3,\omega_4\}$ be the set representing the possible values of X. Let us consider the following possibility distributions representing the opinions of three sources about the exact value of X : $\pi_1[1,0,0,0]$, $\pi_2[1,0.3,0.2,0.5]$, $\pi_3[1,0.8,0.9,1]$.

We can see that $\forall \omega_i$, $\pi_1(\omega_i) \leq \pi_2(\omega_i) \leq \pi_3(\omega_i)$, which means that π_1 is more specific than π_2 which is also more specific than π_3. In this case, it is expected that π_1 is more similar to π_2 than π_3, namely: $s(\pi_1,\pi_2) > s(\pi_1,\pi_3)$.

Let us now permute, for instance, the first and the third elements in the different distributions. We obtain: $\pi'_1[0,0,1,0]$, $\pi'_2[0.2,0.3,1,0.5]$, $\pi'_3[0.9,0.8,1,1]$. Clearly, we have $\forall \omega_i$, $\pi'_1(\omega_i) \leq \pi'_2(\omega_i) \leq \pi'_3(\omega_i)$. Hence, from Property 5, we can conclude that $s(\pi'_1,\pi'_2) \geq s(\pi'_1,\pi'_3)$.

The above six natural properties can be viewed as basic properties of any possibilistic similarity measure. Nevertheless, they are not satisfied as a whole by several existing PSMs as it will be shown in the next section. Note that there are some other evident properties not mentioned here like transitivity and some other not desirable properties in the possibilistic framework like triangle inequality.

4 Property-Based Analysis of Existing Possibilistic Similarity Measures

Several names are given to comparison measures which often belong to two dual measures: *similarity* measures and *dissimilarity* measures. Terms such as closeness, affinity, resemblance, etc. can be considered as expressing the degree of similarity between objects or pieces of information. Likewise, terms such as distance and divergence can be thought of as expressing dissimilarity. We can find many of these words in the literature and generally, they are expressing the same thing with a possibly slight semantic differences.

The choice of the similarity or distance measure is strongly related to the domain value of the objects to compare (real-values, symbolic values, strings, etc.) as well as to the representation of that objects (single value, vector of individual values, sets of values, sets of vectors, etc).

This section reviews some existing possibilistic similarity and distance functions. An example will be given to illustrate each measure and a counter-example will be also provided to show what is (are) the property(ies) violated by that measure. Let us present these measures and show their weaknesses in expressing information similarity between any given two agents (or sensors) who are expressing their opinions (or measures), especially, in the form of possibility distributions.

4.1 Information Closeness

One of the first works, especially dedicated to the problem of measuring information similarity between two possibility distributions was the one of Higashi and Klir in 1983 [16]. Authors have proposed an information variation-based measure which they have called *information closeness* denoted by G. Function G is computed using their U-uncertainty measure [15] (Equation (2)) and it is applicable to any pair of normalized possibility distributions. The less the value of G is, the more the information are similar (G behaves as a distance measure).

Definition 1. *Let π_1 and π_2 be two possibility distributions on the same universe of discourse Ω. The information closeness G between π_1 and π_2 is defined as:*

$$G(\pi_1,\pi_2) = g(\pi_1,\pi_1 \vee \pi_2) + g(\pi_2,\pi_1 \vee \pi_2) \tag{3}$$

where $g(\pi_i,\pi_j) = U(\pi_j) - U(\pi_i)$. $\vee$ is taken as the maximum operator and U is the non-specificity measure given by Equation (2). Consequently, function G can be written as $G(\pi_1,\pi_2) = 2 * U(\pi_1 \vee \pi_2) - U(\pi_1) - U(\pi_2)$.

Example 3. Consider the following distributions π_1, π_2, π_3 and π_4 over $\Omega = \{\omega_1,\ \omega_2,\ \omega_3,\ \omega_4\}$: $\pi_1[1,\ 0.5,\ 0.3,\ 0.7]$, $\pi_2[1,\ 0,\ 0,\ 0]$, $\pi_3[0.9,\ 1,\ 0.3,\ 0.7]$,

$\pi_4[0, 1, 0.3, 0.7]$. Let us compute the order expressing which from the information given by π_2, π_3 and π_4 is closer to π_1. $G(\pi_1, \pi_2) = 1.12$, $G(\pi_1, \pi_3) = 0.52$, $G(\pi_1, \pi_4) = 1.08$. According to G, π_3 is the closest distribution to π_1 and π_4 is closer to π_1 than π_2.

Let π^N be the set of all normalized possibility distributions on Ω.

Let $G_{min} = min_{\pi_i \in \pi^N, \pi_j \in \pi^N}\{G(\pi_i, \pi_j)\}$ and $G_{max} = max_{\pi_i \in \pi^N, \pi_j \in \pi^N}$ $\{G(\pi_i, \pi_j)\}$. One can easily check that $G_{min} = 0$ and $G_{max} = 2 * log_2(|\Omega|) - log_2(|\Omega| - 1)$. G_{min} is obtained when comparing identical possibility distributions whereas G_{max} is obtained when comparing a possibility distribution corresponding to a situation of complete knowledge and its complementary possibility distribution (e.g. $\pi_1[1,0,0,0,0]$ and $\pi_2[0,1,1,1,1]$).

Hence, we can transform G to a similarity measure S_G whose values are in the interval [0,1]:

$$S_G = 1 - \frac{G(\pi_i, \pi_j) - G_{min}}{G_{max}}$$

The similarity measure S_G does not satisfy *Property 4* as it is illustrated by the following counter-example.

Counter-example 1. *Let us consider the following possibility distributions:* $\pi_1[1, 0, 0, 0]$, $\pi_2[0, 1, 1, 1]$, $\pi_3[0, 1, 0, 1]$ *and* $\pi_4[1, 0, 1, 0]$. *Clearly,* $\pi_1 = 1 - \pi_2$ *and* $\pi_3 = 1 - \pi_4$.

Hence, S_G *should take its minimum value when comparing* π_1 *and* π_2 *as well as* π_3 *and* π_4. *Nevertheless, according to* S_G, *we obtain:*

$$S_G(\pi_1, \pi_2) = 1 - \frac{2*log_2(4) - log_2(3)}{2*log_2(4) - log_2(3)} = 1 - 1 = 0$$
$$S_G(\pi_3, \pi_4) = 1 - \frac{2*log_2(4) - 2*log_2(2)}{2*log_2(4) - log_2(3)} = 1 - \frac{2}{2.41} = 0.17.$$

It means that π_3 *and* π_4 *are more similar to each other than* π_1 *and* π_2 *are, which is contrary to what is expected:* $S_G(\pi_3, \pi_4)$ *should be minimal and equal to* $S_G(\pi_1, \pi_2)$.

4.2 *Sangüesa et al. Distance*

In a work by Sangüesa et al. [23] focusing on learning possibilistic causal networks, the authors have proposed a modified version of a distance measure [22] between two possibility distributions for DAG (Directed Acyclic Graph) learning and evaluation. This is done by measuring the distance (which must be minimized) between the possibility distribution implied by a DAG and the one underlying the database. This idea is based on the interpretation of independence as information similarity.

Definition 2. *Given two possibility distributions* π_1 *and* π_2 *on the same universe of discourse* Ω. *The distance between* π_1 *and* π_2 *is defined as the non-specificity of the distribution difference*

$$distance(\pi_1, \pi_2) = U(\pi_d) \tag{4}$$

where $\pi_d(\omega) = |\pi_1(\omega) - \pi_2(\omega)|$ *for each* $\omega \in \Omega$.

Example 4. If we take the same distributions π_1, π_2, π_3 and π_4 of Example 3, we obtain:

$distance(\pi_1,\pi_2) = U([0,0.5,0.3,0.7]) = 1.27,$
$distance(\pi_1,\pi_3) = U([0.1,0.5,0,0]) = 1.1,$
$distance(\pi_1,\pi_4) = U([1,0.5,0,0]) = 0.5.$

Hence according to this measure, π_2 remains the farthest but π_4 becomes the closest to π_1.

Let $distance_{min} = min_{\pi_i \in \pi^N, \pi_j \in \pi^N}\{distance(\pi_i,\pi_j)\}$ and $distance_{max} = max_{\pi_i \in \pi^N, \pi_j \in \pi^N}\{distance(\pi_i,\pi_j)\}$. One can check that $distance_{min} = 0$ and $distance_{max} = log_2(|\Omega|)$.

Hence, we can transform *distance* to a similarity measure $S_{distance}$ whose values are in the interval [0,1]:

$$S_{distance} = 1 - \frac{distance(\pi_i,\pi_j) - distance_{min}}{distance_{max}}$$

$S_{distance}$ is not satisfactory when the distribution difference ($|\pi_1 - \pi_2|$) is sub-normalized (which occurs most of the time). In this case the second term of Equation (2) will be considered. Note that measuring the non-specificity of a sub-normalized distribution π comes down to measure the non-specificity of its normalized distribution π' s.t $\pi'(\omega_i) = \pi(\omega_i) + (1 - max_{\omega\in\Omega}\{\pi(\omega)\})$. Obviously, this normalization scheme is not suited for the proposed similarity function. The following counter-example emphasizes this weakness.

Counter-example 2. *Let us consider the following possibility distributions:*

$\pi_1[1,\ 0,\ 0,\ 0]$, $\pi_2[1,\ 0,\ 0,\ 0]$, $\pi_3[0,\ 1,\ 1,\ 1]$, $\pi_4[1,\ 1,\ 0,\ 0]$.

Clearly, π_2 is the closest possible distribution to π_1 (the best case) while π_3 is the farthest distribution (the worst case). Nevertheless, the results provided by this distance measure do not correspond to these intuitions:

$S_{distance}(\pi_1,\pi_2) = 1 - \frac{2-0}{2} = 0\ (minimum)$
$S_{distance}(\pi_1,\pi_3) = 1 - \frac{2-0}{2} = 0\ (minimum)$
$S_{distance}(\pi_1,\pi_4) = 1 - \frac{0-0}{2} = 1\ (maximum)$

Hence, according to this measure, π_1 and π_2 are minimally similar to each other. This violates Property 3 because $\pi_1 = \pi_2$ and the similarity is minimal.

4.3 Information Divergence

A possibilistic analogy to the probabilistic measure of divergence was proposed by Kroupa [20]. The author has used the Choquet integral [5] as an aggregation operator of the possibility degrees characterizing the, generally, sub-normalized distribution difference ($\pi_d = |\pi_1(\omega_i) - \pi_2(\omega_i)|$, i=1..n) of any two normalized distributions π_1 and π_2.

Definition 3. *Given two possibility distributions π_1 and π_2 on the same universe of discourse Ω, the measure of divergence $D(\pi_1|\pi_2)$ is defined as the discrete Choquet integral of the degrees of π_d:*

$$D(\pi_1|\pi_2) = \sum_{i=1}^{n} \pi_d(\omega_{\sigma(i)})[\Pi_1(A_{\sigma(i)}) - \Pi_1(A_{\sigma(i+1)})] \tag{5}$$

where σ is a permutation of indexes such that $\pi_d(\omega_{\sigma(i)}) \leq \ldots \leq \pi_d(\omega_{\sigma(n)})$ and $A_{\sigma(i)} = \{\omega_{\sigma(i)}, \ldots, \omega_{\sigma(n)}\}$, i=1..n and $A_{\sigma(n+1)} = \emptyset$.

Example 5. Let us reconsider the same distributions of Example 3. The application of the divergence measure gives:

$D(\pi_1|\pi_2) = 0*(1-0.7)+0.3*(0.7-0.7)+0.5*(0.7-0.7)+0.7*(0.7-0) = 0.49$
$D(\pi_1|\pi_3) = 0*(1-1)+0*(1-1)+0.1*(1-0.5)+0.5*(0.5-0) = 0.3$
$D(\pi_1|\pi_4) = 0*(1-1)+0*(1-1)+0.5*(1-1)+1*(1-0) = 1$

Again, we obtain a different order from Example 3 and Example 4. In fact, according to D, π_3 is the closest to π_1 and π_4 is the farthest.

Let $D_{min} = min_{\pi_i \in \pi^N, \pi_j \in \pi^N}\{D(\pi_i|\pi_j)\}$ and $D_{max} = max_{\pi_i \in \pi^N, \pi_j \in \pi^N}\{D(\pi_i|\pi_j)\}$. One can check that $D_{min} = 0$ and $D_{max} = 1$. D_{min} is obtained when we compare identical possibility distributions whereas D_{max} is obtained when we compare maximally conflicting possibility distributions (e.g. $\forall i, j, if\ Inc(\pi_i, \pi_j) = 1\ then\ D(\pi_i, \pi_j) = 1$).

Hence, we can transform D to a similarity measure S_D whose values are in the interval [0,1]:

$$S_D = 1 - D(\pi_i|\pi_j)$$

Clearly, the measure S_D is not symmetric and hence violates Property 2 (for the above example, $S_D(\pi_2|\pi_1) = 1$). $S_D(\pi_2|\pi_1) = 1$ also shows that Property 3 is violated by S_D since the similarity is maximal and $\pi_2 \neq \pi_1$. Moreover, S_D gives the minimum similarity degree (Equal to 0) in the case of maximally conflicting possibility distributions. In other words, S_D is minimal when the distribution difference π_d is normalized. Counter-example 3 emphasizes this limit:

Counter-example 3 *Let us consider the same distributions π_1 and π_4 of Example 3. Let us consider $\pi_5[0, 1, 1, 1]$. $S_D(\pi_1|\pi_5) = S_D(\pi_1|\pi_4) = 0$. We can conclude that this measure is not enough discriminatory since π_4 appears to be closer to π_1 than π_5 was.*

4.4 Minkowski Distance and Its Derivatives

Since possibility distributions are represented by vectors of real values in the unit interval [0,1], and where the position of each individual value (possibility degree) is very important, the well-known Minkowski measure on the set of real numbers IR can be used.

Definition 4. *Given two possibility distributions π_1 and π_2 on $\Omega = \{\omega_1, ..., \omega_n\}$, the non-normalized Minkowski distance is given by:*

$$L_p(\pi_1, \pi_2) = \sqrt[p]{\sum_{i=1}^{n} |\pi_1(\omega_i) - \pi_2(\omega_i)|^p} \tag{6}$$

Particular and interesting cases of Equation (6) are:

- $L_M(\pi_1, \pi_2) = \frac{\sum_{i=1}^{n} |\pi_1(\omega_i) - \pi_2(\omega_i)|}{n}$ *denotes the normalized Manhattan distance (or City-Block distance).*
- $L_E(\pi_1, \pi_2) = \sqrt{\frac{\sum_{i=1}^{n} (\pi_1(\omega_i) - \pi_2(\omega_i))^2}{n}}$ *stands for the normalized Euclidean distance.*
- $L_C(\pi_1, \pi_2) = \max_{i=1}^{n} |\pi_1(\omega_i) - \pi_2(\omega_i)|$ *is known as the normalized Maximum distance (or Chebyshev distance or chessboard distance).*

The above distance measures can be transformed into similarity measures to compare possibility distributions:

$$S_M(\pi_1, \pi_2) = 1 - \frac{\sum_{i=1}^{n} |\pi_1(\omega_i) - \pi_2(\omega_i)|}{n} \tag{7}$$

$$S_E(\pi_1, \pi_2) = 1 - \sqrt{\frac{\sum_{i=1}^{n} (\pi_1(\omega_i) - \pi_2(\omega_i))^2}{n}} \tag{8}$$

$$S_C(\pi_1, \pi_2) = 1 - \max_{i=1}^{n} |\pi_1(\omega_i) - \pi_2(\omega_i)| \tag{9}$$

Example 6. Let us reconsider the following possibility distributions from Example 3: $\pi_1[1, 0.5, 0.3, 0.7]$, $\pi_3[0.9, 1, 0.3, 0.7]$. The application of these measures gives:

$S_M(\pi_1, \pi_3) = 0.85, \quad S_E(\pi_1, \pi_3) = 0.745, \quad S_C(\pi_1, \pi_3) = 0.5.$

We can notice that S_C is not enough discriminatory since it only concentrates on highest differences. For instance, if we take $\pi_i[1, 1, 1]$, $\pi_j[1, 1, 0]$ and $\pi_k[1, 0, 0]$, we obtain $S_C(\pi_i, \pi_j) = S_C(\pi_i, \pi_k) = 0$.

Proposition 1. *S_M, S_E and S_C measures satisfy properties P1-P6.*

S_M, S_E and S_C measures are based on metric distances which, by definition, verify properties of non-negativity, symmetry, identity and triangle inequality (if d is a metric on a set X, then $\forall x, y, z \in X$, $d(x,z) \leq d(x,y) + d(y,z)$). Hence, it is easy to verify that these similarity measures satisfy all the basic properties *P1-P6* proposed in Section 3. For more details see the Proofs in the appendix.

The fact that S_C also satisfies the basic properties, while obviously it is not fully satisfactory as shown by the above example, means that additional properties are needed in order to get a genuine definition.

In addition to the above mentioned possibilistic similarity and distance measures, we also mention a work by Fabris [12] in which the author has proposed a possibilistic counterpart of the well known probabilistic mutual information [25] between two variables X and Y defined on two different domains of discourse. This measure could not be used in our case because we are measuring the similarity between two possibility distributions given on the same domain of discourse (i.e. expressing possibility degrees on the values of the same variable).

5 Additional Properties of a Possibilistic Similarity Measure

The set of properties *P1-P6*, proposed in Section 3, represents basic properties of a possibilistic similarity measure [17]. This set is too minimal and is satisfied by some existing indexes (e.g. S_M, S_E and S_C). However, these properties do not take into account an important factor, namely, the inconsistency degree between the possibility distributions under comparison. Hence, in this section, we will mainly focus on revising and extending these properties to highlight the introduction of inconsistency in measuring possibilistic similarity [18].

5.1 *The Role of Inconsistency in Measuring Possibilistic Similarity*

Let us recall that the inconsistency degree between two possibility distributions π_1 and π_2 is assessed by using the measure *Inc* (see Equation (1)) and is given by $Inc(\pi_1, \pi_2) = Inc(\pi_1 \wedge \pi_2)$. We take the $\wedge$ as the minimum or product operator. When $\pi_1 \wedge \pi_2$ gives a sub-normalized possibility distribution, it indicates that there is a conflict between π_1 and π_2 ($Inc(\pi_1, \pi_2) \in]0,1]$).

This concept of inconsistency (conflict) should be considered when measuring similarity between two normalized possibility distributions. Let us analyze the following example which motivates this assertion.

Example 7. Suppose that a conference chair has to select the best paper among three selected papers (p_1, p_2, p_3) to give an award to its authors. The conference chair decides to make a second reviewing and asks two referees r_1 and r_2 to give their preferences about the papers which, in fact, will be represented in the form of possibility distributions. Let us consider these two situations:

Situation 1: The referee r_1 expresses his full satisfaction for p_3 and fully rejects p_1 and p_2 (i.e. $\pi_1(p_1) = 0$, $\pi_1(p_2) = 0$, $\pi_1(p_3) = 1$) whereas r_2 expresses his full satisfaction for p_2 and fully rejects p_1 and p_3 (i.e. $\pi_2(p_1) = 0$, $\pi_2(p_2) = 1$, $\pi_2(p_3) = 0$). Clearly, p_1 will be rejected but the chair cannot make a decision that fully fits referees' preferences since they are in conflict.

Situation 2: The referee r_1 expresses his full satisfaction for p_1 and p_3 and fully rejects p_2 (i.e. $\pi_1'(p_1) = 1$, $\pi_1'(p_2) = 0$, $\pi_1'(p_3) = 1$) whereas r_2 expresses his full satisfaction for p_1 and p_2 and fully rejects p_3 (i.e. $\pi_2'(p_1) = 1$, $\pi_2'(p_2) = 1$, $\pi_2'(p_3) = 0$). In this case, the conflict disappears and the chair can make a decision that satisfies both reviewers since they agree that p_1 is a good paper.

The fact that possibility degrees 0 and 1 do not play the same role is similar to what is called bipolar information [6, 10, 11] which distinguishes between positive and negative information. In fact, in the bipolar view of preference modeling, negative preferences correspond to what is rejected, considered unacceptable, while positive preferences correspond to what is desired.

Clearly, from the above example, we can see that, in some situations, similarity alone is not sufficient to compare possibility distributions and hence does not allow to make a decision since the expressed preferences in both situations have the same similarity.

In fact, if we apply the similarity measures S_M, S_E and S_C (which, hitherto, represent the only possibilistic similarity indexes (among all above presented measures) satisfying properties *P1-P6*) to compare the possibility distributions (of the above example):

π_1=[0, 0, 1], π_2=[0, 1, 0], π'_1=[1, 0, 1] and π'_2=[1, 1, 0], we obtain:
$S_M(\pi_1,\pi_2)$=$S_M(\pi'_1,\pi'_2)$=0.33,
$S_E(\pi_1,\pi_2)$=$S_E(\pi'_1,\pi'_2)$=0.18,
$S_C(\pi_1,\pi_2)$=$S_C(\pi'_1,\pi'_2)$=0.

To overcome this drawback, we will enrich the proposed properties by some additional ones.

5.2 *Additional Possibilistic Similarity Properties*

The first extension concerns Property 5, where we consider a particular case of strict similarity in case of strict inclusion:

Property 7. Strict inclusion
$\forall \pi_1$, π_2, π_3 s.t. $\pi_1 \neq \pi_2 \neq \pi_3$, if $\pi_1 \leq \pi_2 \leq \pi_3$, then $s(\pi_1,\pi_2) > s(\pi_1,\pi_3)$.
Note that $\pi_1 \neq \pi_2$ and $\pi_1 \leq \pi_2$ imply $\pi_1 < \pi_2$ (strict specificity).

Proposition 2. *Properties P1-P4, P6-P7 imply Property 5.*

Next property analyzes two sources who are proving conflicting pieces of information. Suppose that the two sources, after discovering that they are conflicting, change their opinions in the same way by enhancing the degree of a given situation (with the same value). Then, the similarity will be even larger, if the enhancement leads to a decrease of the amount of conflict:

Property 8. Reaching coherence
Let π_1 and π_2 be two possibility distributions . Let $\omega_i \in \Omega$. Let π'_1 and π'_2 be two normalized possibility distributions s.t.:

i) $\forall j \neq i$, $\pi'_1(\omega_j) = \pi_1(\omega_j)$ and $\pi'_2(\omega_j) = \pi_2(\omega_j)$,
ii) Let α s.t. $\alpha \leq 1 - max(\pi_1(\omega_i), \pi_2(\omega_i))$.

If $\pi'_1(\omega_i) = \pi_1(\omega_i) + \alpha$ and $\pi'_2(\omega_i) = \pi_2(\omega_i) + \alpha$, then:

- If $Inc(\pi'_1, \pi'_2) = Inc(\pi_1, \pi_2)$, then $s(\pi'_1, \pi'_2) = s(\pi_1, \pi_2)$.
- If $Inc(\pi'_1, \pi'_2) < Inc(\pi_1, \pi_2)$, then $s(\pi'_1, \pi'_2) > s(\pi_1, \pi_2)$.

To illustrate Property 8, suppose that we have the following two maximally conflicting possibility distributions: $\pi_1[1,0,0]$ and $\pi_2[0,1,0]$ ($Inc(\pi_1, \pi_2) = 1$). Suppose that the two sources decide to fully support the third situation and give it a possibility degree equal to 1 ($\alpha = 1$). We obtain: $\pi'_1[1,0,1]$ and $\pi'_2[0,1,1]$. Clearly, $Inc(\pi'_1, \pi'_2)$ becomes 0. Hence, $s(\pi'_1, \pi'_2)$ should be greater than $s(\pi_1, \pi_2)$.

The intuition behind next property is the following: consider two sources who provide normalized possibility distributions π_1 and π_2. Assume that there exists a situation ω where they disagree. Now, assume that the second source changes its mind and sets $\pi_2(\omega)$ to be closer to $\pi_1(\omega)$ (for instance, source 2 is convinced by arguments given by source 1, and decides to change the possibility degree given to ω with another degree closer to the one given by source 1, i.e. $\pi_1(\omega)$). Then the new similarity between π_1 and π_2 increases. This is the aim of Property 9. :

Property 9. Mutual convergence
Let π_1 and π_2 be two possibility distributions such that for some ω_i, we have $\pi_1(\omega_i) > \pi_2(\omega_i)$. Let π'_2 be a normalized possibility distribution s.t.:

i) $\pi'_2(\omega_i) \in]\pi_2(\omega_i), \pi_1(\omega_i)]$,
ii) and $\forall j \neq i$, $\pi'_2(\omega_j) = \pi_2(\omega_j)$.

Hence, we obtain: $s(\pi_1, \pi'_2) > s(\pi_1, \pi_2)$.

Corollary 2, concerns the situation when setting the new degree of $\pi_2(\omega_i)$ to be equal to $\pi_1(\omega_i)$.

Corollary 2. Let π_1 and π_2 be two possibility distributions such that for some ω_i, we have $\pi_1(\omega_i) \neq \pi_2(\omega_i)$. Let π'_2 be a normalized possibility distribution s.t.:

i) $\pi'_2(\omega_i) = \pi_1(\omega_i)$,
ii) and $\forall j \neq i$, $\pi'_2(\omega_j) = \pi_2(\omega_j)$.

Hence, we obtain: $s(\pi_1, \pi'_2) > s(\pi_1, \pi_2)$.

Property 10 says that, if we consider two normalized possibility distributions π_1 and π_2 and we increase (resp. decrease) one situation ω_p in π_1 and one situation ω_q in π_2 with a same degree α (leading to π'_1 and π'_2), then the similarity degree between π_1 and π'_1 will be equal to the one between π_2 and π'_2. When decreasing α from π_1 and π_2, the obtained π'_1 and π'_2 should remain normalized.

Property 10. Indifference preserving
Let π_1 and π_2 be two possibility distributions. Let π'_1 and π'_2 be two normalized possibility distributions s.t.

i) $\forall j \neq p$, $\pi'_1(\omega_j) = \pi_1(\omega_j)$ and $\pi'_1(\omega_p) = \pi_1(\omega_p) + \alpha$ (resp. $\pi'_1(\omega_p) = \pi_1(\omega_p) - \alpha$).
ii) $\forall j \neq q$, $\pi'_2(\omega_j) = \pi_2(\omega_j)$ and $\pi'_2(\omega_q) = \pi_2(\omega_q) + \alpha$ (resp. $\pi'_2(\omega_q) = \pi_2(\omega_q) - \alpha$).

Then: $s(\pi_1, \pi'_1) = s(\pi_2, \pi'_2)$.

Let us illustrate property 10. Suppose that we have the following possibility distributions: $\pi_1[1,0,0]$ and $\pi_2[0,1,0]$. The source providing π_1 changes its mind by increasing the possibility degree of the second situation by 1 which leads to $\pi'_1[1,1,0]$. On the other hand, the second source providing π_2 will do the same thing as source one but this time by increasing, with the same value, the possibility degree of the third situation which leads to $\pi'_2[0,1,1]$. Property 10 requires that $s(\pi_1,\pi'_1)$ should be equal to $s(\pi_2,\pi'_2)$.

Next corollary says that if one starts with a normalized possibility distribution π_1, and modify it (without loosing the normalization condition) by decreasing (resp. increasing) only one situation ω_i (leading to π_2), or starts with a same distribution π_1 and only modify, identically, another situation ω_k (leading to π_3), then the similarity degree between π_1 and π_2 is the same as the one between π_1 and π_3.

Corollary 3. Let π_1 be a possibility distribution and α a positive number. Let π_2 be a normalized possibility distribution s.t. $\pi_2(\omega_i) = \pi_1(\omega_i) - \alpha$ (resp. $\pi_2(\omega_i) = \pi_1(\omega_i) + \alpha$) and $\forall j \neq i$, $\pi_2(\omega_j) = \pi_1(\omega_j)$.

Let π_3 be a normalized possibility distribution s.t. for $k \neq i$, $\pi_3(\omega_k) = \pi_1(\omega_k) - \alpha$ (resp. $\pi_3(\omega_k) = \pi_1(\omega_k) + \alpha$) and $\forall j \neq k$, $\pi_3(\omega_j) = \pi_1(\omega_j)$, then: s($\pi_1,\pi_2$)=s($\pi_1,\pi_3$).

Let us illustrate Corollary 3. Suppose that we have the following possibility distribution: $\pi_1[1,0,0]$. The source providing π_1 first changes its mind and gives $\pi_2[1,1,0]$ which increases the possibility degree of the second situation with a degree 1. The source will then realize that it made a mistake by assigning the degree 1 to the second situation instead of the third one. The correction will lead to $\pi_3[1,0,1]$. Corollary 3 requires that $s(\pi_1,\pi_2)$ should be equal to $s(\pi_1,\pi_3)$.

5.3 Derived Propositions

In what follows, we will derive some additional properties that follow from the above defined properties. Proofs of propositions given in this paper can be found in the appendix.

A consequence of Property 7 is that only identity between two distributions imply full similarity, namely:

Proposition 3. *Let s be a possibilistic similarity measure s.t. s satisfies Properties 1-10. Then,* $\forall \pi_i$, π_j, $s(\pi_i,\pi_j)=1$ *iff* $\pi_i=\pi_j$.

This also means that: $\forall \pi_j \neq \pi_i$, $s(\pi_i,\pi_i) < 1$. Clearly, proposition 3 provides a stronger property than property 3.

Similarly, only completely contradictory possibility distributions imply a similarity degree equal to 0. Next proposition provides stronger results than property 4:

Proposition 4. *Let s be a possibilistic similarity measure s.t. s satisfies Properties 1-10. Then,* $\forall \pi_i$, π_j, $s(\pi_i,\pi_j)=0$ *iff* $\forall \omega_p \in \Omega$,

i) $\pi_i(\omega_p) \in \{0,1\}$ *and* $\pi_j(\omega_p) \in \{0,1\}$,
ii) and $\pi_j(\omega_p) = 1 - \pi_i(\omega_p)$

As a consequence of Property 8, discounting the possibility degree of a same situation could lead to a decrease of similarity if the discounting results in an increase of the amount of conflict:

Proposition 5. *Let s be a possibilistic similarity measure satisfying Properties 1-10. Let π_1 and π_2 be two normalized possibility distributions. Let $\omega_i \in \Omega$. Let π'_1 and π'_2 be two normalized possibility distributions s.t.:*

i) $\forall j \neq i$, $\pi'_1(\omega_j) = \pi_1(\omega_j)$ and $\pi'_2(\omega_j) = \pi_2(\omega_j)$,
ii) Let α s.t. $\alpha \leq min(\pi_1(\omega_i), \pi_2(\omega_i))$.

If $\pi'_1(\omega_i) = \pi_1(\omega_i) - \alpha$ and $\pi'_2(\omega_i) = \pi_2(\omega_i) - \alpha$.
Then:

If $Inc(\pi'_1, \pi'_2)$=$Inc(\pi_1, \pi_2)$, then $s(\pi'_1, \pi'_2) = s(\pi_1, \pi_2)$.
If $Inc(\pi'_1, \pi'_2) > Inc(\pi_1, \pi_2)$, then $s(\pi'_1, \pi'_2) < s(\pi_1, \pi_2)$.

In fact, proposition 5 encodes the dual of property 8.

As a consequence of Property 9, starting from a given possibility distribution π_1, we can define a set of possibility distributions that, gradually, converge to the most similar possibility distribution to π_1:

Proposition 6. *Let s be a possibilistic similarity measure satisfying Properties 1-10. Let π_1 and π_2 be two normalized possibility distributions s.t. for some ω_i, $\pi_1(\omega_i) > \pi_2(\omega_i)$. Let π_k (k=3..m) be a set of m possibility distributions. Each π_k is derived in step k from π_{k-1} as follows:*

i) $\pi_k(\omega_i) = \pi_{k-1}(\omega_i) + \alpha_k$ (with $\alpha_k \in]0, \pi_1(\omega_i) - \pi_{k-1}(\omega_i)]$)
ii) and $\forall j \neq i$, $\pi_k(\omega_j) = \pi_{k-1}(\omega_j)$.

Hence, we obtain $s(\pi_1, \pi_2) < s(\pi_1, \pi_3) < s(\pi_1, \pi_4) < ... < s(\pi_1, \pi_m) \leq 1$.

Proposition 7. *Let s be a possibilistic similarity measure satisfying Properties 1-10. $\forall i, j \in \{1,..,m\}$, $\forall \varepsilon \in [0,1]$ and $\forall \omega \in \Omega$, we have:$s(\pi_i, \pi_i^{\omega-\varepsilon}) = s(\pi_j, \pi_j^{\omega-\varepsilon})$. $\pi_i^{\omega-\varepsilon}$ means that we subtract ε from one situation ω of the possibility distribution π_i. We recall that $\forall i, j \in \{1,..,m\}$, the possibility distributions π_i, π_j, $\pi_i^{\omega-\varepsilon}$ and $\pi_j^{\omega-\varepsilon}$ are normalized.*

The following figure illustrates proposition 7:

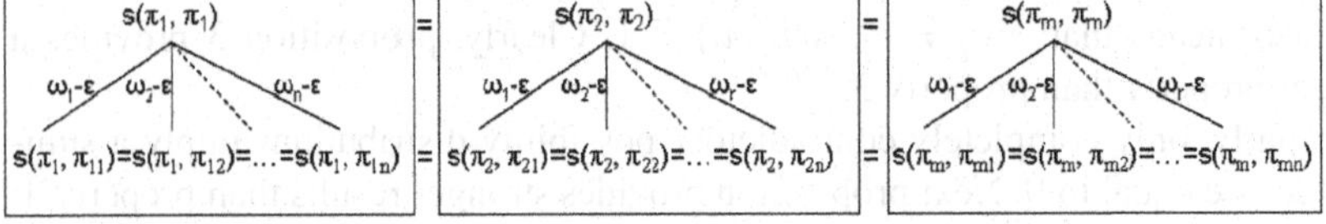

This proposition follows immediately from property 3 and property 10. In fact, let us consider two normalized possibility distributions π_1 and π_2.

The equalities in the first line of all boxes is trivial since the similarity between any identical possibility distributions is maximal and equal to 1 (Property 3).

Let us now analyze one of the boxes of the figure. As we have already explained in the illustration of corollary 3 (Section 5.2), if we subtract a value (here ε) from one situation (for instance ω_1) of a possibility distribution π_1 which leads to π_{11} and do the same thing but we subtract ε from another situation (for instance ω_2) which leads to π_{12}, then $s(\pi_1,\pi_{11}) = s(\pi_1,\pi_{12})$. This applies to all boxes of the figure.

Now, let us look at the equalities between the boxes. Let us take the second box (i.e. π_2) and proceed exactly as we did with π_1. Corollary 3 requires that $s(\pi_2,\pi_{21}) = s(\pi_{21},\pi_{22})$.

Moreover, as illustrated in the explanation of property 10, we should have $s(\pi_1,\pi_{11}) = s(\pi_2,\pi_{21})$ and $s(\pi_1,\pi_{12}) = s(\pi_2,\pi_{22})$.

We can easily check that $s(\pi_1,\pi_{11}) = s(\pi_1,\pi_{12}) = s(\pi_2,\pi_{21}) = s(\pi_2,\pi_{22})$.

The following section provides a similarity measure that satisfies all properties which means that P1-P10 is a coherent set of properties.

6 Information Affinity: The New Inconsistency-Based Possibilistic Similarity Measure

Recall that, in Section 4, we have discarded some similarity functions, namely, S_G, $S_{distance}$ and S_D since they do not satisfy all the basic properties (P1-P6) as shown by the different counter-examples. However, we have provided proofs (in the appendix) showing that S_M, S_E and S_C satisfy basic properties P1-P6.

When considering all the proposed extended properties P1-P10, none of the S_M, S_E and S_C measures satisfy property 8 as shown by the following counter-example:

Counter-example 4 *Let us take again the example of the reviewers given in Example 7 where π_1=[0, 0, 1], π_2=[0, 1, 0], π'_1=[1, 0, 1] and π'_2=[1, 1, 0]. Clearly, we have:*

$Inc(\pi_1,\pi_2) = 1$. *After adding the degree 1 to ω_1 in both π_1 and π_2, the inconsistency has decreased ($Inc(\pi'_1,\pi'_2) = 0$). So, according to property 8, $s(\pi'_1,\pi'_2)$ should be greater than $s(\pi_1,\pi_2)$ which is not the case when taking s as S_M or S_E or S_C. In fact: $S_M(\pi_1,\pi_2)$=$S_M(\pi'_1,\pi'_2)$=0.33, $S_E(\pi_1,\pi_2)$=$S_E(\pi'_1,\pi'_2)$=0.18, $S_C(\pi_1,\pi_2)$= $S_C(\pi'_1,\pi'_2)$=0.*

Considering the weaknesses related to the aforementioned existing possibilistic similarity measures, we will propose a new measure that overcomes these drawbacks, i.e., a measure that will satisfy all the proposed properties *P1-P10*.

Clearly, there are two important criteria in measuring possibilistic similarity, namely, distance and inconsistency. We have chosen to combine them using a sum weighted by κ and λ respectively. The choice of combining these two criteria is justified by the fact that, in the possibilistic framework, a distance measure taken alone does not always allow us to decide about the closest distribution to a given one (as shown by Example 7).

Semantically, the proposed measure takes into account a classical informative distance (Manhattan or Euclidean) along with the well known inconsistency measure.

Definition 5. *Let π_1 and π_2 be two possibility distributions on the same universe of discourse Ω. We define a measure $Aff(\pi_1, \pi_2)$ as follows:*

$$Aff(\pi_1, \pi_2) = 1 - \frac{\kappa * d(\pi_1, \pi_2) + \lambda * Inc(\pi_1, \pi_2)}{\kappa + \lambda} \tag{10}$$

where $\kappa > 0$ and $\lambda > 0$. d denotes one of the L_p normalized metric distances (either Manhattan or Euclidean) between π_1 and π_2. $Inc(\pi_1, \pi_2)$ represents the inconsistency degree between two distributions (see Equation (1)) where $\wedge$ is taken as the product or min conjunctive operators.

Intuitively, using the *min* operator instead of the product means that we give less importance to the inconsistency degree in Equation (10), since $Inc(\pi_1 * \pi_2) > Inc(min(\pi_1, \pi_2))$.

Note that in Equation (10), if $d = L_C$ then P1-P10 are not satisfied. In fact, let us consider the following possibility distributions: $\pi_1[1,1,1]$ and $\pi_2[1,0,0]$. Let us make the possibility degree of the second situation in π_2 equal to the one in π_1 (i.e. by adding a degree 1) which leads to $\pi_2'[1,1,0]$.

Let us take d as the Manhattan distance, $\wedge$ as the minimum conjunctive operator and $\kappa = \lambda = 1$. We obtain $Aff(\pi_1, \pi_2') = Aff(\pi_1, \pi_2) = 0.5$. This violates Property 9 which requires that $Aff(\pi_1, \pi_2') > Aff(\pi_1, \pi_2)$. We can easily check that Property 9 and Property 10 will be also violated if we take $d = L_C$.

Proposition 8. *The Aff measure satisfies all the proposed properties P1-P10.*

Next example shows how the *Aff* measure resolves the problem illustrated by Example 7. We also give another general example that illustrates the *Aff* measure.

Example 8. If we reconsider the example of the referees where π_1=[0, 0, 1], π_2=[0, 1, 0], π_1'=[1, 0, 1] and π_2'=[1, 1, 0]. If we apply *Aff*, we obtain: $Aff(\pi_1,\pi_2)$=0.16 < $Aff(\pi_1',\pi_2')$=0.66 which is conform to what we expect.

Now, let us take π_a=[$\frac{1}{3}$, 1, $\frac{1}{3}$] and π_b=[1, $\frac{1}{3}$, $\frac{1}{3}$]. If we take $\wedge$ as the *min* or the *product* operator, we obtain $Inc(\pi_a, \pi_b)$=1-$\frac{1}{3}$=$\frac{2}{3}$. On the other hand, $d(\pi_a, \pi_b)$= $L_M(\pi_a, \pi_b)$= $\frac{|\frac{1}{3}-1|+|1-\frac{1}{3}|+|\frac{1}{3}-\frac{1}{3}|}{3}$=$\frac{4}{9}$. Thus, $Aff(\pi_a, \pi_b) = 1 - \frac{\frac{4}{9}+\frac{2}{3}}{2} = \frac{5}{9}$.

The following example illustrates some of the proposed additional properties in conjunction with the proposed *Aff* similarity measure.

Example 9. For this example, we will take d as the Manhattan distance, $\wedge$ as the minimum conjunctive operator and $\kappa = \lambda = 1$.

Let us reconsider the example of the reviewers.

- Suppose that we have three reviewers expressing their opinions about three papers by providing three possibility distributions: $\pi_1[0,0,1]$,$\pi_2[0.5,0.5,1]$ and $\pi_3[1,1,1]$.

  Clearly, $\pi_1 < \pi_2 < \pi_3$. We have $Aff(\pi_1, \pi_2) = 0.83 > Aff(\pi_1, \pi_3) = 0.66$. This is confirmed by property 7.

- Let us now consider a fourth reviewer ($\pi_4[1,0,0]$). Clearly, the first and the fourth reviewers are in total conflict ($Inc(\pi_1,\pi_4) = 1$) and the similarity between their opinions is $Aff(\pi_1,\pi_4) = 0.16$. Suppose that both reviewers have changed their minds about the second paper and fully accept it ($\pi_1'[0,1,1]$ and $\pi_4'[1,1,0]$). Clearly, the conflict has disappeared ($Inc(\pi_1',\pi_4') = 0$). In this case, we have $Aff(\pi_1',\pi_4') = 0.66 > Aff(\pi_1,\pi_4) = 0.16$. This is confirmed by property 8.
- Let us now take the opinions of the first and the second reviewers ($\pi_1[0,0,1]$, $\pi_2[0.5,0.5,1]$). Suppose that the first reviewer has changed his opinion about the second paper and has joined the second reviewer ($\pi_1'[0,0.5,1]$). In this case, we have $Aff(\pi_2',\pi_1) = 0.91 > Aff(\pi_2,\pi_1) = 0.83$. This is confirmed by property 9.
- Let us take again the first and the second reviewers ($\pi_1[0,0,1]$,$\pi_2[0.5,0.5,1]$). After reading the revised versions of the papers, the first reviewer has changed his opinion about the second paper (by adding a degree of 0.5) and maintain his opinions on the remaining papers ($\pi_1'[0,0.5,1]$). The second reviewer has also changed his opinion but about the first paper (by adding a degree of 0.5 leading to $\pi_2'[1,0.5,1]$). In this case, we have $Aff(\pi_1',\pi_2') = Aff(\pi_1,\pi_2) = 0.83$. This is confirmed by property 10.

7 Conclusion

In this paper, we have provided an analysis of natural properties that a possibilistic similarity measure should satisfy. In a first stage, we have proposed basic properties of possibilistic similarity functions. Although these basic properties are not met by most of the existing possibilistic similarity measures in the literature, they are too minimal. In fact, as it has been shown, the Minkowski similarity function and some of its derivatives satisfies the proposed properties. Moreover, this analysis led us to the conclusion that inconsistency should be taken into account when comparing possibilistic bodies of evidences.

Consequently, in a second stage, we have revised and extended these axioms and proposed a new similarity measure, so-called, *Information Affinity*, which combines distance and inconsistency in order to assess the similarity degree between two normalized possibility distributions. We have shown that our new measure, that combines a distance-based measure and the inconsistency degree, satisfies both basic and extended set of postulates.

Appendix. Proofs

Proof of Proposition 1. We want to prove that S_M, S_E and S_C satisfy basic properties P1-P6. Let π_1, π_2, π_3 and π_4 be four normalized possibility distributions.

Property 1 - Property 2 - Property 3

Proofs of Properties P1-P3 are immediate since S_M, S_E and S_C are normalized similarity measures derived from the Minkowski metric which by definition satisfy non-negativity, symmetry, upper bound and non-degeneracy properties.

Property 4
If $\forall\, \omega_i, \pi_1(\omega_i) \in \{0,1\}, \pi_2(\omega_i) \in \{0,1\}$ *and* $\pi_2(\omega_i) = 1 - \pi_1(\omega_i)$
$\Rightarrow |\pi_1(\omega_i) - \pi_2(\omega_i)| = 1, \sqrt{(\pi_1(\omega_i) - \pi_2(\omega_i))^2} = 1$ and $max|\pi_1(\omega_i) - \pi_2(\omega_i)| = 1$
$\Rightarrow S_M = 0, S_E = 0$ and $S_C = 0$.

Property 5
If $\forall\, \omega_i\ \pi_1(\omega_i) \le \pi_2(\omega_i) \le \pi_3(\omega_i)$
$\Rightarrow \forall\, \omega_i, |\pi_1(\omega_i) - \pi_2(\omega_i)| \le |\pi_1(\omega_i) - \pi_3(\omega_i)|$
$\Rightarrow \sum_{\omega_i} |\pi_1(\omega_i) - \pi_2(\omega_i)| \le \sum_{\omega_i} |\pi_1(\omega_i) - \pi_3(\omega_i)|$
$\Rightarrow 1 - \sum_{\omega_i} |\pi_1(\omega_i) - \pi_2(\omega_i)| \ge 1 - \sum_{\omega_i} |\pi_1(\omega_i) - \pi_3(\omega_i)|$
$\Rightarrow S_M(\pi_1, \pi_2)| \ge S_M(\pi_1, \pi_3)$.
Similarly for S_E and S_C.

Property 6
The proof of property 6 is immediate.

Let π_1 and π_2 be two normalized possibility distributions and ρ a permutation. Let $\pi_{\rho(1)}, \pi_{\rho(2)}$ be the possibility distributions obtained by permuting elements having the same indexes in π_1 and π_2. The pairwise permutation of the elements has no effect on $S_M \Rightarrow S_M(\pi_1, \pi_2) = S_M(\pi_{\rho(1)}, \pi_{\rho(1)})$(similarly for S_E and S_C).

Proof of Proposition 3. We want to prove that if s satisfies P1-P10, then it satisfies the strong Upper bound property. One direction is immediate since $\pi_1 = \pi_2 \Rightarrow$ s(π_1, π_2)=1 (Property 3). Now, suppose that s(π_1, π_2)=1 and $\pi_1 \neq \pi_2$. $\pi_1 \neq \pi_2 \Rightarrow \exists\ \omega_p$ *s.t.* $\pi_1(\omega_p) \neq \pi_2(\omega_p)$. Suppose that $\pi_1(\omega_p) > \pi_2(\omega_p)$. Let us add α (with $\alpha = \pi_1(\omega_p) - \pi_2(\omega_p)$) to $\pi_2(\omega_p)$ which leads to π_2'.From property 9, we can conclude that s(π_1, π_2')>s(π_1, π_2)=1 (contradiction with property 3). Hence, $\pi_1 = \pi_2$

Proof of Proposition 4. We want to prove that if s satisfies P1-P10, then it satisfies the strong Lower bound property.

One direction is immediate since π_1=1 $- \pi_2$ (with π_1 and π_2 are binary normalized possibility distributions) $\Rightarrow$ s(π_1, π_2)=0 (Property 4).

Let us suppose that:

s(π_1, π_2)=0, π_1 and π_2 are binary possibility distributions and $\pi_1 \neq 1 - \pi_2$.

Let us first show that $\pi_1 = 1 - \pi_2$. Assume that $\pi_1 \neq 1 - \pi_2 \Rightarrow \exists\ \omega_p$ *s.t.* $\pi_1(\omega_p) \neq 1 - \pi_2(\omega_p)$. Suppose that $\pi_1(\omega_p) > \pi_2(\omega_p)$.

Let π_3 s.t. $\pi_3(\omega_p) < \pi_1(\omega_p)$ and $\forall \omega \neq \omega_p, \pi_3(\omega) = \pi_1(\omega)$.

From property 9, we can conclude that s(π_1, π_2)= 0 >s(π_1, π_3) (which is impossible since the lower bound is 0).

Hence, $\pi_1 = 1 - \pi_2$.

Now, let us show that π_1 and π_2 are binary possibility distributions. Indeed, suppose that:

s(π_1, π_2)=0, $\pi_1 = 1 - \pi_2$ and π_1 and π_2 are not binary possibility distributions.

π_1 and π_2 are not binary $\Rightarrow Inc(\pi_1, \pi_2) < 1$.

Let us take all ω_i verifying $min(\pi_1(\omega_i), \pi_2(\omega_i)) = 1 - Inc(\pi_1, \pi_2)$ ($\pi_1(\omega_i) < 1$ and $\pi_2(\omega_i) < 1$).

Let us decrease all these ω_i by α in both π_1 and π_2 and keep the remaining possibility degrees of other ω_j unchanged. According to proposition 5, this will result in an increase of the inconsistency between π_1 and π_2 and hence will result in a decrease of the similarity between π_1 and π_2 which is impossible because $s(\pi_1, \pi_2)$ is already equal to 0.

Proof of Proposition 5

We want to prove that if s satisfies P1-P10, then when decreasing one situation from π_1 and π_2 by a degree α (leading to π_1' and π_2') then if $Inc(\pi_1', \pi_2')$ increases then $s(\pi_1', \pi_2') < s(\pi_1, \pi_2)$.

The proof of this proposition follows immediately from property 8. In fact, it is simply the dual of property 8.

We start by π_1' and π_2' and add (the subtracted) α. Hence, we will recover π_1 and π_2 and $Inc(\pi_1, \pi_2)$ decreases. From property 8, we can conclude that $s(\pi_1, \pi_2) > s(\pi_1', \pi_2')$.

Proof of Proposition 6

The proof of proposition 6 follows immediately from property 9. In fact, we are adding a degree α_k to a given situation (step by step) and hence, the similarity increases at each step.

Proof of Proposition 7

The proof of proposition 7 follows immediately from property 3 and property 10. In fact, let us consider two normalized possibility distributions π_1 and π_2.

$\Rightarrow$ (1) We have: $s(\pi_1, \pi_1)$=$s(\pi_2, \pi_2) = 1$ (Property 3).

Now, let us take two situations ω_p and ω_q from π_1 and ω_k and ω_l from π_2.

Let π_1' and π_1'' be two normalized possibility distributions s.t.:

$\pi_1'(\omega_p) = \pi_1(\omega_p) - \varepsilon$ and $\forall \omega_i \neq \omega_p$, $\pi_1'(\omega_i) = \pi_1(\omega_i)$.
$\pi_1''(\omega_q) = \pi_1(\omega_q) - \varepsilon$ and $\forall \omega_i \neq \omega_q$, $\pi_1''(\omega_i) = \pi_1(\omega_i)$.

Let π_2' and π_2'' be two normalized possibility distributions s.t. $\pi_2'(\omega_k) = \pi_2(\omega_k) - \varepsilon$ and $\forall \omega_i \neq \omega_k$, $\pi_2'(\omega_i) = \pi_2(\omega_i)$.
$\pi_2''(\omega_l) = \pi_2(\omega_l) - \varepsilon$ and $\forall \omega_i \neq \omega_l$, $\pi_2''(\omega_i) = \pi_2(\omega_i)$.

$\Rightarrow$ (2) We have: $s(\pi_1, \pi_1') = s(\pi_1, \pi_1'')$ and $s(\pi_2, \pi_2') = s(\pi_2, \pi_2'')$ (Corollary 3).
$\Rightarrow$ (3) We have also: $s(\pi_1, \pi_1') = s(\pi_2, \pi_2')$ and $s(\pi_1, \pi_1'') = s(\pi_2, \pi_2'')$ (Property 10).
$\Rightarrow$ From (2) and (3) we can conclude that $s(\pi_1, \pi_1') = s(\pi_1, \pi_1'') = s(\pi_2, \pi_2') = s(\pi_2, \pi_2'')$.

Proof of Proposition 8. We want to prove that the *Aff* measure satisfies Properties P1-P10. For the proof, d is either the normalized Manhattan or normalized Euclidean distance. The proof is valid for these two cases. Similarly, $\wedge$ can be either the *minimum* or the *product* operator. Again, the proof is valid for these two cases. Given two possibility distributions π_1 and π_2, we have:

Property 1. Non-negativity:
By definition, $0 \leq d(\pi_1, \pi_2) \leq 1$. Moreover, $0 \leq Inc(\pi_1, \pi_2) \leq 1$ (possibility degrees $\in [0,1]$). Hence $0 \leq \frac{\kappa * d(\pi_1,\pi_2)+\lambda * Inc(\pi_1,\pi_2)}{\kappa+\lambda} \leq 1$, and $0 \leq 1 - \frac{\kappa * d(\pi_1,\pi_2)+\lambda * Inc(\pi_1,\pi_2)}{\kappa+\lambda} \leq 1$ which means $0 \leq Aff(\pi_1, \pi_2) \leq 1$.

Property 2. Symmetry:
Both d and Inc are symmetric.
$Aff(\pi_2, \pi_1) = 1 - \frac{\kappa * d(\pi_2,\pi_1)+\lambda * Inc(\pi_2,\pi_1)}{\kappa+\lambda} = 1 - \frac{\kappa * d(\pi_1,\pi_2)+\lambda * Inc(\pi_1,\pi_2)}{\kappa+\lambda} = (\pi_1, \pi_2)$.

Property 3. Upper bound and Non-degeneracy:
Proof of the first part of property 3 : If $\pi_1 = \pi_2 \Rightarrow Aff(\pi_1, \pi_2) = 1$:
$If\ \pi_1 = \pi_2,\ then\ Aff(\pi_1, \pi_2) = Aff(\pi_1, \pi_1) = 1 - \frac{\kappa * d(\pi_1,\pi_1)+\lambda * Inc(\pi_1,\pi_1)}{\kappa+\lambda} = 1 - \frac{(0+0)}{\kappa+\lambda} = 1$.

The proof of the second part of property 3, namely, $\forall\ \pi_i,\ \pi_j,\ Aff(\pi_i, \pi_j) \leq 1$ has been already given in the proof of property 1.

Property 4. Lower bound
If π_1 and π_2 are binary and maximally contradictory possibility distributions, then: $d(\pi_1, \pi_2)$ is maximal and $Inc(\pi_1, \pi_2)$ is maximal too $\Rightarrow d(\pi_1, \pi_2) = 1$ and $Inc(\pi_1, \pi_2) = 1 \Rightarrow \frac{\kappa * d(\pi_1,\pi_2)+\lambda * Inc(\pi_1,\pi_2)}{\kappa+\lambda} = \frac{\kappa * 1+\lambda * 1}{\kappa+\lambda} = 1 \Rightarrow Aff(\pi_1, \pi_2) = 0$.

Property 5. Large inclusion
If π_1 is more specific than π_2 which is in turn more specific then π_3, then π_1, π_2 and π_3 are fully consistent with each other (they all share at least one state which is fully possible). More formally, $Inc(\pi_1, \pi_2) = Inc(\pi_1, \pi_3) = Inc(\pi_2, \pi_3) = 0$ and $d(\pi_1, \pi_2) \leq d(\pi_1, \pi_3)$. Hence, $1 - \frac{\kappa * d(\pi_1,\pi_2)}{\kappa+\lambda} \geq 1 - \frac{\kappa * d(\pi_1,\pi_3)}{\kappa+\lambda}$ which implies $Aff(\pi_1, \pi_2) \geq Aff(\pi_1, \pi_3)$.

Property 6. Permutation
Let π_1' and π_2' be two possibility distributions obtained by permuting elements having the same indexes in π_1 and π_2. Since we proceed pointwise computation of d and Inc degree by degree, the pairwise permutation of the elements has no effect on d and Inc. So we obtain $d(\pi_1, \pi_2) = d(\pi_1', \pi_2')$ and $Inc(\pi_1, \pi_2) = Inc(\pi_1', \pi_2') \Rightarrow Aff(\pi_1, \pi_2) = Aff(\pi_1', \pi_2')$.

Property 7: Strict inclusion
If π_1 is (strictly) more specific than π_2 which is in turn (strictly) more specific then π_3, then, as we only deal with normalized possibility distributions, it is obvious that all these distributions share one fully possible element $\omega \Rightarrow Inc(\pi_1, \pi_2) = Inc(\pi_1, \pi_3) = Inc(\pi_2, \pi_3) = 0$
$\forall \omega_i,\ \pi_1(\omega_i) \leq \pi_2(\omega_i) \leq \pi_3(\omega_i)$. $\exists\ \omega_k$ s.t. $\pi_1(\omega_k) < \pi_2(\omega_k)$ and $\exists\ \omega_m$ s.t. $\pi_2(\omega_m) < \pi_3(\omega_m)$.

$\Rightarrow$ d(π_1, π_2)<d(π_1, π_3) $\Rightarrow \kappa * d(\pi_1, \pi_2) < \kappa * d(\pi_1, \pi_3)$
$\Rightarrow 1 - \frac{\kappa * d(\pi_1, \pi_2)}{\kappa + \lambda} > 1 - \frac{\kappa * d(\pi_1, \pi_3)}{\kappa + \lambda} \Rightarrow Aff(\pi_1, \pi_2) > Aff(\pi_1, \pi_3)$.

Property 8: Reaching coherence
We have d(π'_1, π'_2)=d(π_1, π_2) since we added the same value α to $\pi_k(\omega_i)_{k \in 1,2}$.
On the other hand, adding α to $\pi_k(\omega_i)_{k \in 1,2}$ may result in a decrease (and not an increase) of the inconsistency. Hence, if Inc(π'_1, π'_2) < Inc(π_1, π_2) then $Aff(\pi'_1, \pi'_2) > Aff(\pi_1, \pi_2)$ and if the inconsistency remains unchanged, then $Aff(\pi'_1, \pi'_2) = Aff(\pi_1, \pi_2)$.

Property 9: Mutual convergence
We have, $\pi_2(\omega_i) \neq \pi_1(\omega_i)$ and $\forall j \neq i$, $\pi'_2(\omega_j) = \pi_2(\omega_j)$. When taking $\pi'_2(\omega_i) = x$ s.t. $x \in]\pi_2(\omega_i), \pi_1(\omega_i)]$, we always have:
$\Rightarrow d(\pi_1, \pi'_2) < d(\pi_1, \pi_2)$.
On the other hand, we have $Inc(\pi_1, \pi'_2) \leq Inc(\pi_1, \pi_2)$
$\Rightarrow \kappa * d(\pi_1, \pi'_2) + \lambda * Inc(\pi_1, \pi'_2) < \kappa * d(\pi_1, \pi_2) + \lambda * Inc(\pi_1, \pi_2)$
$\Rightarrow Aff(\pi_1, \pi'_2) > Aff(\pi_1, \pi_2)$.

Property 10: Indifference preserving
1) If we add α to $\pi_1(\omega_i)$ and keep the other degrees unchanged (which leads to π'_1) and α to $\pi_2(\omega_j)$ and keep the other degrees unchanged (which leads to π'_2)
$\Rightarrow$ d(π_1, π'_1)=d(π_2, π'_2)=$\frac{\alpha}{|\Omega|}$ and Inc(π_1, π'_1)=Inc(π_2, π'_2)=0 (since we only deal with normalized distributions)
$\Rightarrow Aff(\pi_1, \pi'_1) = Aff(\pi_2, \pi'_2)$.
2) The second proof is immediate from 1) if we subtract α.

References

1. Abellan, J., Gomez, M.: Measures of divergence on credal sets. Fuzzy Sets and Systems 157, 1514–1531 (2006)
2. Baets, B.D., Meyer, H.D.: Transitivity-preserving fuzzification schemes for cardinality-based similarity measures. European Journal of Operational Research 160(1), 726–740 (2005)
3. Bouchon-Meunier, B., Rifqi, M., Bothorel, S.: Towards general measures of comparison of objects. Fuzzy sets and systems 84(2), 143–153 (1996)
4. Chan, H., Darwiche, A.: A distance measure for bounding probabilistic belief change. International Journal of Approximate Reasoning 38, 149–174 (2005)
5. Choquet, G.: Theory of capacities. Annales de L'Institut Fourier 54, 131–295 (1953)
6. Dubois, D., Kaci, S., Prade, H.: Bipolarity in reasoning and decision - an introduction. the case of the possibility theory framework. In: Proceedings of the 10th International Conference on Information Processing and Management of Uncertainty in Knowledge-Based Systems (IPMU 2004), pp. 959–966 (2004)

7. Dubois, D., Prade, H.: Possibility theory: An approach to computerized processing of uncertainty. Plenium Press, New York (1988)
8. Dubois, D., Prade, H.: An introductory survey of possibility theory and its recent developments. Journal of Japan Society for Fuzzy Theory and Systems 10, 21–42 (1998)
9. Dubois, D., Prade, H.: Possibility theory in information fusion. In: Proceedings of the Third International Conference on Information Fusion, Paris (2000)
10. Dubois, D., Prade, H.: Information bipolaires. une introduction. Information-Interaction-Intelligence 3(1), 89–106 (2003)
11. Dubois, D., Prade, H.: Bipolar representations in reasoning, knowledge extraction and decision processes. In: Greco, S., Hata, Y., Hirano, S., Inuiguchi, M., Miyamoto, S., Nguyen, H.S., Słowiński, R. (eds.) RSCTC 2006. LNCS (LNAI), vol. 4259, pp. 15–26. Springer, Heidelberg (2006)
12. Fabris, F.: On a measure of possibilistic non-compatibility with an application to data transmission. Journal of Interdisciplinary Mathematics 5(3), 203–220 (2002)
13. Fixsen, D., Mahler, R.P.S.: The modified dempster-shafer approach to classification. IEEE. Trans. Syst. Man and Cybern. 27, 96–104 (1997)
14. Fono, L.A., Gwet, H., Bouchon-Meunier, B.: Fuzzy implication operators for difference operations for fuzzy sets and cardinality-based measures of comparison. European Journal of Operational Research 183, 314–326 (2007)
15. Higashi, M., Klir, G.J.: Measures of uncertainty and information based on possibility distributions. International Journal of General Systems 9(1), 43–58 (1883)
16. Higashi, M., Klir, G.J.: On the notion of distance representing information closeness: Possibility and probability distributions. International Journal of General Systems 9, 103–115 (1983)
17. Jenhani, I., Amor, N.B., Elouedi, Z., Benferhat, S., Mellouli, K.: Information affinity: a new similarity measure for possibilistic uncertain information. In: Mellouli, K. (ed.) ECSQARU 2007. LNCS (LNAI), vol. 4724, pp. 840–852. Springer, Heidelberg (2007)
18. Jenhani, I., Benferhat, S., Elouedi, Z.: Properties analysis of inconsistency-based possibilistic similarity measures. In: Proceedings of IPMU 2008 (2008)
19. Jousselme, A.L., Grenier, D., Bossé, E.: A new distance between two bodies of evidence. Information Fusion 2, 91–101 (2001)
20. Kroupa, T.: Application of the choquet integral to measures of information in possibility theory. International Journal of Intelligent Systems 21(3), 349–359 (2006)
21. Kullback, S., Leibler, R.A.: On information and sufficiency. Annals of Mathematical Statistics 22, 79–86 (1951)
22. Sanguesa, R., Cabos, J., Cortes, U.: Possibilistic conditional independence: a similarity based measure and its application to causal network learning. International Journal of Approximate Reasoning 18, 145–167 (1998)
23. Sanguesa, R., Cortes, U.: Prior knowledge for learning networks in non-probabilistic settings. International Journal of Approximate Reasoning 24, 103–120 (2000)
24. Shafer, G.: A mathematical theory of evidence. Princeton Univ. Press, Princeton (1976)
25. Shannon, C.E.: A mathematical theory of communication. The Bell Systems Technical Journal 27(3), 379–423, 623–656 (1948)
26. Smets, P.: The transferable belief model for quantified belief representation. Handbook of defeasible reasoning and uncertainty management systems 1, 267–301 (1998)
27. Tessem, B.: Approximations for efficient computation in the theory of evidence. Artificial Intelligence 61, 315–329 (1993)
28. Tversky, A.: Features of similarity. Psychological Review 84, 327–352 (1977)

29. Wang, X., Baets, B.D., Kerre, E.: A comparative study of similarity measures. Fuzzy Sets and Systems 73(2), 259–268 (1995)
30. Yager, R.R.: On the specificity of a possibility distribution. Fuzzy Sets and Systems 50, 279–292 (1992)
31. Zadeh, L.A.: Fuzzy sets as a basic for a theory of possibility. Fuzzy Sets and systems 1, 3–28 (1978)
32. Zouhal, L.M., Denoeux, T.: An evidence-theoric *k*-nn rule with paprameter optimization. IEEE Trans. Syst. Man Cybern. 28(2), 263–271 (1998)

Efficient Thresholded Tabulation for Fuzzy Query Answering

Pascual Julián, Jesús Medina, Ginés Moreno, and Manuel Ojeda-Aciego

Abstract. Fuzzy logic programming represents a flexible and powerful declarative paradigm amalgamating fuzzy logic and logic programming, for which there exists different promising approaches described in the literature. In this work we propose an improved fuzzy query answering procedure for the so-called multi-adjoint logic programming approach, which avoids the re-evaluation of goals and the generation of useless computations thanks to the combined use of tabulation with thresholding techniques. The general idea is that, when trying to perform a computation step by using a given program rule R, we firstly analyze if such step might contribute to reach further significant solutions (non-tabulated yet). When it is the case, it is possible to avoid useless computation steps via rule R by using thresholds and filters based on the truth degree of R, as well as a safe, accurate and dynamic estimation of the maximum truth degree associated to its body.

Keywords: Fuzzy Logic Programming, Tabulation, Thresholding.

Pascual Julián
University of Castilla-La Mancha, Department of Information Technologies and Systems, Paseo de la Universidad, 4, 13071 Ciudad Real (Spain)
e-mail: `Pascual.Julian@uclm.es`

Jesús Medina
University of Cádiz, Department of Mathematics, CASEM-Campus Río San Pedro, 11510 Puerto Real —Cádiz— (Spain)
e-mail: `Jesus.Medina@uca.es`

Ginés Moreno
University of Castilla-La Mancha, Department of Computing Systems, Avd. de España s/n, Campus Universitario, 02071 Albacete (Spain)
e-mail: `Gines.Moreno@uclm.es`

Manuel Ojeda-Aciego
University of Málaga, Department of Applied Mathematics, Complejo Tecnológico, Campus de Teatinos, 29071 Málaga (Spain)
e-mail: `aciego@uma.es`

B. Bouchon-Meunier et al. (Eds.) Found. of Reas. under Uncert., STUDFUZZ 249, pp. 125–141.
springerlink.com

1 Introduction

Fuzzy Logic Programming is an interesting and still growing research area that agglutinates efforts to introduce fuzzy logic into Logic Programming. During the last decades, several fuzzy logic programming systems have been developed [8,17,3,7], where the classical inference mechanism of SLD resolution is replaced with a fuzzy variant which is able to handle partial truth and to reason under uncertainty.

This is the case of the extremely flexible framework of *multi-adjoint logic programming* [20,21,22]. Given a multi-adjoint logic program, queries are evaluated in two separate computational phases. Firstly, an *operational* phase in which *admissible steps* (a generalization of the classical *modus ponens* inference rule) are systematically applied by a backward reasoning procedure, in a similar way to classical resolution steps in pure logic programming; until an expression is obtained in which all atoms have been evaluated. Then, this last expression is interpreted in the underlying lattice during an *interpretive* phase [13], providing the computed answer for the given query.

In [5] a non-deterministic tabulation goal-oriented proof procedure was introduced for residuated (a particular case of multi-adjoint) logic programs over complete lattices. The underlying idea of tabulation is, essentially, that atoms of selected tabled predicates as well as their answers are stored in a table. When an identical atom is recursively called, the selected atom is not resolved against program clauses; instead, all corresponding answers computed so far are looked up in the table and the associated answer substitutions are applied to the atom. The process is repeated for all subsequent computed answer substitutions corresponding to the atom.

In [14] a fuzzy partial evaluation framework was introduced for specializing multi-adjoint logic programs. Moreover, it was pointed out that if the proposed partial evaluation process is combined with thresholding techniques, the following benefits can be obtained:

- The *unfolding tree* (i.e., an incomplete search tree used during the partial evaluation process) consumes less computational resources by efficiently pruning unnecessary branches of the tree and, hence, drastically reducing its size.
- Those derivation sequences performed at execution time need less computation steps to reach computed answers.

In this work, we show how the essence of thresholding can be also embedded into a tabulation-based query answering procedure, reinforcing the benefits of both methods in a unified framework. We also provide several kinds of "thresholding filters" which largely help to avoid the generation of redundant and useless computations.

The structure of the rest of the work is as follows. In Section 2 we summarize the main features of multi-adjoint logic programming. Section 3 adapts to the multi-adjoint logic framework the original tabulation procedure for residuated logic programs of [5]. Inspired by [14], the resulting method is refined by using thresholding techniques in Section 4. The benefits of such combination are reinforced in Section 5. In the final section, we draw some conclusions and discuss some lines of future work.

2 Multi-Adjoint Logic Programs

This section is a short summary of the main features of multi-adjoint languages. The reader is referred to [20, 22] for a complete formulation.

We will consider a language, $\mathscr{L}$, containing propositional variables, constants, and a set of logical connectives. In our fuzzy setting, we use implication connectives ($\leftarrow_1, \leftarrow_2, \ldots, \leftarrow_m$) together with a number of aggregators. They will be used to combine/propagate truth values through the rules. The general definition of aggregation operators subsumes conjunctive operators (denoted by $\&_1, \&_2, \ldots, \&_k$), disjunctive operators ($\vee_1, \vee_2, \ldots, \vee_l$), and average and hybrid operators (usually denoted by $@_1, @_2, \ldots, @_n$).

Aggregators are useful to describe/specify user preferences: when interpreted as a truth function they may be considered, for instance, as an arithmetic mean or a weighted sum. For example, if an aggregator @ is interpreted as $[\![@]\!](x,y,z) = (3x + 2y + z)/6$, $x, y, z \in [0,1]$, we are giving the highest preference to the first argument, then to the second, being the third argument the least significant. By definition, the truth function for an n-ary aggregator $[\![@]\!] : L^n \to L$ is required to be increasing in each argument and fulfill $[\![@]\!](\top, \ldots, \top) = \top$, $[\![@]\!](\bot, \ldots, \bot) = \bot$.

The language $\mathscr{L}$ will be interpreted on a *multi-adjoint lattice*, $\langle L, \preceq, \leftarrow_1, \&_1, \ldots, \leftarrow_n, \&_n \rangle$, which is a complete lattice equipped with a collection of adjoint pairs $\langle \leftarrow_i, \&_i \rangle$, where each $\&_i$ is a conjunctor[1] intended to provide a *modus ponens*-rule wrt $\leftarrow_i$. In general, the set of truth values L may be the carrier of any complete bounded lattice but, for simplicity, in the examples of this work we shall select L as the set of real numbers in the interval $[0,1]$.

A *rule* is a formula $A \leftarrow_i \mathscr{B}$, where A is an propositional symbol (usually called the *head*) and $\mathscr{B}$ (which is called the *body*) is a formula built from propositional symbols $B_1, \ldots, B_n$ ($n \geq 0$), truth values of L and conjunctions, disjunctions and aggregations. Rules with an empty body are called *facts*. A *goal* is a body submitted as a query to the system.

Roughly speaking, a *multi-adjoint logic program* is a set of pairs $\langle \mathscr{R}; \alpha \rangle$, where $\mathscr{R}$ is a rule and α is a value of L, which might express the confidence which the user of the system has in the truth of the rule $\mathscr{R}$. Note that the truth degrees in a given program are expected to be assigned by an expert. We will often write "$\mathscr{R}$ with α" instead of $\langle \mathscr{R}; \alpha \rangle$.

2.1 Procedural Semantics

The procedural semantics of the multi–adjoint logic language $\mathscr{L}$ can be thought as an operational phase followed by an interpretive one [13].

In the following, $\mathscr{C}[A]$ denotes a formula where A is a sub-expression (usually a propositional symbol) which occurs in the (possibly empty) context $\mathscr{C}[\,]$, whereas $\mathscr{C}[A/A']$ means the replacement of A by A' in context $\mathscr{C}[\,]$. In the following definition, we always consider that A is the selected propositional symbol in goal $\mathscr{Q}$.

[1] An increasing operator satisfying boundary conditions with the top element.

Definition 1 (Admissible Steps). *Let $\mathcal{Q}$ be a goal, which is considered as a state, and let $\mathcal{G}$ be the set of goals. Given a program $\mathbb{P}$, an* admissible computation *is formalized as a state transition system, whose transition relation $\rightarrow_{AS} \subseteq (\mathcal{G} \times \mathcal{G})$ is the smallest relation satisfying the following* admissible rules*:*

1. $\mathcal{Q}[A] \rightarrow_{AS} \mathcal{Q}[A/v \&_i \mathcal{B}]$ if there is a rule $\langle A \leftarrow_i \mathcal{B}; v\rangle$ in $\mathbb{P}$ and $\mathcal{B}$ is not empty.
2. $\mathcal{Q}[A] \rightarrow_{AS} \mathcal{Q}[A/v]$ if there is a fact $\langle A \leftarrow; v\rangle$ in $\mathbb{P}$.
3. $\mathcal{Q}[A] \rightarrow_{AS} \mathcal{Q}[A/\bot]$ if there is no rule in $\mathbb{P}$ whose head is A.

Note that the third case is introduced to cope with (possible) unsuccessful admissible derivations. We shall use the symbols $\rightarrow_{AS1}$, $\rightarrow_{AS2}$ and $\rightarrow_{AS3}$ to distinguish between computation steps performed by applying one of the specific admissible rules. Furthermore, the application of a specific program rule on a step will be annotated as a superscript of the $\rightarrow_{AS}$ symbol, when considered relevant.

Definition 2. *Let $\mathbb{P}$ be a program and let $\mathcal{Q}$ be a goal. An* admissible derivation *is a sequence $\mathcal{Q} \rightarrow^*_{AS} \mathcal{Q}'$. When $\mathcal{Q}'$ is a formula not containing propositional symbols it is called an* admissible computed answer (a.c.a.) *for that derivation.*

Example 1. Let $\mathbb{P}$ be the following program on the unit interval of real numbers $([0,1], \leq)$.

$$\begin{array}{ll} \mathcal{R}_1 : p \leftarrow_{\mathtt{P}} q \&_{\mathtt{G}} r & with\ 0.8 \\ \mathcal{R}_2 : q \leftarrow_{\mathtt{P}} s & with\ 0.7 \\ \mathcal{R}_3 : q \leftarrow_{\mathtt{L}} r & with\ 0.8 \\ \mathcal{R}_4 : r \leftarrow & with\ 0.7 \\ \mathcal{R}_5 : s \leftarrow & with\ 0.9 \end{array}$$

where the labels P, G and L stand for *Product*, *Gödel* and *Łukasiewicz* connectives.

In the following admissible derivation for the program $\mathbb{P}$ and the goal $p \&_{\mathtt{G}} r$, we underline the selected expression in each admissible step:

$$\begin{array}{l} \underline{p} \&_{\mathtt{G}} r \rightarrow_{AS1}{}^{\mathcal{R}_1} \\ (0.8 \&_{\mathtt{P}} (\underline{q} \&_{\mathtt{G}} r)) \&_{\mathtt{G}} r \rightarrow_{AS1}{}^{\mathcal{R}_2} \\ (0.8 \&_{\mathtt{P}} ((0.7 \&_{\mathtt{P}} \underline{s}) \&_{\mathtt{G}} r)) \&_{\mathtt{G}} r \rightarrow_{AS2}{}^{\mathcal{R}_5} \\ (0.8 \&_{\mathtt{P}} ((0.7 \&_{\mathtt{P}} 0.9) \&_{\mathtt{G}} \underline{r})) \&_{\mathtt{G}} r \rightarrow_{AS2}{}^{\mathcal{R}_4} \\ (0.8 \&_{\mathtt{P}} ((0.7 \&_{\mathtt{P}} 0.9) \&_{\mathtt{G}} 0.7)) \&_{\mathtt{G}} \underline{r} \rightarrow_{AS2}{}^{\mathcal{R}_4} \\ (0.8 \&_{\mathtt{P}} ((0.7 \&_{\mathtt{P}} 0.9) \&_{\mathtt{G}} 0.7)) \&_{\mathtt{G}} 0.7 \end{array}$$

The a.c.a. for this admissible derivation is: $(0.8 \&_{\mathtt{P}} ((0.7 \&_{\mathtt{P}} 0.9) \&_{\mathtt{G}} 0.7)) \&_{\mathtt{G}} 0.7$.

If we exploit all propositional symbols of a goal, by applying admissible steps as much as needed during the operational phase, then it becomes a formula with no propositional symbols which can then be directly interpreted in the multi–adjoint lattice L. We recall from [13] the formalization of this process in terms of the following definition.

Definition 3 (Interpretive Step). *Let $\mathbb{P}$ be a program and $\mathcal{Q}$ a goal. We formalize the notion of* interpretive computation *as a state transition system, whose transition relation $\rightarrow_{IS} \subseteq (\mathcal{G} \times \mathcal{G})$ is defined as the least one satisfying: $\mathcal{Q}[@(r_1, r_2)] \rightarrow_{IS}$*

$\mathcal{Q}[@(r_1,r_2)/[\![@]\!](r_1,r_2)]$, *where* $[\![@]\!]$ *is the truth function of connective* @ *in the lattice* $\langle L, \preceq \rangle$ *associated to* $\mathbb{P}$.

Definition 4. *Let* $\mathbb{P}$ *be a program and* $\mathcal{Q}$ *an a.c.a., that is,* $\mathcal{Q}$ *is a goal not containing propositional symbols. An* interpretive derivation *is a sequence* $\mathcal{Q} \rightarrow^*_{IS} \mathcal{Q}'$. *When* $\mathcal{Q}' = r \in L$, *being* $\langle L, \preceq \rangle$ *the lattice associated to* $\mathbb{P}$, *the value r is called a* fuzzy computed answer (f.c.a.) *for that derivation.*

Example 2. We complete the previous derivation of Example 1 by executing the necessary interpretive steps to obtain the final fuzzy computed answer, 0.504, with respect to lattice $([0,1], \leq)$.

$$(0.8\&_P(\underline{(0.7\&_P0.9)}\&_G0.7))\&_G0.7 \rightarrow_{IS}$$
$$(0.8\&_P\underline{(0.63\&_G0.7)})\&_G0.7 \rightarrow_{IS}$$
$$\underline{(0.8\&_P0.63)}\&_G0.7 \rightarrow_{IS}$$
$$\underline{0.504\&_G0.7} \rightarrow_{IS}$$
$$0.504$$

In this section we have just seen a procedural semantics which provides a means to execute multi-adjoint logic programs. However, there exist a more efficient alternative for obtaining fuzzy computed answers for a given query as occurs with the following tabulation-based proof procedure.

3 The Tabulation Proof Procedure

In what follows, we adapt the original tabulation procedure for propositional residuated logic programs described in [5] to the general case of multi-adjoint logic programs [20]. There are two major problems to address: termination and efficiency. On the one hand, the $T_{\mathbb{P}}$ operator is bottom-up but not goal-oriented; furthermore, the bodies of rules are all recomputed in every step. On the other hand, the usual implementations of Fuzzy Logic Programming languages (e.g. [28]) are goal-oriented, but inherit the problems of non-termination and recomputation of goals. In order to overcome these problems, the tabulation technique has been proposed in the deductive databases and logic programming communities. For instance, in [16] it is proposed an extension of SLD for implementing generalized annotated logic programs that will be used to implement the here defined tabling procedure. Other implementation techniques have been proposed for dealing with uncertainty in logic programming, for instance translation into Disjunctive Stable Models [19], but rely on the properties of specific truth-value domains.

The idea of tabulation (or tabling) is simply to create a table for collecting all the answers to a given goal without repetitions. Every time a goal is invoked it is checked whether there is already a table for that goal. If so, the caller becomes a consumer of the tree, otherwise the construction of a new table is started. All produced answers are kept in the table without repetitions, and are propagated to the pending consumers. The most complete implementation of a full working tabulation system is XSB-Prolog [29] which implements SLG resolution. There is also an extension of SLG for generalized annotated logic programs [24, 16] but differs from the system we present here.

In this section we present a general tabulation procedure for propositional multi-adjoint logic programs. The datatype we will use for the description of the method is that of a *forest*, that is, a finite set of trees. Each one of these trees has a root labeled with a propositional symbol together with a truth-value from the underlying lattice (called the *current value* for the *tabulated* symbol); the rest of the nodes of each of these trees are labeled with an "extended" formula in which some of the propositional symbols have been substituted by its corresponding value. For the description of the adaptation of the tabulation procedure to the framework of multi-adjoint logic programming, we will assume a program $\mathbb{P}$ consisting of a finite number of weighted rules together with a query $?A$. The purpose of the computational procedure is to give (if possible) the greatest truth-value for A that can be inferred from the information in the program $\mathbb{P}$.

3.1 Operations for Tabulation

For the sake of clarity in the presentation, we will introduce the following notation: Given a propositional symbol A, we will denote by $\mathbb{P}(A)$ the set of rules in $\mathbb{P}$ which have head A. The tabulation procedure requires four basic operations: Create New Tree, New Subgoal, Value Update, and Answer Return. The first operation creates a tree for the first invocation of a given goal. New Subgoal is applied whenever a propositional variable in the body of a rule is found without a corresponding tree in the forest, and resorts to the Create New Tree operation. Value update is used to propagate the truth-values of answers to the root of the corresponding tree. Finally, answer return substitutes a propositional variable by the current truth-value in the corresponding tree. We now describe formally the operations:

Rule 1: Create New Tree

Given a propositional symbol A, assume $\mathbb{P}(A) = \{\langle A \leftarrow_j \mathscr{B}_j; \vartheta_j \rangle \mid j = 1,\ldots,m\}$ and construct the tree below, and append it to the current forest. If the forest did not exist, then generate a singleton list with the tree.

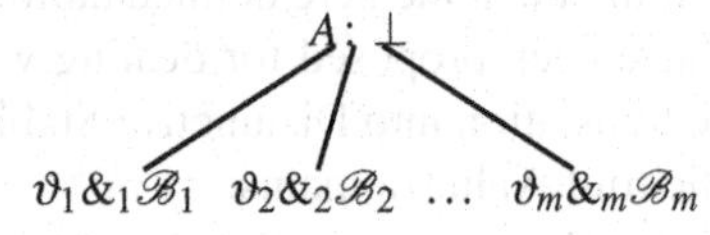

Rule 2: New Subgoal

Select a non-tabulated propositional symbol C occurring in a leaf of some tree (this means that there is no tree in the forest with the root node labeled with C), then create a new tree by directly applying Rule 1, and append it to the forest.

Rule 3: Value Update

If a tree, rooted at $C\colon r$, has a leaf $\mathscr{B}$ with no propositional symbols, and $\mathscr{B} \rightarrow_{IS}{}^{*} s$, where $s \in L$, then update the current value of the propositional symbol C by the value of $\sup_L\{r,s\}$.

Furthermore, once the tabulated truth-value of the tree rooted by C has been modified, for all the occurrences of C in a non-leaf node $\mathscr{B}[C]$ such as the one in the left of the figure below then, update the whole branch substituting the constant u by $\sup_L\{u,t\}$ (where t is the last tabulated truth-value for C, i.e. $\sup_L\{r,s\}$) as in the right of the figure.

$$\mathscr{B}[C] \qquad\qquad \mathscr{B}[C]$$

$$\mathscr{B}[C/u] \qquad\qquad \mathscr{B}[C/\sup_L\{u,t\}]$$

Rule 4: Answer Return

Select in any leaf a propositional symbol C which is tabulated, and assume that its current value is r; then add a new successor node as shown below:

$$\mathscr{B}[C]$$

$$\mathscr{B}[C/r]$$

Once we have presented the rules to be applied in the tabulation procedure, it is worth to recall some facts:

1. The only nodes with several immediate successors are root nodes; the successors correspond to the different rules whose head matches the label of the root node.
2. The leaf of each branch is a conjunction of the truth value of the rule which determined the branch with an instantiation of the body of the rule.
3. The extension of a tree is done only by Rule 4, which applies only to leaves and extends the branch with one new node.
4. The only rule which changes the values of the roots of the trees in the forest is Rule 3 which, moreover, might update the nodes of existing branches.

3.2 A Non-deterministic Procedure for Tabulation

Now, we can state the general non-deterministic procedure for calculating the answer to a given query by using a tabulation technique in terms of the previous rules.

Initial step: Create the initial forest with the *create new tree* rule, applied to the propositional symbol of the query.

Next steps: Non-deterministically select a propositional symbol and apply one of the rules 2, 3, or 4.

Following the steps in [6] it is not difficult to show both that the order of application of the rules is irrelevant, and that the algorithm terminates under very general hypotheses.

Example 3. Consider the following program with mutual recursion and query $?p$:

$\mathcal{R}_1 : p \leftarrow_{\mathrm{P}} q$ with 0.6
$\mathcal{R}_2 : p \leftarrow_{\mathrm{P}} r$ with 0.5
$\mathcal{R}_3 : q \leftarrow$ with 0.9
$\mathcal{R}_4 : r \leftarrow$ with 0.8
$\mathcal{R}_5 : r \leftarrow_{\mathrm{L}} p$ with 0.9

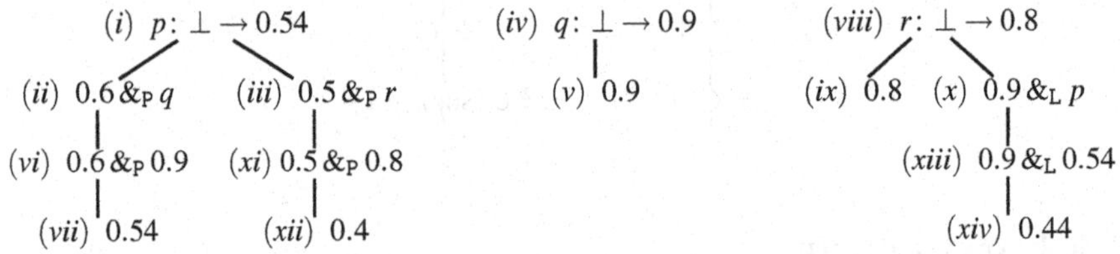

Fig. 1 Example forest for query $?p$

Firstly, the initial tree consisting of nodes $(i),(ii),(iii)$ is generated, see Figure 1. Then *New Subgoal* is applied on q, a new tree is generated with nodes (iv) and (v), and its current value is directly updated to 0.9.

By using this value, *Answer Return* extends the initial tree with node (vi). Now *Value Update* generates node (vii) and updates the current value of p to 0.54.

Then, *New Subgoal* is applied on r, and a new tree is generated with nodes $(viii),(ix)$ and (x). *Value Update* increases the current value to 0.8.

By using this value, *Answer Return* extends the initial tree with node (xi). Now *Value Update* generates node (xii). The current value is not updated since its value is greater than the newly computed one.

Finally, *Answer Return* can be applied again on propositional symbol p in node (x), generating node $(xiii)$. A further application of *Value Update* generates node (xiv) and the forest is terminated, as no rule performs any modification.

4 Combining Tabulation with Thresholding

In this section we will focus on the concept of thresholding, initially proposed in [14] for safely pruning branches when generating *unfolding trees*. The original method was firstly introduced inside the core of a fuzzy *partial evaluation* (PE) framework useful not only for specializing fuzzy programs, but also for generating *reductants* [22]. Reductancts were introduced in the context of multi-adjoint logic programming to cope with a problem of incompleteness that arises for non-linear lattices. For instance, given two non-comparable elements a,b in $\langle L, \preceq \rangle$; assume that for a goal A there are only two facts whose heads are A, namely $\langle A \leftarrow ; a \rangle$ and $\langle A \leftarrow ; b \rangle$; both a and b are correct answers and, moreover, by definition of correct answer [22], the supremum $\sup_L\{a,b\}$, is also a correct answer which cannot be computed. The problem above can be solved by extending the original program with a special rule $\langle A \leftarrow \sup_L\{a,b\}; \top \rangle$, the so-called *reductant*.

The above discussion shows that a multi-adjoint logic program, interpreted inside a partially ordered lattice, needs to contain all its reductants in order to guarantee the completeness property of a sequence of admissible computations. This obviously increases both the size and execution time of the final *"completed"* program. However, this negative effects can be highly diminished if the proposed reductants have been partially evaluated before being introduced in the target program: the computational effort done (just once) at generation time is avoided (many times) at execution time.

Fortunately, if queries are evaluated following the tabulation method proposed before, reductants are not required to be included in a program (which obviously would increase both the size and execution time of the final completed program) because their effects are efficiently achieved by the direct use of *Rule 3: Value Update*, as the reader can easily check. Anyway, even when reductants are not mandatory in the tabulation method described in Section 3, it is important to recast some useful ideas introduced in [14], where a refined notion of reductant (called PE-reductant) was given by using partial evaluation techniques with thresholding. *Partial evaluation* [18, 9, 1] is an automatic program transformation technique aiming at the optimization of a program with respect to parts of its input: hence, it is also known as *program specialization*. It is expected that the *partially evaluated* (or *residual*) program could be executed more efficiently than the original program. This is because the residual program is able to save some computations, at execution time, that were done only once at PE time. To fulfill this goal, PE uses symbolic computation as well as some techniques provided by the field of program transformation [4, 25, 2], specially the so called *unfolding* transformation (essentially, the replacement of a call by its definition body).

Following this path, the idea is to unfold goals, as much as possible, using the notion of unfolding rule developed in [12, 13] for multi-adjoint logic programs, in order to obtain an optimized version of the original program. In [14], the construction of such "unfolding trees" was improved by pruning some useless branches or, more exactly, by avoiding the use (during unfolding) of those program rules whose weights do not surpass a given "threshold" value. For this enhanced definition of unfolding tree we have that:

1. Nodes contain information about an upper bound of the truth degree associated to their goal;
2. A set of threshold values is dynamically set to limit the generation of useless nodes.

This last feature provides great chances to reduce the unfolding tree shape, by stopping unfolding of those nodes whose upper bound for the truth-valued component falls below a threshold value α.

4.1 Rules for Tabulation with Thresholding

In what follows, we will see that the general idea of thresholding can be combined with the tabulation technique shown in the previous section, in order to provide more efficient query answering procedures. Specifically, we will discard the previous descriptions of *Rule 1: Create New Tree* and *Rule 2: New Subgoal*, and instead of them, we propose new definitions:

Rule 1: Root Expansion

Given a tree with root $A: r$ in the forest, and a program rule $\langle A \leftarrow_i \mathscr{B}; \vartheta \rangle$ not consumed before, such that $\vartheta \not\leq r$, append the new child $\vartheta \&_i \mathscr{B}$ to the root of the tree and mark the program rule as consumed.

Rule 2: New Subgoal/Tree

Select a non-tabulated propositional symbol C occurring in a leaf of some tree (this means that there is no tree in the forest with the root node labeled with C), then create a new tree with a single node, the root $C: \bot$, and append it to the forest.

There are several remarks to do regarding the new definitions of Rules 1 and 2. Firstly, notice that the creation of new trees is now performed in Rule 2, instead of Rule 1, which justifies its new name. On the other hand, the new Rule 1, does not create a new tree by expanding (one level) all the possible children of the root. Instead of it, the *Root Expansion* rule has a *lazy* behaviour: each time it is fired, it expands the tree by generating at most one new leaf, if and only if this new leaf might contribute in further steps to reach greater truth degrees than the current one heading the tree. In this sense, the truth degree attached to the root of the tree, acts as a threshold for deciding which program rules can be used for generating new nodes in the tree. Note also that this threshold is dynamically updated by rule *Value Update*: the more it grows, the less chances for *Root Expansion* to create new children of the root.

The new non-deterministic procedure for tabulation with thresholding is as follows:

Initial step: Create an initial tree by using the rule *new subgoal/tree* on the query.
Next steps: Non-deterministically select a propositional symbol and apply one of the rules 1, 2, 3, or 4.

In order to show the correctness of the new tabulation procedure, we have just to note that, in the *Root Expansion* rule, when we generate a leaf $\vartheta \&_i \mathscr{B}$ for a root node $A: r$, the value generated by the leaf will always be less than ϑ, independently of the truth degree eventually computed for the subgoal $\mathscr{B}$. So, we can safely discard at run-time the use of those program rules (or facts) whose weight ϑ falls below the threshold value r. Otherwise, we would generate useless nodes which never would increase the truth degree of the root.

4.2 A Deterministic Procedure for Tabulation with Thresholding

The main goal of thresholding is to reduce the number and size of trees in the forest. This way, although the order of application of the rules is irrelevant because they generate the same solutions, the refinements introduced by thresholding might produce different forests depending on how and when rules are applied. In this section we provide some heuristics in order to minimize as much as possible the complexity of the generated forest.

To begin with, we assume now that the procedure starts with a forest containing a single tree with root $A: \bot$, being A the propositional query we plan to answer.

Obviously, the *Root Expansion* rule has a crucial role in this sense: the more lazily it is applied, the less chances it has to generate new nodes. So, we assign it the lowest priority in our deterministic procedure. For a similar reason, it is also important to increase the threshold at the root of a tree as fast as possible. In order to do this, we propose:

1. Assign maximum priority to *Value Update* and *Answer Return*.
2. When program rules are consumed by *Root Expansion* in a top-down way, we assume that facts textually appear before rules with body, and program rules are distributed in a descending ordering w.r.t. their weights, whenever possible.

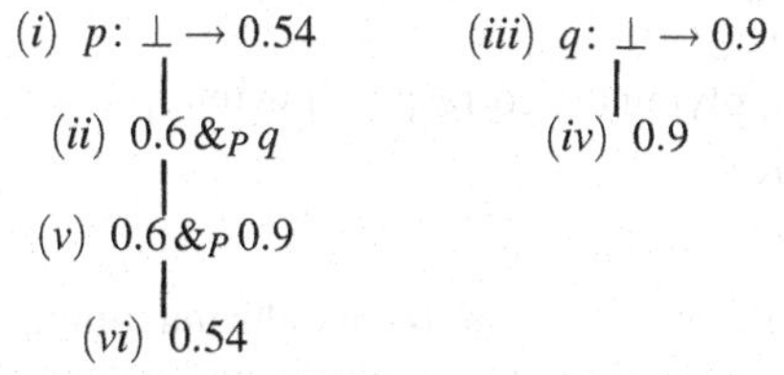

Fig. 2 Example threshold forest for p

Notice for instance, the distribution of the rules in Example 3, which accomplish with the ordering we have just commented. The proposed strategy applied to the example avoids the construction of a number of nodes, see Figure 2, which evidences the benefits of combining tabulation with thresholding.

The answer to the query example with this optimized procedure is as follows: the initial tree consisting of nodes $(i), (ii)$ is generated. Then *New Subgoal/Tree* is applied on q, a new tree is generated with nodes (iii) and (iv), and its current value is directly updated to 0.9.

By using this value, *Answer Return* extends the initial tree with node (v). Now *Value Update* generates node (vi) and updates the current value of p to 0.54.

Now, *Root Expansion* prevents using the rule with body r, since its weight is smaller than the currently computed for p. Hence, the forest is terminated.

5 Reinforcing Thresholding

As we have shown in the previous section, thresholding can be seen as an improvement performed on the core of the basic tabulation proof procedure. The general idea is that all nodes whose value of the body cannot surpass the current value of the root node can be safely removed, or directly, not generated. The thresholding technique described in Section 4 was based on the truth degree of each program rule tried to expand the root of a given tree. However, there is at least two more opportunities for performing thresholding, thus avoiding the unnecessary expansion of trees, as we will see in this section.

A sound rule for determining the maximum value of the body of a program rule, might consist in substituting all the propositional variables occurring in it by the top element of the lattice, $\top$. It is easy to see that this second kind of *filter* can reduce the search space if it is appropriately implemented inside the *Root Expansion Rule*. This idea was initially proposed as a further refinement of the original tabulation method for propositional, residuated logic programs of [5]. In the multi-adjoint logic setting, we also find a recent precedent: the same test was used in the PE-based reductant calculus proposed in [14], when collecting leaves of residual unfolding trees. As we are interested in formalizing the same idea inside our *thresholded tabulation* method for multi-adjoint logic programs, we will consider the following descriptions:

- Let $\mathscr{R} = \langle A \leftarrow_i \mathscr{B}; \vartheta \rangle$ be a program rule.
- Let $\mathscr{B}'$ be an expression with no atoms, obtained from body $\mathscr{B}$ by replacing each occurrence of a propositional symbol by $\top$.
- Let $v \in L$ be the result of interpreting (by applying the corresponding interpretive steps) $\mathscr{B}'$ under a given lattice, i.e. $\mathscr{B}' \rightarrow_{IS}^{*} v$.
- Then, $Up\_body(\mathscr{R}) = v$.

Apart from the truth degree ϑ of a program rule $\mathscr{R} = \langle A \leftarrow_i \mathscr{B}; \vartheta \rangle$ and the maximum truth degree of its body $Up\_body(\mathscr{R})$, in the multi-adjoint logic setting, we can consider a third kind of filter for reinforcing thresholding. The idea is to combine the two previous measures by means of the adjoint conjunction $\&_i$ of the implication $\leftarrow_i$ in rule $\mathscr{R}$. Now, we define the *maximum truth degree of a program rule*, symbolized by function *Up_rule*, as: $Up\_rule(\mathscr{R}) = \vartheta \&_i (Up\_body(\mathscr{R}))$.

Putting all pieces together, we propose the new improved version of the root expansion rule as follows:

Rule 1: Root Expansion

Given a tree with root $A : r$ in the forest, if there is at least a program rule $\mathscr{R} = \langle A \leftarrow_i \mathscr{B}; \vartheta \rangle$ not consumed before and verifying the three conditions below, append the new child $\vartheta \&_i \mathscr{B}$ to the root of the tree.

- Condition 1. $\vartheta \nleq r$.
- Condition 2. $Up\_body(\mathscr{R}) \nleq r$.
- Condition 3. $Up\_rule(\mathscr{R}) \nleq r$.

There are some remarks to do about our definition.

1. The more *filters* for thresholding we use, the more efficient the method becomes, since the number of nodes in trees can be drastically diminished. Note that by avoiding the generation of a single node, the method implicitly avoids the generation of all its possible descendants as well.
2. On the other hand, the time required to properly evaluate the filters is largely compensated by the effects explained in the previous item.
3. Anyway, in order to perform an efficient evaluation of filters, it must be taken into account that a condition only is checked if none of the previous ones fails. In particular, the only situation in which the three filters are completely evaluated appears only when the first two ones do not fail.

In order to illustrate the advantages of our improved method, consider that in our running example, we replace the second program rule $\mathcal{R}_2 : p \leftarrow_{\mathrm{P}} r \; with \; 0.5$ by $\mathcal{R}'_2 : p \leftarrow_{\mathrm{P}} (r \&_{\mathrm{P}} 0.9) \; with \; 0.55$. It is important to note that with the old version (previous section) of the *Root Expansion Rule*, we could not obtain thresholding benefits, due to the new truth degree 0.55 of $\mathcal{R}'_2$. Note also, that this value verifies the first condition of the new *Root Expansion Rule* when building the forest of Figure 2. So, we proceed by evaluating the second one, which is also satisfied since $Up\_body(\mathcal{R}'_2) = 1 * 0.9 = 0.9 \nleq 0.54$. Fortunately, the third condition fails, since $Up\_rule(\mathcal{R}'_2) = 0.55 * 0.9 = 0.495 < 0.54$, which avoids future expansions of the tree and in our case, the process finishes generating exactly the same forest of Figure 2.

5.1 Thresholded Tabulation and Termination

As far as we have seen before, the original procedural semantics of the multi-adjoint logic framework described in Section 2, exhibits a non terminating behaviour in all our previous examples, whereas tabulation has successfully avoided this problem in an elegant, practicable way. Moreover, we have also shown that thresholding has offered great opportunities to pure tabulation for improving its efficiency, by allowing the safe pruning of many useless trees/branches on computational forests. However, the following question arises now: can thresholding safely cut infinite computations that could not be avoided by other simpler operational mechanisms such as the ones proposed in Sections 2 (based on admissible/interpretive steps) and 3 (based on pure tabulation techniques)?

In a propositional logic context, in order to answer this query, we have observed that it is mandatory to consider programs having infinitely many propositional symbols in their signature which should eventually be tabulated: in other words, the Herbrand universe associated to a given fuzzy program must be infinite in order to introduce risks of non termination when executing goals by using a pure tabulation process[2]. This is just the case of our following example.

[2] Obviously, if the number of trees to be built at tabulation time according Section 3 is infinite (for infinite propositional symbols too), the operational semantics described in Section 2 would also produce non-ending branches on the associated computational trees.

Example 4. Consider the following infinite program $\mathbb{P}$:

$$
\begin{array}{llll}
\mathscr{R}_1: & p & \leftarrow_{\mathrm{P}} & & \text{with } 0.9 \\
\mathscr{R}_2: & p & \leftarrow_{\mathrm{P}} & q_1 & \text{with } 0.4 \\
\mathscr{R}_3: & q_1 & \leftarrow_{\mathrm{P}} & q_2 & \text{with } 0.4 \\
\mathscr{R}_4: & q_2 & \leftarrow_{\mathrm{P}} & q_3 & \text{with } 0.4 \\
\mathscr{R}_5: & q_3 & \leftarrow_{\mathrm{P}} & q_4 & \text{with } 0.4 \\
 & & \vdots & &
\end{array}
$$

Here, if we try to solve goal p by using admissible/interpretive steps as described in Section 2, we immediately find the unique solution 0.9 after using the first rule of $\mathbb{P}$ by means of a simple $\rightarrow_{AS2}$ step. Unfortunately, the system enters into a loop when generating the following infinite admissible/interpretive derivation:

$$
\begin{array}{l}
\underline{p} \rightarrow_{AS1}{}^{\mathscr{R}_2} (0.4\&_{\mathrm{P}}(\underline{q_1})) \rightarrow_{AS1}{}^{\mathscr{R}_3} (0.4\&_{\mathrm{P}}(0.4\&_{\mathrm{P}}(\underline{q_2}))) \\
\quad \rightarrow_{AS1}{}^{\mathscr{R}_4} (0.4\&_{\mathrm{P}}(0.4\&_{\mathrm{P}}(0.4\&_{\mathrm{P}}(\underline{q_3})))) \rightarrow_{AS1}{}^{\mathscr{R}_5} \ldots
\end{array}
$$

A similar infinite behaviour emerges when using the (non-thresholded yet) tabulation method introduced in Section 3, where an infinite forest is intended to be built by generating trees rooted with propositional symbols p, q_1, q_2, q_3, $\cdots$

Fortunately, by using the thresholding techniques reported in this work, a very small forest with just a single tree for p is needed to reach the single solution 0.9. And still better, such a tree is not only finite, but it also has a tiny shape: after creating its root and its unique leaf, our first thresholding filter avoids its infinite expansion since it is not allowed to open a new branch (which would generate infinitely many extensions of the forest) by using the second program rule, since its associated truth degree 0.4 does not surpass the best value 0.9 currently found.

Although finding propositional examples with an infinite Herbrand universe seem rather artificial, it is important to note that in first-order this fact is the most common case, since a finite signature containing just a single non-constant function symbol generates infinitely many atoms. For instance, the following finite first-order program $\mathbb{P}'$ has clear correspondences with the previous infinite propositional program $\mathbb{P}$ (note the associations among propositional symbols p, q_1, q_2, q_3, …, and first order atoms $p'(X)$, $q(X)$, $q(s(X))$, $q(s(s(X)))$, …, respectively):

$$
\begin{array}{lllll}
\mathscr{R}_1: & p'(X) & \leftarrow & & \text{with } 0.9 \\
\mathscr{R}_2: & p'(X) & \leftarrow_{\mathrm{P}} & q'(X) & \text{with } 0.4 \\
\mathscr{R}_3: & q'(X) & \leftarrow_{\mathrm{P}} & q'(s(X)) & \text{with } 0.4
\end{array}
$$

We are currently working on the extension of our results to first-order multi-adjoint logic programming (more on this will be said in the final section), but regarding future improvements via thresholding of termination results regarding fuzzy tabulation techniques, we strongly believe that modern termination results on pure logic tabulation methods (see [26], and its counterpart related to partial evaluation techniques reported in [27]), admit much more affordable characterizations in a fuzzy setting thanks to the possibility of flexible and intelligent manipulation of truth degrees. Some experiences, for the specific problem of computing reductants by using methods based on fuzzy partial evaluation with thresholding, are described in [15, 14].

6 Conclusions and Further Research

In this work, an extended and enhanced version of [10, 11], we were concerned with efficient query answering procedures for propositional multi-adjoint logic programs. We have shown that, by using a fuzzy variant of tabulation (specially tailored for the multi-adjoint logic approach) it is possible to avoid the repeated evaluation of redundant goals. Moreover, in the same fuzzy setting, we have also combined tabulation with thresholding, thus safely avoiding other kind of non-redundant, but useless computations.

- Thresholding has been naturally embedded into the core of the tabulation method by reformulating in a lazy way the rule which expands the root node of trees.
- By proposing a deterministic strategy which assigns priorities to each "tabulation rule", it is possible to increase the efficiency of the whole method.
- We exploit three kinds of "thresholding filters" for stopping the creation of new tree nodes and maximally reducing the search space.
- Such filters (based on the truth degree of program rules, an upper bound estimation of the truth degrees of their bodies, and a suitable combination of both values), specially the first and third one, have been specially formulated for the multi-adjoint logic approach, and cannot be applied to other settings not based in weighted rules (such as pure logic programming and residuated logic programming).

Nowadays, we are working in two practical extensions of our approach:

1. In order to cover more realistic programs than the ones reported in this paper, we are enriching our technique to cope with the first-order case. In this sense, we plan to take advantage from the experience acquired in [6] when lifting to this more general case the original tabulation proof procedure for propositional residuated logic programs [5].
2. Regarding implementation issues, our efforts are devoted to incorporate the proposed technique inside the kernel of the FLOPER environment (see, for instance, [23] and visit `http://www.dsi.uclm.es/investigacion/dect/FLOPERpage.htm`). Our tool offers several programming resources regarding the *multi-adjoint logic approach*, including two operational procedures for debugging/tracing and executing goals. The first way is based on a direct translation of fuzzy logic programs into Prolog code in order to safely execute these final programs (via classical SLD-resolution steps) inside any standard Prolog interpreter in a completely transparent way for the final user. The second alternative implements the notion of admissible step seen in Definition 1, in order to generate declarative traces based on unfolding trees with any level of depth. We think that the inclusion of a third operational semantics supporting the thresholded tabulation technique studied so far, will give us great opportunities for highlighting the practical benefits of our approach and providing experimental results.

For the future, beyond query answering procedures, we also plan to study the role that tabulation combined with thresholding might play in program transformation techniques such as partial evaluation and fold/unfold, in order to efficiently specialize and optimize multi-adjoint logic programs.

Acknowledgements. The second and fourth authors would like to thank the wide support of the EU (FEDER), and the Spanish Science and Education Ministry (MEC) (under grant TIN 2006-15455-C03-01) and the Junta de Andalucía (under grant P06-FQM-02049) that partially financed this research work. Also the first and third authors want to show their gratitude with the EU (FEDER), and the Spanish Science and Education Ministry (MEC), which partially supported this work under grant TIN 2007-65749, and with the Castilla-La Mancha Regional Administration, which partially supported this work under grant PII1I09-0117-4481.

References

1. Alpuente, M., Falaschi, M., Julián, P., Vidal, G.: Specialization of Lazy Functional Logic Programs. In: Proc. of the ACM SIGPLAN Conf. on Partial Evaluation and Semantics-Based Program Manipulation, PEPM 1997. Sigplan Notices, vol. 32(12), pp. 151–162. ACM Press, New York (1997)
2. Alpuente, M., Falaschi, M., Moreno, G., Vidal, G.: Rules + Strategies for Transforming Lazy Functional Logic Programs. Theoretical Computer Science 311(1-3), 479–525 (2004)
3. Baldwin, J.F., Martin, T.P., Pilsworth, B.W.: Fril- Fuzzy and Evidential Reasoning in Artificial Intelligence. John Wiley & Sons, Inc., Chichester (1995)
4. Burstall, R., Darlington, J.: A Transformation System for Developing Recursive Programs. Journal of the ACM 24(1), 44–67 (1977)
5. Damásio, C., Medina, J., Ojeda-Aciego, M.: A tabulation proof procedure for residuated logic programming. In: Proc. of the European Conference on Artificial Intelligence. Frontiers in Artificial Intelligence and Applications, vol. 110, pp. 808–812 (2004)
6. Damásio, C., Medina, J., Ojeda-Aciego, M.: Termination of logic programs with imperfect information: applications and query procedure. Journal of Applied Logic 5(3), 435–458 (2007)
7. Guadarrama, S., Muñoz, S., Vaucheret, C.: Fuzzy Prolog: A new approach using soft constraints propagation. Fuzzy Sets and Systems 144(1), 127–150 (2004)
8. Ishizuka, M., Kanai, N.: Prolog-ELF Incorporating Fuzzy Logic. In: Joshi, A.K. (ed.) Proceedings of the 9th International Joint Conference on Artificial Intelligence (IJCAI 1985), Los Angeles, CA, pp. 701–703. Morgan Kaufmann, San Francisco (1985)
9. Jones, N., Gomard, C., Sestoft, P.: Partial Evaluation and Automatic Program Generation. Prentice-Hall, Englewood Cliffs (1993)
10. Julián, P., Medina, J., Moreno, G., Ojeda-Aciego, M.: Combining tabulation and thresholding techniques for executing multi-adjoint logic programs. In: Proc. of the 12th Intl Conf on Information Processing and Management of Uncertainty in Knowledge-based Systems (IPMU 2008), Málaga, Spain, pp. 505–512 (2008)
11. Julián, P., Medina, J., Moreno, G., Ojeda-Aciego, M.: Thresholded tabulation in a fuzzy logic setting. Electronic Notes in Theoretical Computer Science 248, 115–130 (2009)
12. Julián, P., Moreno, G., Penabad, J.: On Fuzzy Unfolding. A Multi-adjoint Approach. Fuzzy Sets and Systems 154, 16–33 (2005)
13. Julián, P., Moreno, G., Penabad, J.: Operational/Interpretive Unfolding of Multi-adjoint Logic Programs. Journal of Universal Computer Science 12(11), 1679–1699 (2006)
14. Julián, P., Moreno, G., Penabad, J.: Efficient reductants calculi using partial evaluation techniques with thresholding. Electronic Notes in Theoretical Computer Science 188, 77–90 (2007)
15. Julián, P., Moreno, G., Penabad, J.: An Improved Reductant Calculus using Fuzzy Partial Evaluation Techniques. Fuzzy Sets and Systems 160, 162–181 (2009)

16. Kifer, M., Subrahmanian, V.: Theory of generalized annotated logic programming and its applications. Journal of Logic Programming 12, 335–367 (1992)
17. Li, D., Liu, D.: A fuzzy Prolog database system. John Wiley & Sons, Inc., Chichester (1990)
18. Lloyd, J., Shepherdson, J.: Partial Evaluation in Logic Programming. Journal of Logic Programming 11, 217–242 (1991)
19. Lukasiewicz, T.: Fixpoint characterizations for many-valued disjunctive logic programs with probabilistic semantics. In: Eiter, T., Faber, W., Truszczyński, M. (eds.) LPNMR 2001. LNCS (LNAI), vol. 2173, pp. 336–350. Springer, Heidelberg (2001)
20. Medina, J., Ojeda-Aciego, M., Vojtáš, P.: Multi-adjoint logic programming with continuous semantics. In: Eiter, T., Faber, W., Truszczyński, M. (eds.) LPNMR 2001. LNCS (LNAI), vol. 2173, pp. 351–364. Springer, Heidelberg (2001)
21. Medina, J., Ojeda-Aciego, M., Vojtáš, P.: A procedural semantics for multi-adjoint logic programming. In: Brazdil, P.B., Jorge, A.M. (eds.) EPIA 2001. LNCS (LNAI), vol. 2258, pp. 290–297. Springer, Heidelberg (2001)
22. Medina, J., Ojeda-Aciego, M., Vojtáš, P.: Similarity-based Unification: a multi-adjoint approach. Fuzzy Sets and Systems 146, 43–62 (2004)
23. Morcillo, P., Moreno, G.: Programming with Fuzzy Logic Rules by using the FLOPER Tool. In: Bassiliades, N., Governatori, G., Paschke, A. (eds.) RuleML 2008. LNCS, vol. 5321, pp. 119–126. Springer, Heidelberg (2008)
24. Swift, T.: Tabling for non-monotonic programming. Annals of Mathematics and Artificial Intelligence 25(3-4), 201–240 (1999)
25. Tamaki, H., Sato, T.: Unfold/Fold Transformations of Logic Programs. In: Tärnlund, S. (ed.) Proc. of Second Int'l Conf. on Logic Programming, pp. 127–139 (1984)
26. Verbaeten, S., De Schreye, D., Sagonas, K.: Termination proofs for logic programs with tabling. ACM Trans. Comput. Logic 2(1), 57–92 (2001)
27. Vidal, G.: Quasi-Terminating Logic Programs for Ensuring the Termination of Partial Evaluation. In: Proc. of the ACM SIGPLAN 2007 Workshop on Partial Evaluation and Program Manipulation (PEPM 2007), pp. 51–60. ACM Press, New York (2007)
28. Vojtáš, P., Paulík, L.: Soundness and completeness of non-classical extended SLD-resolution. In: Herre, H., Dyckhoff, R., Schroeder-Heister, P. (eds.) ELP 1996. LNCS, vol. 1050, pp. 289–301. Springer, Heidelberg (1996)
29. Warren, D.S., et al.: The XSB system version 3.1 volume 1: Programmer's manual. Tech. Rep. Version released on August, 30, Stony Brook University, USA (2007), `http://xsb.sourceforge.net/`

Linguistic Summaries of Time Series: On Some Additional Data Independent Quality Criteria

Janusz Kacprzyk and Anna Wilbik

Abstract. We further extend our approach on the linguistic summarization of time series (cf. Kacprzyk, Wilbik and Zadrożny) in which an approach based on a calculus of linguistically quantified propositions is employed, and the essence of the problem is equated with a linguistic quantifier driven aggregation of partial scores (trends). Basically, we present here some reformulation and extension of our works mainly by including a more complex evaluation of the linguistic summaries obtained. In addition to the basic criterion of a degree of truth (validity), we also use here as the additional criteria a degree of imprecision, specificity, fuzziness and focus. However, for simplicity and tractability, we use in the first shot the degrees of truth (validity) and focus, which usually reduce the space of possible linguistic summaries to a considerable extent, and then – for a usually much smaller set of linguistic summaries obtained – we use the remaining three degrees of imprecision, specificity and fuzziness for making a final choice of appropriate linguistic summaries. We show an application to the absolute performance type analysis of daily quotations of an investment fund.

1 Introduction

Financial data analysis is one of the most important application areas of advanced data mining and knowledge discovery tools and techniques. For instance, in a report

Janusz Kacprzyk,
Systems Research Institute, Polish Academy of Sciences,
ul. Newelska 6, 01-447 Warsaw, Poland
PIAP – Industrial Research Institute for Automation and Measurements,
Al. Jerozolimskie 202, 02-486 Warsaw, Poland
e-mail: kacprzyk@ibspan.waw.pl

Anna Wilbik
Systems Research Institute, Polish Academy of Sciences,
Newelska 6, 01-447 Warsaw, Poland
e-mail: wilbik@ibspan.waw.pl

B. Bouchon-Meunier et al. (Eds.) Found. of Reas. under Uncert., STUDFUZZ 249, pp. 143–166.
springerlink.com

presented by G. Piatetsky-Shapiro's KDNuggets (`http://www.kdnuggets.com`) on top data mining applications in 2008, the first two positions are, in the sense of yearly increase:

- *Investment/Stocks*, up from 3% of respondents in 2007 to 14% of respondents in 2008% (350% increase),
- *Finance*, up form 7.2% in 2007 to 16.8% in 2008 (108% increase).

This general trend will presumably continue over the next years, maybe decades, in view of a world wide financial and economic difficulties that are expected to continue well after 2009.

This paper is a continuation of our previous works (cf. Kacprzyk, Wilbik, Zadrożny [19, 20, 21, 22, 24, 26] or Kacprzyk, Wilbik [12, 13, 14]) which deal with the problem of how to effectively and efficiently support a human decision maker in making decisions concerning investments. We deal mainly with investment (mutual) funds. Clearly, decision makers are here concerned with possible future gains/losses, and their decisions is related to what might happen in the future.

However, our aim is not the forecasting of the future daily prices, which could have been eventually used directly for a purchasing decision. Instead, in our works, we follow a decision support paradigm (Fig. 1), that is we try to provide the decision maker with some information that can be useful for his/her decision on whether and how many units of funds to purchase. We do not intend to replace the human decision maker.

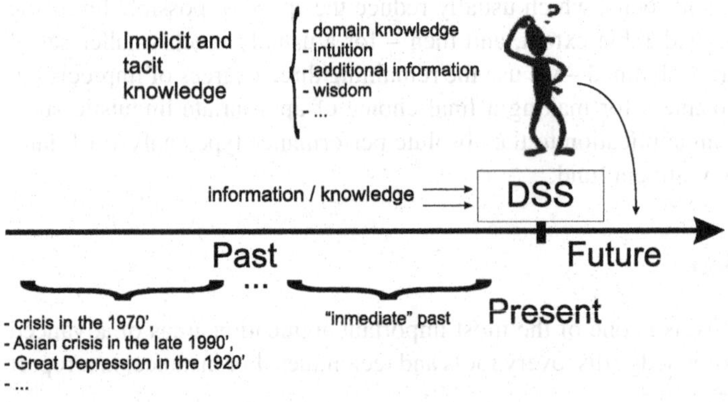

Fig. 1 Decision support paradigm

This problem is very complex. First of all, there may be two general approaches. The first one, which may seem to be the most natural is to provide means to derive a price forecast for an investment unit so that the decision maker could "automatically" purchase what has been forecast, and as much as he/she could wish and/or afford. Unfortunately, the success in such a straightforward approach has been much less than expected. Basically, statistical methods employed usually for this purpose are primitive in the sense that they just extrapolate the past and do not use domain knowledge, intuition, some inside information, etc. A natural solution may be to try to just support the human decision maker in making those investment decisions by providing him/her with some additional useful information, and not getting involved in the actual investment decision making.

Various philosophies in this respect are possible. Basically, from our perspective, the following one will be followed. In all investment decisions the future is what really counts, and the past is irrelevant. But, the past is what we know, and the future is (completely) unknown. Behavior of the human being is to a large extent driven by his/her (already known) past experience. We usually assume that what happened in the past will also happen (to some, maybe large extent) in the future. This is basically, by the way, the very underlying assumption behind the statistical methods too!

This clearly indicates that the past can be employed to help the human decision maker find a good solution. We present here a method to subsume the past, the past performance of an investment (mutual) fund, by presenting results in a very human consistent way, using natural language statements.

We will apply our method to mutual funds quotations, as those time series are easily available, and almost everyone can invest money in a mutual fund. However if one looks at an information leaflet, one may always notice a disclaimer stating that "Past performance is no indication of future returns" which is true. However, on the other hand, in a well known posting "Past Performance Does Not Predict Future Performance" [3], they state something that may look strange in this context, namely: "...according to an Investment Company Institute study, about 75% of all mutual fund investors mistakenly use short-term past performance as their primary reason for buying a specific fund". But, in an equally well known posting "Past performance is not everything" [4], they state: "...disclaimers apart, as a practice investors continue to make investments based on a scheme's past performance. To make matters worse, fund houses are only too pleased to toe the line by actively advertising the past performance of their schemes leading investors to conclude that it is the single-most important parameter (if not the most important one) to be considered while investing in a mutual fund scheme".

As strange as this apparently is, we may ask ourselves why it is so. Again, in a well known posting "New Year's Eve: Past performance is no indication of future return" [2], they say "...if there is no correlation between past performance and future return, why are we so drawn to looking at charts and looking at past performance? I believe it is because it is in our nature as human beings ... because we don't know what the future holds, we look toward the past ...".

And, continuing along this line of reasoning, we can find many other examples of similar statements supporting our position. For instance, in [34], the author says: "...Does this mean you should ignore past performance data in selecting a mutual fund? No. But it does mean that you should be wary of how you use that information ...While some research has shown that consistently good performers continue to do well at a better rate than marginal performers, it also has shown a much stronger predictive value for consistently bad performers ...*Lousy performance in the past is indicative of lousy performance in the future...*". And, further: in [7], we have:

"... there is an important role that past performance can play in helping you to make your fund selections. While you should disregard a single aggregate number showing a fund's past long-term return, you can learn a great deal by studying the *nature of its past returns*. Above all, look for consistency.". In [36], we find: "While past performance does not necessarily predict future returns, it can tell you how volatile a fund has been". In the popular "A 10-step guide to evaluating mutual funds" [1], they say in the last, tenth, advise: "Evaluate the funds performance. Every fund is benchmarked against an index like the BSE Sensex, Nifty, BSE 200 or the CNX 500 to cite a few names. Investors should compare fund performance over varying time frames vis-a-vis both the benchmark index and peers. Carefully evaluate the funds performance across market cycles particularly the downturns".

Therefore we think, that linguistic summaries may be easily understood by the humans and present them briefly the performance of the mutual fund, and this knowledge may be later incorporated while making up decisions.

Here we extend our previous works on linguistic summarization of time series (cf. Kacprzyk, Wilbik, Zadrożny [19, 20, 21, 22, 24, 26] or Kacprzyk, Wilbik [12, 13, 14]), mainly towards a more complex evaluation of results. Generally the basic criterion for the evaluation of linguistic summaries is a degree of truth (used initially by us in first our papers [19, 20, 21, 26]) as it was originally proposed in the static context by Yager [39]. However, later Kacprzyk and Yager [27] and Kacprzyk, Yager and Zadrożny [28, 29] and Kacprzyk and Zadrożny [10, 11] introduced some additional quality criteria. One of those was a degree of imprecision.

In this paper we will discuss the degree of imprecision, as well as two other measures similar in spirit, namely the degree of specificity and the degree of fuzziness. The purpose of this is to provide an additional mechanism for the selection of proper linguistic data summaries. Basically, there may often be a situation that two linguistic data summaries have the same degree of truth, and are therefore considered to be equally good while employing the basic criterion of a degree of truth. However, one of them has a higher degree of imprecision and/or fuzziness, and may be therefore considered to provide information that is too general to be useful for the user.

2 Linguistic Data Summaries

Under the term *linguistic data (base) summary* we understand a (usually short) sentence (or a few sentences) that captures the very essence of the set of data, that is numeric, large, and because of its size is not comprehensible for human being.

In Yager's basic approach [39], and later papers on this topic, as well as here the following notation is used:

- $Y = \{y_1, y_2, \ldots, y_n\}$ is the set of objects (records) in the database D, e.g., a set of employees;
- $A = \{A_1, A_2, \ldots, A_m\}$ is the set of attributes (features) characterizing objects from Y, e.g., a salary, age in the set of employees.

A linguistic summary includes:

- a summarizer P, i.e. an attribute together with a linguistic value (fuzzy predicate) defined on the domain of attribute A_j (e.g. *low* for attribute *salary*);
- a quantity in agreement Q, i.e. a linguistic quantifier (e.g. *most*);
- truth (validity) $\mathcal{T}$ of the summary, i.e. a number from the interval $[0,1]$ assessing the truth (validity) of the summary (e.g. 0.7);
- optionally, a qualifier R, i.e. another attribute together with a linguistic value (fuzzy predicate) defined on the domain of attribute A_k determining a (fuzzy) subset of Y (e.g. *young* for attribute *age*).

Thus, a linguistic summary may be exemplified by

$$\mathcal{T}(\textit{most} \text{ of employees earn } \textit{low} \text{ salary}) = 0.7 \tag{1}$$

or in richer (extended) form, including a qualifier (e.g. *young*), by

$$\mathcal{T}(\textit{most} \text{ of } \textit{young} \text{ employees earn } \textit{low} \text{ salary}) = 0.82 \tag{2}$$

Thus, basically the core of a linguistic summary is a linguistically quantified proposition in the sense of Zadeh [45] which for (1) may be written as

$$Qy's \text{ are } P \tag{3}$$

and for (2) may be written as

$$QRy's \text{ are } P \tag{4}$$

Then the truth (validity), $\mathcal{T}$, of a linguistic summary directly corresponds to the truth value of (3) and (4). This may be calculated using either original Zadeh's calculus of quantified propositions (cf. [45]) or other interpretations of linguistic quantifiers. In the former case the truth values of (3) and (4) are calculated, respectively, as

$$\mathcal{T}(Qy's \text{ are } P) = \mu_Q\left(\frac{1}{n}\sum_{i=1}^{n}\mu_P(y_i)\right) \tag{5}$$

$$\mathcal{T}(QRy's \text{ are } P) = \mu_Q\left(\frac{\sum_{i=1}^{n}\mu_P(y_i)\wedge\mu_R(y_i)}{\sum_{i=1}^{n}\mu_R(y_i)}\right) \tag{6}$$

where $\wedge$ is the minimum operation (more generally it can be another appropriate operator, notably a t-norm), and Q is a fuzzy set representing the linguistic quantifier in the sense of Zadeh [45], i.e. regular, nondecreasing and monotone:

(a) $\mu_Q(0) = 0$,
(b) $\mu_Q(1) = 1$, and
(c) if $x > y$, then $\mu_Q(x) \geq \mu_Q(y)$;

It may be exemplified by *most* given by

$$\mu_Q(x) = \begin{cases} 1 & \text{for } x \geq 0.8 \\ 2x - 0.6 & \text{for } 0.3 < x < 0.8 \\ 0 & \text{for } x \leq 0.3 \end{cases} \tag{7}$$

Other methods of calculating $\mathscr{T}$ can be used here, notably those based on OWA (ordered weighted averaging) operators (cf. Yager [40, 42] and Yager and Kacprzyk [44]), and the Sugeno and Choquet integrals (cf. Bosc and Lietard [8] or Grabisch [9]).

3 Linguistic Summaries of Trends

In our first approach we summarize the trends (segments) extracted from time series. Therefore as the first step we need to extract the segments. We assume that segment is represented by a fragment of straight line, because such segments are easy for interpretation.

There are many algorithms for the piecewise linear segmentation of time series data, including e.g. on-line (sliding window) algorithms, bottom-up or top-down strategies (cf. Keogh [30, 31]). In our works [16, 17, 18, 19, 21, 22, 23, 24, 25] we used a simple on-line algorithm, a modification of the Sklansky and Gonzalez one [37].

We consider the following three features of (global) trends in time series:

1. dynamics of change,
2. duration, and
3. variability.

By *dynamics of change* we understand the speed of change of the consecutive values of time series. It may be described by the slope of a line representing the trend, represented by a linguistic variable.

Duration is the length of a single trend, and is also represented by a linguistic variable.

Variability describes how "spread out" a group of data is. We compute it as a weighted average of values taken by some measures used in statistics: (1) the range, (2) the interquartile range (IQR), (3) the variance, (4) the standard deviation, and (5) the mean absolute deviation (MAD). This is also treated as a linguistic variable.

For practical reasons for all we use a fuzzy granulation (cf. Bathyrshin et al. [5, 6]) to represent the values by a small set of linguistic labels as, e.g.: quickly increasing, increasing, slowly increasing, constant, slowly decreasing, decreasing, quickly decreasing. These values are equated with fuzzy sets.

For clarity and convenience we employ Zadeh's [46] protoforms for dealing with linguistic summaries [11]. A protoform is defined as a more or less abstract prototype (template) of a linguistically quantified proposition. We have two types of protoforms of linguistic summaries of trends:

- a short form:

$$\text{Among all segments, } Q \text{ are } P \tag{8}$$

e.g.: "Among all segments, *most* are *slowly increasing*".

- an extended form:

$$\text{Among all } R \text{ segments, } Q \text{ are } P \tag{9}$$

e.g.: "Among all *short* segments, *most* are *slowly increasing*".

The protoforms are very convenient for various reasons, notably: they make it possible to devise general tools and techniques for dealing with a variety of statements concerning different domains and problems, and their form is often easily comprehensible to domain specialists.

In static context Kacprzyk and Yager [27], Kacprzyk, Yager and Zadrożny [28, 29], and Kacprzyk and Zadrożny [10, 11] proposed several additional quality criteria, except from the basic one, the truth value. One of those was, among others, degree of imprecision. We will discuss it here, as well as two other measures similar in spirit, namely degree of specificity and degree of fuzziness.

Generating the set of summaries requires checking many possible summaries and may be time consuming. However we follow a simplified approach, in that, we use a two-level procedure:

1. we reduce the search space of possible linguistic summaries – for this purpose we use the truth value and the degree of focus, and then
2. we additionally use the remaining degrees of imprecision, specificity and fuzziness.

This heuristic method makes it possible to generate good summaries in computationally reasonable time.

3.1 Truth Value

The truth value (a degree of truth or validity), introduced by Yager in [39], is the basic criterion describing the degree of truth (in $[0,1]$) to which a linguistically quantified proposition equated with a linguistic summary is true.

Using Zadeh's calculus of linguistically quantified propositions [45] it is calculated in dynamic context using the same formulas as in the static case. Thus, the truth value is calculated for the simple and extended form as, respectively:

$$\mathscr{T}(\text{Among all } y\text{'s, } Q \text{ are } P) = \mu_Q\left(\frac{1}{n}\sum_{i=1}^{n}\mu_P(y_i)\right) \tag{10}$$

$$\mathscr{T}(\text{Among all } Ry\text{'s, } Q \text{ are } P) = \mu_Q\left(\frac{\sum_{i=1}^{n}\mu_R(y_i)\wedge\mu_P(y_i)}{\sum_{i=1}^{n}\mu_R(y_i)}\right) \tag{11}$$

where $\wedge$ is the minimum operation (more generally it can be another appropriate operator, notably a t-norm). In Kacprzyk, Wilbik and Zadrożny [23] results obtained by using different t-norms were compared. Various t-norms can be in principle used in Zadeh's calculus but clearly their use may result in different results of the linguistic quantifier driven aggregation. It seems that the minimum operation is a good choice since it can be easily interpreted and the numerical values correspond to the intuition.

3.2 Degree of Focus

The very purpose of a degree of focus is to limit the search for best linguistic summaries by taking into account some additional information in addition to the degree of truth (validity). The extended form of linguistic summaries (9) does limit by itself the search space as the search is performed in a limited subspace of all (most) trends that fulfill an additional condition specified by qualifier R. The very essence of the degree of focus is to give the proportion of trends satisfying property R to all trends extracted from the time series. It provides a measure that, in addition to the basic degree of truth (validity), can help control the process of discarding non-promising linguistic summaries. The details are described in Kacprzyk and Wilbik's paper [15].

The degree of focus is similar in spirit to a degree of covering [14], however it measures how many trends fulfill property R. That is, we focus our attention on such trends, fulfilling property R. The degree of focus makes obviously sense for the extended form summaries only, and is calculated as:

$$d_{foc}(\text{Among all } Ry, \text{'s } Q \text{ are } P) = \frac{1}{n}\sum_{i=1}^{n} \mu_R(y_i) \tag{12}$$

In our context, the degree of focus describes how many trends extracted from a given time series fulfill qualifier R in comparison to all extracted trends. If the degree of focus is high, then we can be sure that such a summary concerns many trends, so that it is more general. However, if the degree of focus is low, we may be sure that such a summary describes a (local) pattern seldom occurring.

As we wish to discover a more general, global relationship, we can eliminate linguistic summaries, that concern a small number of trends only. The degree of focus may be used to eliminate the whole groups of extended form summaries for which qualifier R limits the set of possible trends to, for instance, 5%. Such summaries, although they may be very true, will not be representative.

3.3 Degree of Imprecision

A *degree of imprecision*, introduced by Kacprzyk and Yager in [27] and Kacprzyk, Yager and Zadrożny [28], describes how imprecise the fuzzy predicates used in the summary are. This measure does not depend on the data to be summarized, but only on the form of a summary and the definition of linguistic values.

The degree of imprecision of a single fuzzy set A_i, defining the linguistic value of a summarizer, is calculated as

$$im(A_i) = \frac{card\{x \in X_i : \mu_{A_i} > 0\}}{card X_i} \tag{13}$$

In our summaries to define membership functions of the linguistic values we use trapezoidal functions since they are sufficient in most applications [47]. Moreover, they can be very easily interpreted and defined by a user not familiar with fuzzy sets and fuzzy logic, as shown in Figure 2. To represent a fuzzy set with a trapezoidal membership function we need to store four numbers only, a, b, c and d. The use of such a form of a fuzzy set is a compromise between a so-called cointension and computational complexity (cf. Zadeh [47]).

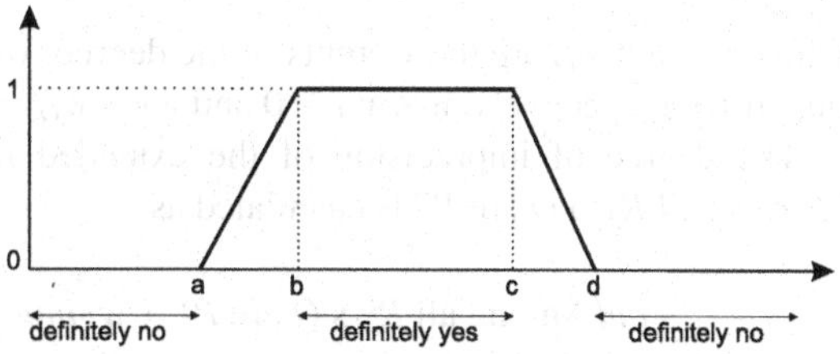

Fig. 2 A trapezoidal membership function of a set

In a case of trapezoidal membership functions, defined as above, the degree of imprecision of a fuzzy set A_i is calculated as:

$$im(A_i) = \frac{d-a}{range(X_i)} \tag{14}$$

where $range(X_i)$ is the range of values taken by the feature considered.

Then, these values – calculated for each fuzzy set A_i belonging to the summarizer – are aggregated using the geometric mean. The degree of imprecision of the summary, or in fact of summarizer P, is therefore calculated as

$$im_P = \sqrt[n]{\prod_{i=1}^{n} im(A_i)} \tag{15}$$

where n is the number of fuzzy predicates in summarizer P which are defined as fuzzy sets A_i.

This degree focuses on the summarizer only. Similarly we can introduce the two additional measures, a degree of imprecision of a qualifier and that of a quantifier, as it was proposed in [35].

Hence, the degree of imprecision of a qualifier is calculated as

$$im_R = \sqrt[n]{\prod_{i=1}^{n} im(A_i)} \tag{16}$$

where n is the number of fuzzy predicates in qualifier R which are defined as fuzzy sets A_i.

And the degree of imprecision of a quantifier is calculated as

$$im_Q = im(Q) \tag{17}$$

We can aggregate those three measures using the the weighted average. Then the degree of imprecision of a simple form of the linguistic summary “Among all y’s Q are P” is calculated as

$$im(\text{Among all } y\text{'s } Q \text{ are } P) = w_P im_P + w_Q im(Q) \tag{18}$$

where w_P and w_Q are the weights of the degrees of imprecision of summarizer and quantifier, respectively. $w_P, w_Q \geq 0$ and $w_P + w_Q = 1$.

The degree of imprecision of the extended form of the linguistic summary “Among all Ry’a Q are P” is calculated as

$$im(\text{Among all } Ry\text{'s } Q \text{ are } P) = w_P im_P + w_R im_R + w_Q im(Q) \tag{19}$$

where w_P, w_Q and w_R are the weights of the degrees of imprecision of summarizer, quantifier and qualifier, respectively. $w_P, w_Q, w_R \geq 0$ and $w_P + w_Q + w_R = 1$.

In the fuzzy set theory there are other concepts capturing the notion of uncertainty, like e.g. specificity or fuzziness.

3.4 Degree of Specificity

The concept of specificity provides a measure of the amount of information contained in a fuzzy subset or possibility distribution. The specificity measure evaluates the degree to which a fuzzy subset points to one and only one element as its member, cf. Yager [43]. It is closely related to the inverse of the cardinality of a fuzzy set. Klir (cf. Klir and Wierman [32] or Klir and Yuan [33]) has proposed the notion of nonspecificity.

We will now consider the original Yager’s proposal [43] in which the specificity measures a degree to which a fuzzy subset contains one and only one element. The measure of specificity is a measure $Sp : I^X \longrightarrow I$, $I \in [0,1]$ if it has the following properties:

- $Sp(A) = 1$ if and only if $A = \{x\}$, (is a singleton set),
- $Sp(\varnothing) = 0$,
- $\frac{\partial Sp(A)}{\partial a_1} > 0$ and $\frac{\partial Sp(A)}{\partial a_j} \leq 0$ for all $j \geq 2$, where A is a fuzzy subset over X and a_j is the j-th largest membership grade in A.

Yager [38] proposed a measure of specificity as

$$Sp(A) = \int_0^{\alpha_{max}} \frac{1}{card(A_\alpha)} d\alpha \tag{20}$$

where α_{max} is the largest membership grade in A, A_α is the α-level set of A, (i.e. $A_\alpha = \{x : A(x) \geq \alpha\}$) and $cardA_\alpha$ is the number of elements in A_α.

Let X be a continuous space, e.g. a real interval. Yager [41] proposed a general class of specificity measures in the continuous domain as

$$Sp(A) = \int_0^{\alpha_{max}} F(\mu(A_\alpha))d\alpha \qquad (21)$$

where α_{max} is the maximum membership grade in A, F is a function $F : [0,1] \longrightarrow [0,1]$ such that $F(0) = 1$, $F(1) = 0$ and $F(x) \leq F(y) \leq 0$ for $x > y$, μ is a fuzzy measure (cf. e.g. Grabisch [9]) and A_α is the α-level set.

If F is defined as $F(z) = 1 - z$, measure μ of an interval $[a,b]$ is defined as $\mu([a,b]) = b - a$, and the space is normalized to $[0,1]$, then the degree of specificity of the fuzzy set A is calculated as

$$Sp(A) = \alpha_{max} - \text{area under } A \qquad (22)$$

If the fuzzy set A has a trapezoidal membership function, as e.g. shown in Figure 2, then

$$Sp(A) = 1 - \frac{c + d - (a + b)}{2} \qquad (23)$$

In most applications, both the fuzzy predicates P and R are assumed to be of a rather simplified, atomic form referring to just one attribute. They can be extended to cover more sophisticated summaries involving some confluence of various attribute values as, e.g, "slowly decreasing and short" trends. To combine more then one attribute values we will use t-norms (for instance, the minimum or product) for conjunction and a corresponding s-norm (for instance, the maximum or probabilistic sum, respectively) for disjunction.

We can aggregate the degrees of specificity of a summarizer, qualifier and quantifier using the weighted average. Then the degree of specificity of the simple form of the linguistic summary "Among all y's Q are P" is calculated as

$$im(\text{Among all } y\text{'s } Q \text{ are } P) = w_P Sp(P) + w_Q Sp(Q) \qquad (24)$$

where w_P and w_Q are the weights of the degrees of specificity of the summarizer and quantifier, respectively. $w_P, w_Q \geq 0$ and $w_P + w_Q = 1$.

The degree of specificity of the extended form of the linguistic summary "Among all Ry's Q are P" is calculated as

$$im(\text{Among all } Ry\text{'s } Q \text{ are } P) = w_P Sp(P) + w_R Sp(R) + w_Q Sp(Q) \qquad (25)$$

where w_P, w_Q and w_R are the weights of the degrees of specificity of summarizer, quantifier and qualifier, respectively. $w_P, w_Q, w_R \geq 0$ and $w_P + w_Q + w_R = 1$.

If we consider the approach proposed by Klir and his collaborators (cf. Klir and Wierman [32] or Klir and Yuan [33]) then the nonspecificity measure from fuzzy sets theory is defined using the so-called Hartley function. For a finite, nonempty (crisp) set, A, we measure this amount using a function from the class of functions

$$U(A) = c \log_b |A|, \qquad (26)$$

where $|A|$ denotes the cardinality of A, b and c are positive constants, $b,c \geq 1$ (usually, $b = 2$ and $c = 1$). This function is applicable to finite sets only but it can be modified for infinite sets of $\mathbb{R}$ as follows: $U(A) = \log[1 + \mu(A)]$, where $\mu(A)$ is the measure of A defined by the Lebesque integral of the characteristic function of A. When $A = [a,b]$, then $\mu(A) = b - a$ and $U([a,b]) = \log[1 + b - a]$.

For any nonempty fuzzy set A defined on a finite universal set X, function $U(A)$ has the form

$$U(A) = \frac{1}{h(A)} \int_0^{h(A)} \log_2 |A^\alpha| d\alpha, \tag{27}$$

where $|A^\alpha|$ is the cardinality of the α-cut of A and $h(A)$ – the height of A. If A is a normal fuzzy set, then $h(A) = 1$.

If a nonempty fuzzy set is defined in $\mathbb{R}$ and the α-cuts are infinite sets (e.g., intervals of real numbers), then:

$$U(A) = \frac{1}{h(A)} \int_0^{h(A)} \log[1 + \mu(A^\alpha)] d\alpha, \tag{28}$$

For convenience, the values of nonspecificity are normalized.

Then the degree of specificity of "Among all y's, Q are P" may be:

$$d_s(\text{Among all } y\text{'s } Q \text{ are } P) = 1 - U(P) \tag{29}$$

and the degree of specificity of "Among all Ry's, Q are P" may be:

$$d_s(\text{“Among all } Ry\text{'s }, Q \text{ are } P\text{”}) = 1 - (U(P) \wedge U(R)) \tag{30}$$

where $U(P)$ is the degree of nonspecificity of the summarizer P, given by (28), $U(R)$ is the degree of nonspecificity of the qualifier R, and $\wedge$ is a t-norm (minimum or product).

We must emphasize the distinction between specificity and fuzziness. Fuzziness is generally related to the lack of clarity, relating to the membership of some set, whereas specificity is related to the lack of exact knowledge of some attribute.

3.5 Degree of Fuzziness

A degree of fuzziness describes a degree of imprecision (which may well be equated with fuzziness) of the linguistic predicates in the summary. In general, a measure of fuzziness of a fuzzy set is a function $f : \mathscr{F} \longrightarrow \mathbb{R}^+$, where $\mathscr{F}$ denotes the family of all fuzzy subsets of X. In other words, for each fuzzy set A, this function assigns a nonnegative real number $f(A)$ that expresses a degree to which the boundary of A is not sharp.

The function f must satisfy the following three requirements (cf. Klir and Yuan [33]):

1. $f(A) = 0$ iff A is a crisp set.
2. $f(A)$ attains its maximum value iff $A(x) = 0.5$ for all $x \in X$
3. $f(A) \leq f(B)$ when set A is undoubtly sharper than set B:

- $A(x) \leq B(x)$ when $B(x) \leq 0.5$ for all $x \in X$, or
- $A(x) \geq B(x)$ when $B(x) \geq 0.5$ for all $x \in X$.

One way to measure the fuzziness of A is by using a distance (metric) between its membership function and the membership function of its nearest crisp set defined as: a *nearest crisp set* of a fuzzy set A is a set $\underline{A} \subset X$ given by its characteristic function:

$$\mu_{\underline{A}} = \begin{cases} 0 & \mu_A(x) \leq 0.5 \\ 1 & \mu_A(x) > 0.5 \end{cases} \tag{31}$$

Then, using different distance function we can obtain different measures, for instance:

- the linear degree of fuzziness:

$$\delta(A) = \frac{2}{n} \sum_{x \in X} |\mu_A(x_i) - \mu_{\underline{A}}(x_i)| \tag{32}$$

- the quadratic degree of fuzziness:

$$\eta(A) = \frac{2}{n} \sqrt{\sum_{x \in X} (\mu_A(x_i) - \mu_{\underline{A}}(x_i))^2} \tag{33}$$

- the vector degree of fuzziness

$$\nu(A) = \frac{2}{n} \sum_{x \in X} \mu_{\underline{A} \cap \neg A}(x_i) \tag{34}$$

Another way of measuring the (degree of) fuzziness of a fuzzy set is to measure a (degree of) lack of distinction between a fuzzy set and its complement. Of course, also here we can choose different forms of the fuzzy complements and distance functions.

If we choose the standard complement and the Hamming distance, we have:

$$f(A) = \sum_{x \in X} (1 - |2A(x) - 1|) \tag{35}$$

where the range of f is $[0, |X|]$, $f(A) = 0$ iff A is a crisp set and $A = |X|$ when $A(x) = 0.5$ for all $x \in X$.

The above form is only valid for fuzzy sets defined in finite universes of discourse. However we can modify it to fuzzy sets defined in $\mathbb{R}$, the set of real numbers: if $X = [a, b]$, then

$$f(A) = \int_a^b (1 - |2A(x) - 1|)dx = b - a - \int_a^b |2A(x) - 1|dx \tag{36}$$

and this form of $f(.)$ will be used here.

If the set A has a trapezoidal membership function, as e.g. shown in Figure 2, then

$$f(A) = \frac{b + d - (a + c)}{2} \tag{37}$$

In general, the summarizer and the qualifier may involve more than one attribute value. To combine them we will use a t-norm (for instance, the minimum or product) for conjunction and a corresponding s-norm (for instance, the maximum or probabilistic sum, respectively) for the disjunction.

The degree of fuzziness of "Among all y's, Q are P" is:

$$d_f(\text{Among all } y\text{'s } Q \text{ are } P) = f(P) \wedge f(Q) \tag{38}$$

where $f(P)$ is the degree of fuzziness of the summarizer P, $f(Q)$ is the degree of fuzziness of the quantifier Q, and $\wedge$ is a t-norm (minimum or product).

The degree of fuzziness of "Among all Ry's, Q are P" is:

$$d_f(\text{Among all } Ry\text{'s } Q \text{ are } P) = f(P) \wedge f(R) \wedge f(Q) \tag{39}$$

where $f(P)$ is the degree of fuzziness of the summarizer P, $f(R)$ is the degree of fuzziness of the qualifier R, $f(Q)$ is the degree of fuzziness of the quantifier Q, and $\wedge$ is a t-norm (minimum or product).

The degree of fuzziness is not of high importance in evaluation of the summaries. However we discussed it for completeness.

4 Numerical Experiments

The method proposed in this paper was tested on data on quotations of an investment (mutual) fund that invests at least 50% of assets in shares listed at the Warsaw Stock Exchange. Data shown in Figure 3 were collected from January 2002 until the end of March 2009 with the value of one share equal to PLN 12.06 in the beginning of the period to PLN 21.82 at the end of the time span considered (PLN stands for the Polish Zloty). The minimal value recorded was PLN 9.35 while the maximal one during this period was PLN 57.85. The biggest daily increase was equal to PLN 2.32, while the biggest daily decrease was equal to PLN 3.46.

It should be noted that the example shown below is meant to illustrate the method proposed by analyzing the absolute performance of a given investment fund. We do not deal here with a presumably more common way of analyzing an investment fund by relating its performance to a benchmark (or benchmarks) exemplified by an average performance of a group of (similar) funds, a stock market index or a synthetic index reflecting, for instance, the bond versus stock allocation.

Using the modified Sklansky and Gonzalez algorithm (cf. [37]) and $\varepsilon = 0.25$ we obtained 422 extracted trends. The shortest trend took 1 time unit only, while the longest one – 71. The histograms for duration, dynamics of change and variability are shown in Figure 4.

We have applied different granulations, namely with 3, 5 and 7 labels for each feature (dynamics of change, duration and variability). Minimal accepted truth value was 0.6 and the degree of focus threshold was 0.1. The degree of focus, and the method of effective and efficient generating summaries is described in Kacprzyk and Wilbik's paper [15].

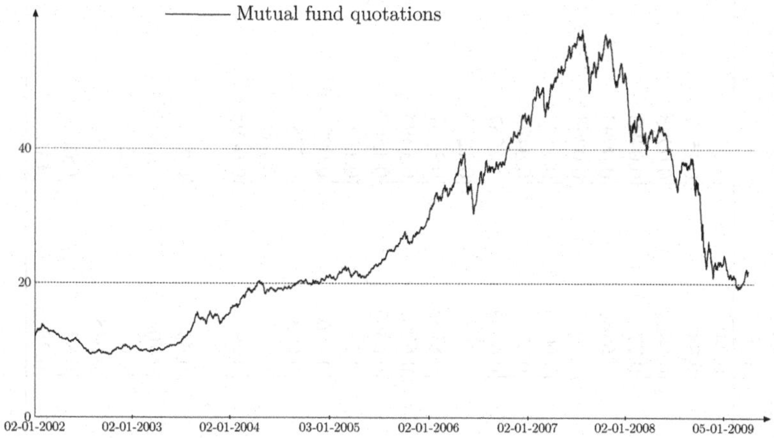

Fig. 3 Mutual fund quotations

When we have used 3 labels for dynamics of change (decreasing, constant and increasing), 3 labels for duration (short, medium length and long) and 3 labels for variability (low, moderate and high), then we have obtained the summaries shown in Table 1.

The linguistic summaries are sorted according to the truth values, and later by the values of degree of focus. The simple form summaries are before the extended ones with the same truth value. The summaries here have high values of specificity, indicating that they may be potentially useful for the user. Only a few summaries have the degree of imprecision greater than 0.5, and they should be analyzed with care. The values of degree of fuzziness are small, only for 3 summaries they exceed the value of 0.2.

Let us now slightly modify the used properties. We add linguistic labels *A*, *B*, *C*. Their membership functions together with the membership function of the fuzzy set with label *low* are depicted in Fig. 5.

The values of the degree of imprecision, specificity and fuzziness of a single fuzzy set are shown in Table 2.

Let us now analyze some of the summaries obtained.

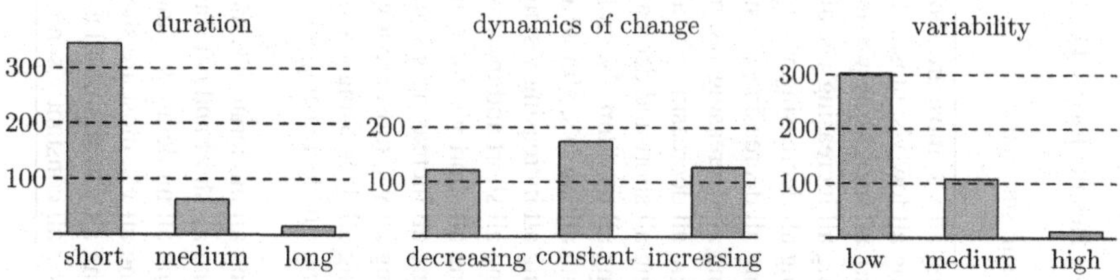

Fig. 4 Histograms of duration, dynamics of change and variability

Table 1 Results for 3 labels

linguistic summary	truth value	degree of focus	degree of imprecision	degree of specificity	degree of fuzziness
Among all y's, most are short	1		0.385	0.745	0.135
Among all low y's, most are short	1	0.7227	0.39	0.73	0.1567
Among all increasing y's, most are short	1	0.2984	0.4047	0.6867	0.0993
Among all increasing y's, almost all are short	1	0.2984	0.2713	0.77	0.0493
Among all decreasing y's, most are short	1	0.2880	0.4047	0.6867	0.0993
Among all decreasing y's, most are short and low	1	0.2880	0.4371	0.6067	0.0993
Among all decreasing y's, most are low	1	0.2880	0.5147	0.6067	0.1593
Among all decreasing y's, almost all are short	1	0.2880	0.2713	0.77	0.0493
Among all short and decreasing y's, most are low	1	0.2842	0.4254	0.6067	0.1567
Among all medium y's, most are constant	1	0.1308	0.3393	0.765	0.1253
Among all low y's, almost all are short	0.9674	0.7227	0.2567	0.8133	0.1067
Among all increasing y's, most are low	0.9610	0.2984	0.5147	0.6067	0.1593
Among all short and increasing y's, most are low	0.9588	0.2946	0.4254	0.6067	0.1567
Among all short y's, most are low	0.9483	0.8341	0.39	0.73	0.1567
Among all increasing y's, most are short and low	0.9386	0.2984	0.4371	0.6067	0.0993
Among all y's, most are low	0.8455		0.55	0.625	0.225
Among all decreasing y's, almost all are low	0.8122	0.2880	0.3813	0.69	0.1093
Among all decreasing y's, almost all are short and low	0.7916	0.2880	0.3038	0.69	0.0493
Among all moderate y's, most are short	0.7393	0.2483	0.4567	0.6967	0.2233
Among all short and constant y's, most are low	0.7325	0.2565	0.4028	0.7033	0.1567
Among all moderate y's, most are constant	0.7024	0.2483	0.4893	0.67	0.2353
Among all y's, most are short and low	0.6915		0.4337	0.625	0.135
Among all y's, almost all are short	0.6706		0.185	0.87	0.06
Among all constant y's, most are short	0.6405	0.4136	0.3127	0.7833	0.1087

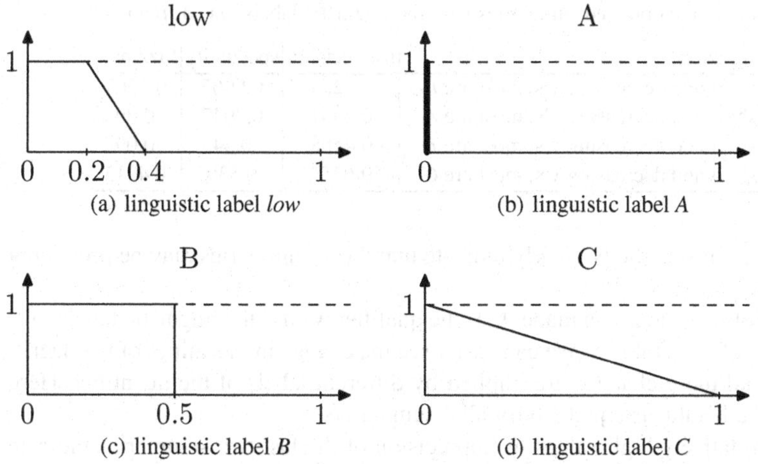

Fig. 5 Illustration of the membership functions for the linguistic labels *low*, *A*, *B* and *C*

Here, in Table 3, we may observe that higher values of the degree of specificity (or lower of the degree of imprecision) may result in lower truth values. The low or high values of fuzziness do not considerably affect the results in this case.

Those big differences in the values of degrees of imprecision, specificity and fuzziness are visible also in evaluation of more complex linguistic summaries, e.g. Table 4. The values of the degree of truth and the values of the degree of focus are not shown in Table 4 as they are the same for all 4 summaries and equal to 1.0 and 0.2842, respectively.

The high values of imprecision of the last two summaries indicate, that they are general, and should be analyzed with special care. On the other hand, high values

Table 2 The values of imprecision, specificity and fuzziness of fuzzy sets representing linguistic labels *low*, *A*, *B* and *C*

lingistic label	imprecision	specificity	fuzziness
low	0.4	0.7	0.2
A	0.006	0.994	0
B	0.5	0.5	0
C	1	0.497	0.994

Table 3 Some of the obtained summaries with the linguistic labels *low*, *A*, *B* and *C*

linguistic summary	truth value	imprecision	specificity	fuzziness
Among all *y*'s, most are *low*	0.8455	0.55	0.625	0.225
Among all *y*'s, most are *A*	0.5280	0.353	0.772	0.125
Among all *y*'s, most are *B*	1	0.6	0.525	0.125
Among all *y*'s, most are *C*	1	0.85	0.5235	0.622

Table 4 Some of the obtained summaries with the linguistic labels *low*, *A*, *B* and *C*

linguistic summary	imprecision	specificity	fuzziness
Among all short and decreasing *y*'s, most are *low*	0.4254	0.6067	0.1567
Among all short and decreasing *y*'s, most are *A*	0.2941	0.7047	0.09
Among all short and decreasing *y*'s, most are *B*	0.4588	0.54	0.09
Among all short and decreasing *y*'s, most are *C*	0.6254	0.539	0.4213

of the degree of specificity clearly indicate that those summaries may be promising and useful.

Similar observations can made if, as the qualifier we use the linguistic labels *low*, *A*, *B* and *C*, cf. e.g. Table 5. We can easily see the change in the values of the degree of focus, and these changes are implied by different labels of the quantifier. Here again the truth value is equal 1.0 for all 4 summaries.

Again, relatively high values of imprecision of the last two summaries indicate that they are general, and should be analyzed with special care. Very high values of the degree of specificity of the two first summaries indicate that those summaries may be promising and useful.

For 5 labels for the dynamics of change (quickly decreasing, decreasing, constant, increasing and quickly increasing), 5 labels for the duration (very short, short, medium length, long and very long) and 5 labels for the variability (very low, low, moderate, high and very high) we have obtained the summaries shown in Table 6.

Similarly, the linguistic summaries are sorted first according to the truth values, and later according to the values of the degree of focus. The summaries here have high values of the degree of specificity, indicating that they may be potentially useful for the user. Values of the degree of specificity are higher than the values in the case with 3 linguistic labels. Values of the degree of imprecision as well as the ones of the degree of focus are smaller than in the case with 3 linguistic labels.

We also have used 7 labels for dynamics of change (quickly decreasing, decreasing, slowly decreasing, constant, slowly increasing, increasing and quickly increasing), 7 labels for duration (very short, short, rather short, medium length, rather long, long and very long) and 7 labels for variability (very low, low, rather low, moderate, rather high, high and very high) to describe the segments. In this case the summaries obtained are shown in Table 7.

As previously the linguistic summaries are sorted first according to the truth values, and later by the values of the degree of focus. There are only 4 summaries that "globally" describe the situation. Those are: one of a simple form and 3 with high

Table 5 Some of the obtained summaries with the linguistic labels *low*, *A*, *B* and *C*

linguistic summary	focus	imprecision	specificity	fuzziness
Among all *low* *y*'s, most are short	0.7227	0.39	0.73	0.1567
Among all *A* *y*'s, most are short	0.5640	0.2587	0.828	0.09
Among all *B* *y*'s, most are short	0.8910	0.4233	0.6633	0.09
Among all *C* *y*'s, most are short	0.8353	0.59	0.6623	0.4213

Table 6 Results for 5 labels

linguistic summary	truth value	degree of focus	degree of imprecision	degree of specificity	degree of fuzziness
Among all very short *y*'s, most are very low	1	0.7180	0.3167	0.7867	0.1233
Among all very low *y*'s, most are very short	1	0.6141	0.3167	0.7867	0.1233
Among all very low *y*'s, almost all are very short	1	0.6141	0.1833	0.87	0.0733
Among all increasing *y*'s, most are very short	1	0.1903	0.2967	0.7993	0.1087
Among all quickly decreasing *y*'s, most are very short	1	0.1484	0.3613	0.73	0.0933
Among all quickly decreasing *y*'s, most are very short and very low	1	0.1484	0.378	0.6933	0.0933
Among all quickly decreasing *y*'s, most are very low	1	0.1484	0.4113	0.6933	0.126
Among all quickly decreasing *y*'s, almost all are very short	1	0.1484	0.228	0.8133	0.04933
Among all quickly decreasing *y*'s, almost all are very short and very low	1	0.1484	0.2447	0.7767	0.0493
Among all quickly decreasing *y*'s, almost all are very low	1	0.1484	0.278	0.7767	0.076
Among all very short and quickly decreasing *y*'s, most are very low	1	0.1464	0.3431	0.6933	0.1233
Among all decreasing *y*'s, most are very short	1	0.1434	0.2967	0.7993	0.1087
Among all very short and decreasing *y*'s, most are very low	1	0.1275	0.3279	0.7627	0.1233
Among all quickly increasing *y*'s, most are very short	1	0.1101	0.3613	0.73	0.0933
Among all quickly increasing *y*'s, most are very short and very low	1	0.1101	0.378	0.6933	0.0933
Among all quickly increasing *y*'s, most are very low	1	0.1101	0.4113	0.6933	0.126
Among all quickly increasing *y*'s, almost all are very short	1	0.1101	0.228	0.8133	0.0493
Among all quickly increasing *y*'s, almost all are very short and very low	1	0.1101	0.2447	0.7767	0.04933
Among all quickly increasing *y*'s, almost all are very low	1	0.1101	0.278	0.7767	0.076
Among all very short and quickly increasing *y*'s, most are very low	1	0.1100	0.3431	0.6933	0.1233
Among all decreasing *y*'s, almost all are very short	0.9446	0.1434	0.1633	0.8827	0.0587
Among all short *y*'s, most are constant	0.8999	0.1979	0.3193	0.78	0.1153
Among all low *y*'s, most are constant	0.8872	0.1471	0.3893	0.7367	0.1687
Among all decreasing *y*'s, most are very low	0.8585	0.1434	0.3467	0.7627	0.1353
Among all decreasing *y*'s, most are very short and very low	0.8477	0.1434	0.3133	0.7627	0.1087
Among all *y*'s, most are very short	0.8360		0.375	0.755	0.135
Among all very short and increasing *y*'s, most are very low	0.7720	0.1611	0.3279	0.7627	0.1233
Among all very short and constant *y*'s, most are very low	0.7572	0.1857	0.3306	0.7533	0.1233
Among all increasing *y*'s, almost all are very short	0.7325	0.1903	0.1633	0.8827	0.0587
Among all moderate *y*'s, most are constant	0.6674	0.1987	0.4227	0.7033	0.1687
Among all *y*'s, most are very low	0.6282		0.45	0.7	0.175

Table 7 Results for 7 labels

linguistic summary	truth value	degree of focus	degree of imprecision	degree of specificity	degree of fuzziness
Among all very low y's, most are very short	1	0.5984	0.3	0.795	0.1067
Among all very low y's, almost all are very short	1	0.5984	0.1667	0.8783	0.0567
Among all quickly decreasing y's, most are very short	1	0.1484	0.3613	0.73	0.0933
Among all quickly decreasing y's, most are very short and very low	1	0.1484	0.3735	0.7017	0.0933
Among all quickly decreasing y's, most are very low	1	0.1484	0.3947	0.7017	0.1093
Among all quickly decreasing y's, almost all are very short	1	0.1484	0.228	0.8133	0.0493
Among all quickly decreasing y's, almost all are very short and very low	1	0.1484	0.2402	0.785	0.0493
Among all quickly decreasing y's, almost all are very low	1	0.1484	0.2613	0.785	0.0593
Among all very short and quickly decreasing y's, most are very low	1	0.1464	0.3264	0.7017	0.1067
Among all increasing y's, most are very short	1	0.1345	0.2873	0.8087	0.1087
Among all slowly decreasing y's, most are very short	1	0.1124	0.278	0.818	0.1087
Among all quickly increasing y's, most are very short	1	0.1101	0.3613	0.73	0.0933
Among all quickly increasing y's, most are very short and very low	1	0.1101	0.3735	0.7017	0.0933
Among all quickly increasing y's, most are very low	1	0.1101	0.3947	0.7017	0.1093
Among all quickly increasing y's, almost all are very short	1	0.1101	0.228	0.8133	0.0493
Among all quickly increasing y's, almost all are very short and very low	1	0.1101	0.2402	0.785	0.0493
Among all quickly increasing y's, almost all are very low	1	0.1101	0.2613	0.785	0.0593
Among all very short and quickly increasing y's, most are very low	1	0.1100	0.3264	0.7017	0.1067
Among all very short y's, most are very low	0.9847	0.7180	0.3	0.795	0.1067
Among all y's, most are very short	0.8360		0.375	0.755	0.135
Among all very short and increasing y's, most are very low	0.7938	0.11533	0.3083	0.7803	0.1067
Among all increasing y's, almost all are very short	0.7881	0.1345	0.154	0.892	0.0587
Among all slowly increasing y's, most are very short	0.7591	0.1372	0.278	0.818	0.1087
Among all slowly decreasing y's, most are very low	0.6361	0.1124	0.3113	0.7897	0.1187

values of the degree of focus. The degree of focus for the other summaries is smaller than 15% so that they describe patterns more locally occurring. The summaries here have high values of the degree of specificity, indicating that they may be potentially useful for the user. Values of the degree of imprecision as well as the ones of the degree of fuzziness are small. These small values of the degree of imprecision are implied by a fine granulation used.

Let us note that the degree of imprecision and the degree of specificity are to some extent related, notably large values of the degree of imprecision are associated with small values of the degree of specificity and vice versa. The degree of fuzziness describes a different aspect. So it is possible to have a summary with a very small value of the degree of specificity (i.e. big of the degree of imprecision) which may have either a very small or big value of the degree of fuzziness. However, if a summary has a very high degree of specificity, then its degree of fuzziness is low.

5 Concluding Remarks

We extended our approach to the linguistic summarization of time series based on a calculus of linguistically quantified propositions used for a linguistic quantifier driven aggregation of partial scores (trends). We presented a reformulation and extension of our works mainly by including a more complex evaluation of the linguistic summaries obtained. In addition to the degree of truth (validity), we additionally used a degree of imprecision, specificity, fuzziness and focus. However, for simplicity and numerical tractability, we used in the first shot the degrees of truth (validity) and focus, to reduce the space of possible linguistic summaries, and then – for a usually much smaller set of linguistic summaries obtained – we used the remaining three degrees of imprecision, specificity and fuzziness for making a final choice of appropriate linguistic summaries. So this does not guarantee the optimality, our experience however suggests that it makes possible to generate good summaries in computationally reasonable time. A more formalized approach of this heuristic method to find best summaries will be presented in next papers. We showed an application to the absolute performance type analysis of daily quotations of an investment fund. The results obtained give more insight into the nature of the time series of quotations analyzed, and may be very useful for supporting decision makers.

References

1. A 10-step guide to evaluating mutual funds, `http://www.personalfn.com/detail.asp?date=5/18/2007&story=2`
2. New year's eve: past performance is no indication of future return, `http://stockcasting.blogspot.com/2005/12/new-years-evepast-performance-is-no.html`
3. Past performance does not predict future performance, `http://www.freemoneyfinance.com/2007/01/past_performanc.html`

4. Past performance is not everything, http://www.personalfn.com/detail.asp?date=9/1/2007&story=3
5. Batyrshin, I.: On granular derivatives and the solution of a granular initial value problem. International Journal Applied Mathematics and Computer Science 12(3), 403–410 (2002)
6. Batyrshin, I., Sheremetov, L.: Perception based functions in qualitative forecasting. In: Batyrshin, I., Kacprzyk, J., Sheremetov, L., Zadeh, L.A. (eds.) Perception-based Data Mining and Decision Making in Economics and Finance. Springer, Heidelberg (2006)
7. Bogle, J.C.: Common Sense on Mutual Funds: New Imperatives for the Intelligent Investor. Wiley, New York (1999)
8. Bosc, P., Lietard, L., Pivet, O.: Quantified statements and database fuzzy queries. In: Bosc, P., Kacprzyk, J. (eds.) Fuzziness in Database Management Systems. Springer, Heidelberg (1995)
9. Grabisch, M.: Fuzzy integral as a flexible and interpretable tool of aggregation. In: Bouchon-Meunier, B. (ed.) Aggregation and Fusion of Imperfect Information, pp. 51–72. Physica–Verlag, Heidelberg (1998)
10. Kacprzyk, J., Zadrożny, S.: Fuzzy linguistic data summaries as a human consistent, user adaptable solution to data mining. In: Gabrys, B., Leiviska, K., Strackeljan, J. (eds.) Do Smart Adaptive Systems Exist?, pp. 321–339. Springer, Heidelberg (2005)
11. Kacprzyk, J., Zadrożny, S.: Linguistic database summaries and their protoforms: toward natural language based knowledge discovery tools. Information Sciences 173, 281–304 (2005)
12. Kacprzyk, J., Wilbik, A.: An extended, specificity based approach to linguistic summarization of time series. In: Proceedings of the 12th International Conference Information Processing and Management of Uncertainty in Knowledge-based Systems, pp. 551–559 (2008)
13. Kacprzyk, J., Wilbik, A.: Linguistic summarization of time series using linguistic quantifiers: augmenting the analysis by a degree of fuzziness. In: Proceedings of 2008 IEEE World Congress on Computational Intelligence, pp. 1146–1153. IEEE Press, Los Alamitos (2008)
14. Kacprzyk, J., Wilbik, A.: A new insight into the linguistic summarization of time series via a degree of support: Elimination of infrequent patterns. In: Dubois, D., Lubiano, M.A., Prade, H., Gil, M.A., Grzegorzewski, P., Hryniewicz, O. (eds.) Soft Methods for Handling Variability and Imprecision, pp. 393–400. Springer, Heidelberg (2008)
15. Kacprzyk, J., Wilbik, A.: Towards an efficient generation of linguistic summaries of time series using a degree of focus. In: Proceedings of the 28th North American Fuzzy Information Processing Society Annual Conference – NAFIPS 2009 (2009)
16. Kacprzyk, J., Wilbik, A., Zadrożny, S.: Capturing the essence of a dynamic behavior of sequences of numerical data using elements of a quasi-natural language. In: Proceedings of the 2006 IEEE International Conference on Systems, Man, and Cybernetics, pp. 3365–3370. IEEE Press, Los Alamitos (2006)
17. Kacprzyk, J., Wilbik, A., Zadrożny, S.: A linguistic quantifier based aggregation for a human consistent summarization of time series. In: Lawry, J., Miranda, E., Bugarin, A., Li, S., Gil, M.A., Grzegorzewski, P., Hryniewicz, O. (eds.) Soft Methods for Integrated Uncertainty Modelling, pp. 186–190. Springer, Heidelberg (2006)

18. Kacprzyk, J., Wilbik, A., Zadrożny, S.: Linguistic summaries of time series via a quantifier based aggregation using the Sugeno integral. In: Proceedings of 2006 IEEE World Congress on Computational Intelligence, pp. 3610–3616. IEEE Press, Los Alamitos (2006)
19. Kacprzyk, J., Wilbik, A., Zadrożny, S.: Linguistic summarization of trends: a fuzzy logic based approach. In: Proceedings of the 11th International Conference Information Processing and Management of Uncertainty in Knowledge-based Systems, pp. 2166–2172 (2006)
20. Kacprzyk, J., Wilbik, A., Zadrożny, S.: On some types of linguistic summaries of time series. In: Proceedings of the 3rd International IEEE Conference Intelligent Systems, pp. 373–378. IEEE Press, Los Alamitos (2006)
21. Kacprzyk, J., Wilbik, A., Zadrożny, S.: Linguistic summaries of time series via an owa operator based aggregation of partial trends. In: Proceedings of the FUZZ-IEEE 2007 IEEE International Conference on Fuzzy Systems, pp. 467–472. IEEE Press, Los Alamitos (2007)
22. Kacprzyk, J., Wilbik, A., Zadrożny, S.: Linguistic summarization of time series by using the choquet integral. In: Melin, P., Castillo, O., Aguilar, L.T., Kacprzyk, J., Pedrycz, W. (eds.) IFSA 2007. LNCS (LNAI), vol. 4529, pp. 284–294. Springer, Heidelberg (2007)
23. Kacprzyk, J., Wilbik, A., Zadrożny, S.: Linguistic summarization of time series under different granulation of describing features. In: Kryszkiewicz, M., Peters, J.F., Rybiński, H., Skowron, A. (eds.) RSEISP 2007. LNCS (LNAI), vol. 4585, pp. 230–240. Springer, Heidelberg (2007)
24. Kacprzyk, J., Wilbik, A., Zadrożny, S.: Mining time series data via linguistic summaries of trends by using a modified sugeno integral based aggregation. In: IEEE Symposium on Computational Intelligence and Data Mining, CIDM 2007, pp. 467–472. IEEE Press, Los Alamitos (2007)
25. Kacprzyk, J., Wilbik, A., Zadrożny, S.: On linguistic summaries of time series via a quantifier based aggregation using the sugeno integral. In: Melin, O.C.P., Kacprzyk, J., Pedrycz, W. (eds.) Hybrid Intelligent Systems Analysis and Design, pp. 421–439. Springer, Heidelberg (2007)
26. Kacprzyk, J., Wilbik, A., Zadrożny, S.: Linguistic summarization of time series using a fuzzy quantifier driven aggregation. Fuzzy Sets and Systems 159(12), 1485–1499 (2008)
27. Kacprzyk, J., Yager, R.R.: Linguistic summaries of data using fuzzy logic. International Journal of General Systems 30, 33–154 (2001)
28. Kacprzyk, J., Yager, R.R., Zadrożny, S.: A fuzzy logic based approach to linguistic summaries of databases. International Journal of Applied Mathematics and Computer Science 10, 813–834 (2000)
29. Kacprzyk, J., Yager, R.R., Zadrożny, S.: Fuzzy linguistic summaries of databases for an efficient business data analysis and decision support. In: Zurada, J., Abramowicz, W. (eds.) Knowledge Discovery for Business Information Systems, pp. 129–152. Kluwer, Boston (2001)
30. Keogh, E., Chu, S., Hart, D., Pazzani, M.: An online algorithm for segmenting time series. In: Proceedings of the 2001 IEEE International Conference on Data Mining (2001)
31. Keogh, E., Chu, S., Hart, D., Pazzani, M.: Segmenting time series: A survey and novel approach. In: Last, M., Kandel, A., Bunke, H. (eds.) Data Mining in Time Series Databases. World Scientific Publishing, Singapore (2004)

32. Klir, G.J., Wierman, M.J. (eds.): Uncertainty-Based Information, Elements of Generalized Information Theory. Physica-Verlag (1999)
33. Klir, G.J., Yuan, B. (eds.): Fuzzy Stes and Fuzzy Logic, Theory and Applications. Prentice Hall, Englewood Cliffs (1995)
34. Myers, R.: Using past performance to pick mutual funds. Nation's Business (October 1997), http://www.findarticles.com/p/articles/mi_m1154/is_n10_v85/ai_19856416
35. Niewiadomski, A. (ed.): Methods for the Linguistic Summarization of Data: Aplications of Fuzzy Sets and Their Extensions. Academic Publishing House EXIT (2008)
36. U.S. Securities and Exchange Commission. Mutual fund investing: Look at more than a fund's past performance, http://www.sec.gov/investor/pubs/mfperform.htm
37. Sklansky, J., Gonzalez, V.: Fast polygonal approximation of digitized curves. Pattern Recognition 12(5), 327–331 (1980)
38. Yager, R.R.: Measuring tranquility and anxiety in decision making: An application of fuzzy sets. International Journal of General Systems 8, 139–146 (1982)
39. Yager, R.R.: A new approach to the summarization of data. Information Sciences 28, 69–86 (1982)
40. Yager, R.R.: On ordered weighted averaging aggregation operators in multicriteria decision making. IEEE Transactions on Systems, Man and Cybernetics, SMC-18, 183–190 (1988)
41. Yager, R.R.: On the specificity of a possibility distribution. Fuzzy Sets and Systems 50, 279–292 (1992)
42. Yager, R.R.: Quantifier guided aggregation using OWA operators. International Journal of Intelligent Systems 11, 49–73 (1996)
43. Yager, R.R.: On measures of specificity. In: Kaynak, O., Zadeh, L.A., Türksen, B., Rudas, I.J. (eds.) Computational Intelligence: Soft Computing and Fuzzy-Neuro Integration with Applications, pp. 94–113. Springer, Berlin (1998)
44. Yager, R.R., Kacprzyk, J. (eds.): The Ordered Weighted Averaging Operators: Theory and Applications. Kluwer, Boston (1997)
45. Zadeh, L.A.: Toward a theory of fuzzy information granulation and its centrality in human reasoning and fuzzy logic. Fuzzy Sets and Systems 9(2), 111–127 (1983)
46. Zadeh, L.A.: A prototype-centered approach to adding deduction capabilities to search engines – the concept of a protoform. In: Proceedings of the Annual Meeting of the North American Fuzzy Information Processing Society (NAFIPS 2002), pp. 523–525 (2002)
47. Zadeh, L.A.: Computation with imprecise probabilities. In: IPMU 2008, Torremolinos, Malaga, June 22-27 (2008)

Preference Modelling Using the Level-Dependent Choquet Integral with Respect to Łukasiewicz Filters

Martin Kalina, Dana Hliněná, and Pavol Král'

1 Introduction

We would like to deal with the problem how to identify the best n objects within a small database consisting of N $(n < N)$ objects with respect to the given set of criteria assuming only one decision maker, i.e., we restrict ourselves to the multicriteria mono personal decision making problems called preference modelling where the preference structure is based on a query of the decision maker. We prefer to use a fuzzy preference structure because of its ability to deal with a possible uncertainty and inconsistency of a decision maker. For more details on fuzzy preference modelling we refer the reader to [9]. For recent development on this topic see also [5,29]. We focus on the fuzzy preference relation derived from the set of utility functions for quantitative criteria and the appropriate linguistic scales for ordinal qualitative criteria. Nominal qualitative criteria are omitted. Unfortunately such fuzzy preference relation does not meet any kind of transitivity in general. Even more the related fuzzy incomparability relation can be non-empty. On the other hand, assumed criteria can interact and their weights can be value dependent, i.e., the relative importance of criteria depends on the query and their weights can be different for different levels of values and query.

Martin Kalina
Dept. of Mathematics, Slovak Uni. of Technology, Radlinského 11,
813 68 Bratislava, Slovakia
e-mail: kalina@math.sk

Dana Hliněná
Dept. of Mathematics, FEEC Brno Uni. of Technology Technická 8,
616 00 Brno, Czech Republic,
e-mail: hlinena@feec.vutbr.cz

Pavol Král'
Institute of Mathematics and Computer Science, UMB and MÚ SAV, Ďumbierska 1,
974 11 Banská Bystrica, Slovakia
e-mail: pavol.kral@umb.sk

B. Bouchon-Meunier et al. (Eds.) Found. of Reas. under Uncert., STUDFUZZ 249, pp. 167–188.
springerlink.com

The lack of transitivity, non-empty incomparability relation, and interaction between criteria may cause serious problems to identify the required number of elements for the given query. So we need to find some suitable aggregation procedure (dependent on the actual query of the decision maker) to deal with interacting criteria with value dependent weights such that the procedure leads to a fuzzy preference structure with fuzzy preference relation satisfying some kind of transitivity and, if possible, an empty fuzzy incomparability relation. For simplicity we will assume that all available information is included in the query and in the database. We do not presume any additional information about criterial importance provided by the decision maker. In our opinion a possible solution can be based on a special type of a so-called level-dependent Choquet integral. We will present here a partial solution of this problem.

An alternative approach to this problem is presented in [7] where fuzzy rule base with conditional and unconditional fuzzy rules is assumed. Roughly speaking it means that we can obtain a k-tuple of values from the unit interval representing the coherence between the object and the query. Values included in the k-tuple are then aggregated using some appropriate aggregation procedure, e.g. we can eliminate values which are not sufficiently large etc. The main advantage of this procedure is its simplicity and that we always obtain the ordered set of objects. But there are also several drawbacks. We need a special querying language, for example FuzzySQL (see [7]), the weights for partial criteria cannot be derived from the fuzzy rule base, such procedure does not include the interaction between criteria and does not take into account the fact that the importance of a criterion can be dependent on the given query.

The paper is organized as follows: in Section 2 we present some basic properties of the evaluators (especially T_L-, S_L-evaluators), fuzzy preference structures and study the connection between fuzzy preference and incomparability relations and T_L-evaluators; in Section 3 we present the generalization of the so-called level-dependent Choquet integral, Section 4 is devoted to the description of our proposed algorithm and finally Section 5 illustrates our approach using rather simple, but practically oriented example.

A substantial part of this article was presented by the authors at IPMU 2008, see [14].

2 Preliminaries

First of all we will recall some well-known definitions used in the rest of our paper.

Definition 1. *(e.g. [19]) A triangular norm (t-norm for short) on the unit interval* $[0,1]$ *is a commutative, associative mapping* $T : [0,1]^2 \to [0,1]$ *which is increasing in both places and for which* $T(x,1) = x$*, for all* $x \in [0,1]$.

Remark 1. Note that, if T is a t-norm, then its dual t-conorm $S : [0,1]^2 \to [0,1]$ is given by

$$S(x,y) = 1 - T(1-x, 1-y).$$

Remark 2. Note that each t-norm T and each t-conorm S are associative, i.e., though they are defined as binary operations, they can be uniquely extended into ternary, and by induction also into n-ry, operations by

$$T(x,y,z) = T(T(x,y),z) = T(x,T(y,z))$$
$$S(x,y,z) = S(S(x,y),z) = S(x,S(y,z)).$$

In the rest of our paper we will restrict ourselves to the Łukasiewicz t-norm,

$$T_L(x,y) = \max\{0, x+y-1\}$$

and Łukasiewicz t-conorm,

$$S_L(x,y) = \min\{1, x+y\}.$$

In our considerations we need to construct a comparison of elements from a given set which is based on comparison of numerical evaluations of elements, i.e., we will use evaluators defined on some at most countable fixed set $X \neq \emptyset$. Then we denote by $\mathscr{M}$ the system of all functions $f : X \to [0,1]$. Hence $(\mathscr{M}, \wedge, \vee, \top, \bot)$ is a lattice with top and bottom elements $\top$ and $\bot$, equal to constants 1 and 0, respectively. In [4], evaluators have been defined on the system $\mathscr{M}$.

Remark 3. In the decision-making context the set X represents the set of criteria.

Definition 2. *([4]) A function $\varphi : \mathscr{M} \to [0,1]$ is called an evaluator on $\mathscr{M}$ if*

1. $\varphi(\top) = 1$, $\varphi(\bot) = 0$,
2. for all $f, g \in \mathscr{M}$, if $f \leq g$ then $\varphi(f) \leq \varphi(g)$.

In fact, evaluators can be defined for an arbitrary bounded lattice. However, we restrict our considerations to the lattice $\mathscr{M}$.

For each evaluator φ, we can define its dual evaluator $\bar{\varphi}$ by

$$\bar{\varphi}(f) = 1 - \varphi(1-f).$$

The special type of evaluators, T_L-evaluators and S_L-evaluators, were first proposed in [3]. The more detailed description of T_L- and/or S_L-evaluators can be found in [4].

Definition 3. *([3]) An evaluator $\varphi : \mathscr{M} \to [0,1]$ is said to be a T_L-evaluator on $\mathscr{M}$ if it satisfies the formula*

$$\varphi(f \wedge g) \geq T_L(\varphi(f), \varphi(g)). \tag{1}$$

Definition 4. *([3]) An evaluator $\varphi : \mathscr{M} \to [0,1]$ is said to be an S_L-evaluator on $\mathscr{M}$ if it satisfies the formula*

$$\varphi(f \vee g) \leq S_L(\varphi(f), \varphi(g)). \tag{2}$$

There is a connection between T_L-evaluators and S_L-evaluators and Łukasiewicz filters and Łukasiewicz ideals, respectively.

Definition 5. *([18]) Let $X \neq \emptyset$ be a given at most countable set. Then a set-function $\mathscr{F} : 2^X \to [0,1]$ is said to be a Łukasiewicz filter on X if and only if the following are satisfied:*

1. $\mathscr{F}(X) = 1, \mathscr{F}(\emptyset) = 0$,
2. $(\forall A, B \subseteq X)(A \subseteq B \Rightarrow \mathscr{F}(A) \leq \mathscr{F}(B))$,
3. for all $A, B \subseteq X$ the following holds
$$\mathscr{F}(A \cap B) \geq T_L(\mathscr{F}(A), \mathscr{F}(B)).$$

Łukasiewicz ideals represent a complementary notion to Łukasiewicz filters.

Definition 6. *([13]) Let $X \neq \emptyset$ be a given at most countable set. A set-function $\mathscr{I} : 2^X \to [0,1]$ is said to be a Łukasiewicz ideal on X if and only if:*

1. $\mathscr{I}(\emptyset) = 1, \mathscr{I}(X) = 0$,
2. If $A \subseteq B \subseteq X$, then $\mathscr{I}(A) \geq \mathscr{I}(B)$,
3. for all $A, B \subseteq X$ the following is satisfied
$$T_L(\mathscr{I}(A), \mathscr{I}(B)) \leq \mathscr{I}(A \cup B).$$

It is easy to show that every Łukasiewicz filter $\mathscr{F} : 2^X \to [0,1]$ is connected to a T_L-evaluator (see Lemma 1), and every Łukasiewicz ideal $\mathscr{I} : 2^X \to [0,1]$ is connected to an S_L-evaluator (see Lemma 2). Łukasiewicz filters which are self-dual, will be called Łukasiewicz ultrafilters. Generalizing the results of [3], we get the following assertions.

Lemma 1. ([15]) *Let $\varphi : \mathscr{M} \to [0,1]$ be a T_L-evaluator. Then there exists a Łukasiewicz filter $\mathscr{F} : 2^X \to [0,1]$ such that, for each $A \in 2^X$,*
$$\mathscr{F}(A) = \varphi(1_A).$$

Lemma 2. ([15]) *Let $\varphi : \mathscr{M} \to [0,1]$ be an S_L-evaluator. Then there exists a Łukasiewicz ideal $\mathscr{I} : 2^X \to [0,1]$ such that, for each $A \in 2^X$,*
$$\mathscr{I}(A) = 1 - \varphi(1_A).$$

Obviously, the dual to each T_L-evaluator is an S_L-evaluator and vice versa. By Definitions 3 and 4 we straightforwardly get the following:

Lemma 3. ([15]) *Let $\varphi : \mathscr{M} \to [0,1]$ be a self-dual T_L-evaluator (S_L-evaluator). Then it is also an S_L-evaluator (a T_L-evaluator).*

On the other hand, there are evaluators which are T_L- and S_L-evaluators at the same time. However, they are not self-dual (see [15]).

Example 1. ([15]) Let $X = \{a,b,c,d,e\}$. First, we define a Łukasiewicz filter $\mathscr{F}$: $X \to [0,1]$ by

$$\mathscr{F}(A) = \begin{cases} 0, & \text{if } |A| = 0, \\ 0.25, & \text{if } |A| = 1, \\ 0.5, & \text{if } |A| = 2, \\ 0.5, & \text{if } |A| = 3, \\ 0.75, & \text{if } |A| = 4, \\ 1, & \text{if } |A| = 5, \end{cases}$$

where $|A|$ means the cardinality of A. For functions $f : X \to [0,1]$ we denote

$$core(f) = \{A \subseteq X; (\forall x \in A) f(x) = 1\}$$

and

$$supp(f) = \{B \subseteq X; (\forall x \in B) f(x) > 0\}.$$

Evaluator $\varphi : \mathscr{M} \to [0,1]$, defined as

$$\varphi(f) = \mathscr{F}(core(f)),$$

is both T_L- and S_L-evaluator. However, the dual to φ is

$$\bar{\varphi}(f) = \mathscr{F}(supp(f)).$$

Example 2. ([15]) Let $\mathscr{U} : 2^X \to [0,1]$ be a Łukasiewicz ultrafilter (i.e. a self-dual Łukasiewicz filter). Then $\varphi : \mathscr{M} \to [0,1]$, defined by

$$\varphi(f) = \frac{1}{2}\left(\mathscr{U}(supp(f) + core(f))\right),$$

is a self-dual T_L- and S_L-evaluator.

Example 3. Another example of a self-dual T_L- and S_L-evaluator arises if we take a probability distribution on the set X (which is a Łukasiewicz ultrafilter) and define $\varphi : \mathscr{M} \to [0,1]$ by

$$\varphi(f) = \mu(\{x \in X; f(x) \geq 0.5\}).$$

Another important notion in our considerations is preference structure. The preference structure is a basic concept of preference modelling. In a classical preference structure (PS), a decision-maker makes three decisions for any pair (a,b) from the set A of all alternatives. His or her decision defines a triplet P,I,J of crisp binary relations on A:

1. a is preferred to b if and only if $P(a,b) = 1$ (strict preference).
2. a and b are indifferent if and only if $I(a,b) = 1$ (indifference).
3. a and b are incomparable if and only if $J(a,b) = 1$ (incomparability).

A preference structure (PS) on a set A is a triplet (P,I,J) of binary relations on A, P: $A \times A \to \{0,1\}$, I: $A \times A \to \{0,1\}$, J: $A \times A \to \{0,1\}$, such that

(ps1) I is reflexive, P and J are antireflexive.
(ps2) P is asymmetric, I and J are symmetric.

(ps3) $P \cap I = P \cap J = I \cap J = \emptyset$.
(ps4) $P \cup I \cup J \cup P^t = A \times A$ where $P^t(x,y) = P(y,x)$.

A preference structure can be characterized by the reflexive relation $R = P \cup I$ called the large preference relation. It can be easily proved that

$$P = R \cap (R^t)^c, I = R \cap R^t, J = R^c \cap (R^t)^c,$$

where $R^c(a,b)$ is the complement of $R(a,b)$. This allows us to construct a preference structure (P,I,J) from a reflexive binary operation R only.

Decision-makers are often uncertain, even inconsistent, in their judgements. In such cases, the restriction to two-valued relations has been an important drawback in their practical use. A natural demand led researchers to introducing of a fuzzy preference structure (FPS). The original idea of using numbers between zero and one to describe the strength of links between two alternatives goes back to Menger [21]. The introducing of fuzzy relations enables us to express degrees of preference, indifference and incomparability. Of course, the attempts to simply replace the notion used in the definition of (PS) by their fuzzy equivalents have brought some problems.

To define (FPS) it is necessary to consider some fuzzy connectives. We shall consider a continuous De Morgan triple (T,S,N) consisting of a continuous t-norm T, continuous t-conorm S and a strong negator N satisfying formula $T(x,y) = N(S(N(x),N(y)))$. The main problem lies in the fact that the completeness condition (ps4) can be written in many forms, e.g.:

$$(P \cup P^t)^c = I \cup J, P = (P^t \cup I \cup J)^c,$$

$$P \cup I = (P^t \cup J)^c.$$

Let (T,S,N) be a De Morgan triplet. A fuzzy preference structure (FPS) on a set A is a triplet (P,I,J) of binary fuzzy relations on A such that:

(f1) I is reflexive, P and J are antireflexive. $I(a,a) = 1$, $P(a,a) = J(a,a) = 0$.
(f2) P is T-asymmetric, i.e. $T(P(a,b),P(b,a)) = 0$, and I, J are symmetric.
(f3) $T(P,I) = T(P,J) = T(I,J) = 0$.
(f4) $(\forall (a,b) \in A)\, S(P,P^t,I,J) = 1$ or $N(S(P,I)) = S(P^t,J)$ or other completeness conditions.

In [2, 28] it was shown that the concept of a fuzzy preference structure is only meaningful provided that the de Morgan triplet involved, contains a continuous Archimedean triangular norm having zero divisors. Moreover, any fuzzy preference structure with respect to a de Morgan triplet containing a continuous non-Archimedean triangular norm having zero divisors can be transformed into a fuzzy preference structure with respect to the standard Łukasiewicz triplet. This is why we will use the triplet (T_L,S_L,N), where $N(x) = 1-x$ in this paper.

As it has been already mentioned in our decision-making process it is important to get some kind of transitivity of fuzzy preference relation P. Due to the used triplet (T_L,S_L,N) it is natural to assume the T_L-transitivity of fuzzy preference relation P.

The T_L-transitivity of fuzzy preference relation P means; if a_1, a_2, a_3 are some objects, then the following holds:

$$T_L(P(a_1,a_2),P(a_2,a_3)) \leq P(a_1,a_3). \tag{3}$$

Assuming an evaluator φ, we define the fuzzy preference relation P as aggregation of partial fuzzy preference relations P_i given criterion by criterion

$$P(a,b) = \varphi\left(P_1(a_1,a_2),P_2(a_1,a_2),\ldots,P_n(a_1,a_2)\right), \tag{4}$$

where n is the number of criteria. If φ is a T_L-evaluator we get exactly the T_L-transitivity of P (formula (3)).

Hence we get the following theorem:

Theorem 1. ([15]) *Let P be a fuzzy preference relation given by formula (4). Then P is T_L-transitive if and only if $\varphi : \mathscr{M} \to [0,1]$ is a T_L-evaluator.*

We can define the indifference relation I, again criterion by criterion, by

$$I_i(a_1,a_2) = 1 - S_L(P_i(a_1,a_2),P_i(a_2,a_1))$$

and $I(a_1,a_2) = \varphi(I_1(a_1,a_2),\cdots,I_n(a_1,a_2))$. The incomparability relation J is defined by

$$J(a_1,a_2) = 1 - S_L(P(a_1,a_2),P(a_2,a_1),I(a_1,a_2)). \tag{5}$$

We will denote P_φ, I_φ and J_φ the preference, indifference and incomparability relations with respect to the evaluator φ.

For the proofs of Theorems 2-5 see [15].

Theorem 2. ([15]) *Assume φ is a T_L-evaluator, which is bounded from below by a self-dual T_L-evaluator $\tilde{\varphi}$. Then the relation J_φ (relation of incomparability) is empty.*

The reader can find more on the connection between Łukasiewicz filters and fuzzy preference relations in [11].

The interaction of criteria means that if we have two different criteria, c_1 and c_2 with their weights w_1 and w_2, respectively, then the weight of the set of criteria $\{c_1,c_2\}$ is different from the sum $w_1 + w_2$. An axiomatic approach to the measurement of the amount of interaction among criteria was given by Kojadinovic in [20].

The interaction of criteria has been considered through non-additive integrals such as Choquet integral. The interaction indices with respect to this integral were introduced by Murofushi and Soneda ([23]) with respect to only couples of criteria, and by Mesiar ([22]) with respect to all possible subsets of criteria. For the first time this kind of integral, i.e. integral with respect to non-additive measure, was defined by Vitali (1925, [27]) and then by Choquet (1953-54, [6]). Another important paper on this topic was that of J. Šipoš (1979, [25]). We will restrict our interest to the discrete case. The definition of the Choquet integral with respect to a fuzzy measure is the following:

Definition 7. *([15]) Let $\mu : 2^X \to [0,1]$ be a fuzzy measure, which is continuous from below. Then the following mapping, $(C)\int : \mathcal{M} \to [0,1]$, is called the Choquet integral with respect to μ:*

$$(C)\int f\,d\mu = \int_0^1 \mu(\{z; f(z) \geq x\})\,dx$$

for each $f \in \mathcal{M}$, where the right-hand integral is in the sense of Riemann.

Łukasiewicz filters are special fuzzy measures, i.e., we may integrate with respect to them. Choosing some particular Łukasiewicz filter means choosing weights of individual criteria, but also of any system of criteria (which means their interaction).

Example 4. ([15]) Let us have three criteria: $X = \{x_1, x_2, x_3\}$. We construct a Łukasiewicz filter $\mathcal{F}$ in choosing weights for each criterion according to its importance, but also for each couple of criteria, see Table 1. Of course, the complete set of criteria, X, has its weight equal to one. The system of weights has to fulfill conditions of Definition 5:

Table 1 Weights of criteria

singletons	weights	couples	weights
x_1	0.1	x_1, x_2	0.5
x_2	0.3	x_1, x_3	0.7
x_3	0.5	x_2, x_3	0.9

The connection between T_L- and/or S_L-evaluators and Choquet integrals is given by the following three theorems.

Theorem 3. ([15]) *Let $\mathcal{F}$ be an arbitrary Łukasiewicz filter on X. Then the Choquet integral with respect to $\mathcal{F}$ is a T_L-evaluator.*

Theorem 4. ([15]) *Let $\mathcal{U}$ be an arbitrary Łukasiewicz ultrafilter on X. Then the Choquet integral with respect to $\mathcal{U}$ is both a T_L- and S_L-evaluator.*

In Example 2 we have constructed a self-dual T_L- and S_L-evaluator. Another construction method uses Choquet integral:

Theorem 5. ([15]) *Let $\mathcal{U} : 2^X \to [0,1]$ be a Łukasiewicz ultrafilter. We denote by $\varphi : \mathcal{M} \to [0,1]$ the following evaluator:*

$$\varphi(f) = (C)\int f\,d\mathcal{U}.$$

Then φ is a self-dual T_L- and S_L-evaluator.

T_L- and/or S_L-evaluators can be constructed also using other types of fuzzy integrals, especially Shilkret ([24]) and Sugeno ([26]) integrals, but also in some other cases. The reader can find more on this topic in [12].

3 Generalized Choquet Integral

As it has been already mentioned the Choquet integral can be used to model the interaction between criteria. However, in our situation we need the weights of criteria and their interaction simultaneously according to level at which they are fulfilled. Criteria with lower level of fulfillment could have other weights than criteria with higher level of fulfillment because the importance of criteria changes according the level they achieved. This problem can be solved using a so-called level-dependent Choquet integral, introduced by S. Greco, S. Giove and B. Matarazzo (see [10]).

Definition 8. *([10]) Let us consider a set of criteria $X = \{1,2,...,m\}$. We define a generalized capacity as function $\mu^G : 2^X \times [0,1] \to [0,1]$ such that*

1) for all $t \in [0,1]$ and $A \subset B \subset X$, $\mu^G(A,t) \leq \mu^G(B,t)$,
2) for all $t \in [0,1], \mu^G(\emptyset,t) = 0$ and $\mu^G(X,t) = 1$,
3) (The regularity property) for all $t \in [0,1]$ and for all $A \subset X, \mu^G(A,t)$ is continuous with respect to t almost everywhere.

Definition 9. *([10]) We define the level-dependent Choquet integral of a function $f : X \to [0,1]$ with respect to the generalized capacity μ^G as follows*

$$Ch^G(f,\mu^G) = \int_0^1 \mu^G(A(f,t),t)\,dt$$

where the right-hand-side is the Lebesgue integral and

$$A(f,t) = \{x \in X; f(x) \geq t\}.$$

Let us remark that if f achieves k different positive values $z_1 < z_2 < \cdots < z_k$ then the level-dependent Choquet integral can always be written as

$$Ch^G(f,\mu^G) = \sum_{i=1}^{k} \int_{z_{i-1}}^{z_i} \mu^G(A(f,t),t)\,dt,$$

where we have put $z_0 = 0$. To illustrate how the level-dependent Choquet integral works, we give the following example.

Example 5. Let $X = \{x_1,x_2,x_3,x_4,x_5\}$ and $f : X \to [0,1]$ be a function given by the formula $f(x_i) = \frac{1}{i}$, for $i = 1,2,3,4,5$. Further, let us take the following generalized capacity $\mu^G : 2^X \times [0,1] \to [0,1]$, defined as follows

$$\mu^G(A,t) = \begin{cases} \frac{|A|}{5}, & \text{if } 0.4 < t \leq 1, \\ \left(\frac{|A|}{5}\right)^2, & \text{if } 0 \leq t \leq 0.4, \end{cases}$$

where $|A|$ is the cardinality of A. Then

$$Ch^G(f,\mu^G) = \frac{1}{5}\cdot 1 + \left(\frac{1}{4}-\frac{1}{5}\right)\left(\frac{4}{5}\right)^2 + \left(\frac{1}{3}-\frac{1}{4}\right)\left(\frac{3}{5}\right)^2 + \left(0.4-\frac{1}{3}\right)\left(\frac{2}{5}\right)^2 + \left(\frac{1}{2}-0.4\right)\frac{2}{5} + \left(1-\frac{1}{2}\right)\frac{1}{5} \doteq 0.455.$$

Roughly speaking, in the above computation we have split the area to be counted into two parts - below 0.4 and above 0.4. In each of these areas we have used the classical Choquet integral with respect to the corresponding (fuzzy) measure.

Remark 4. The level-dependent Choquet integral makes a completely different kind of interaction of criteria possible. Though their weights might be additive, they interact in changing the values of corresponding weights.

Observe that it is not a problem to generalize the level-dependent Choquet integral to the case when X is a countable set (and with additional property of measurability of μ^G even to uncountable set X). The following theorems can be found in [15].

Theorem 6. ([15]) *Let $\mu^G : 2^X \times [0,1] \to [0,1]$ be a generalized capacity such that for each $t \in [0,1]$ $\mu^G(\cdot,t) : 2^X \to [0,1]$ is a Łukasiewicz filter. Then*

$$Ch^G(f,\mu^G) = \int_0^1 \mu^G(A(f,t),t)\,dt$$

is a T_L-evaluator.

Theorem 7. ([15]) *Let $\mu^G : 2^X \times [0,1] \to [0,1]$ be a generalized capacity such that for each $t \in [0,1]$ $\mu^G(\cdot,t) : 2^X \to [0,1]$ is a Łukasiewicz ultrafilter. Then*

$$Ch^G(f,\mu^G) = \int_0^1 \mu^G(A(f,t),t)\,dt$$

is a T_L- and S_L-evaluator at the same time.

Other possible modification of the Choquet integral lies in transforming the values of the function to be integrated. The next two theorems concern this kind of modification.

Theorem 8. ([15]) *Let us denote for each $x \in X$ $\eta_x : [0,1] \to [0,1]$ some isotone transformation with 0 and 1 as fixed points. For each $f \in \mathscr{M}$ we set $\tilde{\eta}(f)(x) = \eta_x(f(x))$. Further, let $\mathscr{F} : 2^X \to [0,1]$ be a Łukasiewicz filter. Then $\varphi : \mathscr{M} \to [0,1]$ defined by*

$$\varphi(f) = (C)\int \tilde{\eta}(f)\,d\mathscr{F} \tag{6}$$

is a T_L-evaluator.

Theorem 9. ([15]) *Let us denote for each* $x \in X$ $\eta_x : [0,1] \to [0,1]$ *some isotone transformation with* 0 *and* 1 *as fixed points. For each* $f \in \mathscr{M}$ *we set* $\tilde{\eta}(f)(x) = \eta_x(f(x))$. *Further, let* $\mathscr{U} : 2^X \to [0,1]$ *be a Łukasiewicz ultrafilter. Then* $\varphi : \mathscr{M} \to [0,1]$ *defined by*

$$\varphi(f) = (C) \int \tilde{\eta}(f) d\mathscr{U} \tag{7}$$

is a T_L*- and* S_L*-evaluator.*

Level-dependent Choquet integrals are not necessarily self-dual evaluators (see [15]). A sufficient condition for the level-dependent Choquet integral to be a self-dual evaluator, is using the same Łukasiewicz ultrafilter G_α for all couples of levels $(\alpha, 1-\alpha)$ where $\alpha \in [0, 0.5]$.

4 Construction of Preference Structures

Let us consider the following situation. We want to find the best fitting decision with respect to preferences of a querying subject without, in general, any additional information about the querying subject than the query itself. The starting point of our algorithm is the construction of a partial preference relation given by a query for each criterion. The preference structure for a selected criterion is often derived from the corresponding utility function of the subject. We consider that the true utility function is unknown. It is obvious that the preference relation must include as much information as possible about the unknown utility function for a criterion. Although the utility function is fully determined by the querying subject and in general unknown, in many cases we know some of its general properties regardless of the querying subject. Let us assume for simplicity that, for each criterion, the query can be transformed either to real numbers or to triangular fuzzy numbers. From the query the triplet (c_1, c_2, c_3) of real numbers can be derived, where c_1, c_3 represent the maximal and minimal acceptable values for the criterion, respectively, and c_2 is the value explicitly specified by the querying subject as desired (not necessarily optimal). Then for the selected criterion we can often identify if the utility function should be increasing or decreasing, optimal value of a criterion and minimal or maximal acceptable value of that criterion. For example, in the case of price we can assume that utility function is decreasing, the optimal but not practically reachable value is 0 (c_1), the values derived from the query represents the maximal acceptable price (c_3) and the desired price (c_2), and objects with values up to the desired price are definitely substantially better than other objects (from the point of view of price). Let us start with quantitative criteria. Then the corresponding utility function with respect to the query can be defined as follows:

Definition 10. *Let* c *be the quantitative criterion with the range* $[z_1, z_2] \subset \mathbb{R}$, *and* $c(a) \in [z_1, z_2]$ *be the value of a criterion for an object* a. *Assume corresponding decreasing utility function with the domain* $[z_1, z_2] \subset \mathbb{R}$. *Let* $c = (c_1, c_2, c_3)$ *be values representing the query,* $c_1, c_2, c_3 \in [z_1, z_2]$, $c_1 = z_1$ *and* $\alpha \in [0,1]$ *be a parameter related to the querying subject. Then the corresponding* α*-utility function* $u^d_{c,\alpha}$: $\mathbb{R} \to [0,1]$ *is given as*

$$u^d_{c,\alpha}(c(a)) = \begin{cases} f(c(a)) & \textit{for } c(a) \in [z_1, c_2[, \\ \alpha & \textit{for } c(a) = c_2, \\ g(c(a)) & \textit{for } c(a) \in]c_2, c_3], \\ 0 & \textit{elsewhere}, \end{cases} \tag{8}$$

where f, g are decreasing continuous functions such that

$$\lim_{c(a)\to c_2^-} f(c(a)) = \lim_{c(a)\to c_2^+} g(c(a)) = \alpha, \quad f(z_1) = 1, \quad g(z_2) = 0.$$

In the previous definition we assume that the optimal value of a criterion (c_1) is identical with the lower bound of the domain of the criterion c. Then functions f, g and constant α are selected in order to better characterize the querying person. For example in the case of price f, g can be strictly decreasing linear functions and α can be 0.9.

Definition 11. *Let c be the quantitative criterion with the range $[z_1, z_2] \subset \mathbb{R}$, and $c(a) \in [z_1, z_2]$ be the value of a criterion for an object a. Assume an increasing corresponding utility function with the domain $[z_1, z_2] \subset \mathbb{R}$. Let $c = (c_1, c_2, c_3)$ be values representing the query, $c_1, c_2, c_3 \in [z_1, z_2]$, $c_3 = z_2$ and $\alpha \in [0,1]$ be related to the querying subject. Then the corresponding α-utility function $u^i_{c,\alpha} : \mathbb{R} \to [0,1]$ is given as*

$$u^i_{c,\alpha}(c(a)) = \begin{cases} f(c(a)) & \textit{for } c(a) \in [c_1, c_2[, \\ \alpha & \textit{for } c(a) = c_2, \\ g(c(a)) & \textit{for } c(a) \in]c_2, z_2], \\ 0 & \textit{elsewhere}, \end{cases} \tag{9}$$

where f, g are increasing continuous functions such that

$$\lim_{c(a)\to c_2^-} f(c(a)) = \lim_{c(a)\to c_2^+} g(c(a)) = \alpha, \quad f(z_1) = 0, \quad g(z_2) = 1.$$

The fuzzy preference for two objects a_1, a_2 can be derived from the above mentioned utility functions in the following way:

Definition 12. *Let c be the quantitative criterion with the range $[z_1, z_2] \subset \mathbb{R}$, and $c(a_i) \in [z_1, z_2]$, for $i = 1, 2$, be the values of a criterion for objects a_1, a_2, respectively. Assume a decreasing utility function with the domain $[z_1, z_2] \subset \mathbb{R}$. Let $c = (c_1, c_2, c_3)$ be values representing the query, $c_1, c_2, c_3 \in [z_1, z_2]$, and $\alpha \in [0,1]$ be related to the querying subject. Then the corresponding fuzzy preference for the given objects a_1, a_2 is given by the following formula:*

$$FP_c(a_1, a_2) = \begin{cases} \max\left\{ \left(u^d_{c,\alpha}(c(a_1)) - u^d_{c,\alpha}(c(a_2))\right), 0 \right\} & \textit{if } c(a_1) < c(a_2) < c_2 \\ & \textit{or } c_2 > c(a_1) > c(a_2), \\ 1 & \textit{if } c(a_1) < c_2 < c(a_2), \\ 0 & \textit{elsewhere}. \end{cases}$$

Definition 13. *Let c be the quantitative criterion with the range* $[z_1,z_2] \subset \mathbb{R}$, *and* $c(a_i) \in [z_1,z_2]$, *for* $i = 1,2$, *be the values of a criterion for objects* a_1,a_2, *respectively. Assume an increasing utility function with the domain* $[z_1,z_2] \subset \mathbb{R}$. *Let* c_2 *be a value explicitly specified by the querying subject,* $c_2 \in [z_1,z_2]$, *and* $\alpha \in [0,1]$ *be parameter related to the querying subject. Then the corresponding fuzzy preference is given by the following formula:*

$$FP_c(a_1,a_2) = \begin{cases} \max\left\{\left(u^i_{c,\alpha}(c(a_1)) - u^i_{c,\alpha}(c(a_2))\right),0\right\} & \text{if } c(a_1) > c(a_2) > c_2 \\ & \text{or } c(a_2) < c(a_1) < c_2, \\ 1 & \text{if } c(a_2) < c_2 < c(a_1), \\ 0 & \text{elsewhere.} \end{cases}$$

If we assume ordinal qualitative data, the possible values are described using an appropriate linguistic scale. We will assume only the triangular fuzzy numbers forming a partition in the sense of Ruspini. The considered support is expressed using the unit interval, i.e., if the utility function is increasing 0 is the total absence of optimality of a selected criterion, 1 is the perfect match with the optimal value. (Similarly for a decreasing utility function 0 is the perfect match with the optimal value and 1 is the total absence of optimality for the criterion in question.) In this case the fuzzy preference can be based on distances between fuzzy numbers.

Definition 14. *Let c be the ordinal qualitative criterion where the minimal value of the assumed scale is optimal. Denote by* $c(a_i) \in [z_1,z_2]$ *for* $i = 1,2$ *the values of a criterion for objects* a_1,a_2, *respectively. Let* $c = (c_1,c_2,c_3)$ *be a triangular fuzzy number representing the query,* $c_1,c_2,c_3 \in [0,1]$. *Then the corresponding fuzzy preference is given by the following formula:*

$$FP_c(a_1,a_2) = \begin{cases} d(c(a_1),c(a_2)) & \text{if } c(a_1) < c(a_2) < c \text{ or } c < c(a_1) > c(a_2), \\ 1 & \text{if } c(a_1) < c < c(a_2), \\ 0 & \text{elsewhere.} \end{cases},$$

where d is a distance between fuzzy sets.

Definition 15. *Let c be the ordinal qualitative criterion where the optimum is the maximal value of the assumed scale. Denote by* $c(a_i) \in [z_1,z_2]$ *for* $i = 1,2$ *the values of a criterion for objects* a_1,a_2, *respectively. Let* $c = (c_1,c_2,c_3)$ *be a triangular fuzzy number representing the query,* $c_1,c_2,c_3 \in [0,1]$. *Then the corresponding fuzzy preference is given by the following formula:*

$$FP_c(a_1,a_2) = \begin{cases} d(c(a_1),c(a_2)) & \text{if } c(a_1) > c(a_2) > c \text{ or } c(a_2) < c(a_1) < c, \\ 1 & \text{if } c(a_2) < c < c(a_1), \\ 0 & \text{elsewhere,} \end{cases}$$

where d is a distance between fuzzy sets.

The constructed fuzzy preferences with respect to all criteria can be used to identify the interaction between pairs of criteria and the relative importance of criteria. The main idea is very simple. The similarity of preference relations means the similarity of corresponding criteria and vice versa. More precisely a similarity of preference relations takes values from the unit interval where 1 means that two criteria are closely related (identical with respect to a generated ordering) to each other and 0 means that two criteria should be assumed independent of each other. Using similarity between fuzzy preferences we can construct the similarity of criteria (for an alternative view see [16, 17]).

Definition 16. *Let Sim be the similarity between fuzzy preferences FP_{c_1}, FP_{c_2} for criteria c_1, c_2, respectively. We say that criteria c_1, c_2 are similar at the level β if we have that*

$$Sim(FP_{c_1}, FP_{c_2}) \geq \beta.$$

We would like to group criteria with the similarity at a certain level β, therefore the similarity of preference relations should fulfil some kind of transitivity, e.g., T_L-transitivity. Assuming T_L-transitive similarity relation of preference relation, we obtain T_L-transitive similarity of criteria at β and T_L-transitive similarity of criteria forms a T_L-partition (see [1]). It is evident that at $\beta = 0$ all criteria belong to the same group and at $\beta = 1$ only criteria with completely identical corresponding preference structures are grouped together. If we assume all levels β for which the grouping of criteria is changed, we obtain the levels of the assumed level-dependent Choquet integral. Let us assume for simplicity that all criteria in the same group have equal weights and all groups at certain level have the same weight. From the previous section it is obvious that the level-dependent Choquet integral in our consideration always exists and it is a T_L-evaluator. Using the level-dependent Choquet integral we obtain a fuzzy preference relation describing objects in our database. Due to T_L-transitivity we are able to identify at least n objects which are better then other objects. Pairs of objects with the corresponding preference less than given threshold β are considered as indifferent objects. Then we obtain groups of objects consisting of at least n objects. The rest of the database is omitted. There are two problems related to this procedure. The first one is related to the assumption of T_L- transitivity. This fact can cause some cycles in our preference structure. Objects (groups) corresponding to cycles are indifferent for us, i.e., we cannot separate them using our preference relation. The second one is connected to the incomparability relation. Because of Theorem 2 we know that the level-dependent Choquet integral leads to empty incomparability relation if it is bounded from below by a self dual T_L evaluator. It is easy to show that such Choquet integral is not bounded from below by a self-dual T_L-evaluator in general, and this means that the relation of incomparability is not empty.

Example 6. Let us assume five criteria, x_1, x_2, x_3, x_4, x_5, and two objects (alternatives), a_1 and a_2. In the following table we give the strict preferences of a_1 to a_2, of a_2 to a_1, and of indifferences between a_1 and a_2, respectively, according to particular criteria.

Table 2 Strict preferences and indifferences with respect to particular criteria

	x_1	x_2	x_3	x_4	x_5
$P_i(a_1,a_2)$	0.3	0	1	0.7	0
$P_i(a_1,a_2)$	0	0.3	0	0	0.7
$I_i(a_1,a_2)$	0.7	0.7	0	0.3	0.3

Table 3 Weights of criteria at corresponding levels

	x_1	x_2	x_3	x_4	x_5
$\alpha \leq 0.3$	1/6	1/6	1/3	1/6	1/6
$0.3 < \alpha \leq 0.7$	1/4	1/4	1/4	1/8	1/8
$0.7 < \alpha \leq 1$	1/5	1/5	1/5	1/5	1/5

The weights from Table 3 (depending on particular levels) define the generalized capacity, let us denote it by μ^G. We count now the strict preferences and the indifference using the level-dependent Choquet integral with respect to μ^G.

$$\begin{aligned}
P(a_1,a_2) &= Ch^G(P_i(a_1,a_2),\mu^G) = 0.3\left(\frac{1}{6}+\frac{1}{3}+\frac{1}{6}\right)+0.4\left(\frac{1}{4}+\frac{1}{8}\right)+0.3\frac{1}{5}\\
&= 0.41\\
P(a_2,a_1) &= Ch^G(P_i(a_2,a_1),\mu^G) = 0.3\left(\frac{1}{6}+\frac{1}{6}\right)+0.4\frac{1}{8} = 0.15\\
I(a_1,a_2) &= Ch^G(I_i(a_1,a_2),\mu^G) = 0.3\left(\frac{1}{6}+\frac{1}{6}+\frac{1}{6}+\frac{1}{6}\right)+0.4\left(\frac{1}{4}+\frac{1}{4}\right) = 0.4,
\end{aligned}$$

and hence $J(a_1,a_2) = 1 - S_L(P(a_1,a_2),P(a_2,a_1),I(a_1,a_2)) = 0.04$. This means that in this example the lack of comparability (i.e., the incomparability) is not high. But in general it may cause some troubles.

It means that the aggregation could result to a partial order, where some elements are, to a certain level, incomparable. In order to have an empty incomparability relation we have to add one additional step to our procedure, a modification of the Choquet integral. From Example 6 and from the last section it is evident that there is a problem with weights on levels under 0.5 which should be completely determined by weights on levels above 0.5. It means that for $\alpha \geq 0.5$ we construct weights as was mentioned above and for each $\alpha < 0.5$ we put the corresponding weights equal to those of the dual level $1-\alpha$. If we repeat the first step of our procedure using the newly defined weights we obtain the preference structure with empty incomparability relation. So we are able to select at least n objects related to query. In general these objects can be separated into several groups. We have a comparison between groups but objects within groups are indifferent.

In general, we may get yet another problem. Namely, if the group of objects is too large, we can possibly get too much levels (values) of criteria to be distinguished. In such a case, at least in the first step we can transform the values of criteria (utilizing Theorems 8 and 9) to simplify our model.

Table 4 Values of criteria for our system of objects

	a_1	a_2	a_3	a_4	a_5	a_6	a_7	a_8	a_9	a_{10}
x1	7	6,5	6,5	6,2	5,5	7,3	7,5	7,8	4,5	4,5
x2	11,9	10,4	11,3	14,7	8,8	20,2	26,7	9	8,9	9,3
x3	70	68	88	77	51	105	110	50	66	38
x4	medium	high	low	medium	medium	medium	high	low	low	low
x5	medium	good	medium	good	good	good	good	low	low	low
x6	medium	medium	medium	medium	high	low	low	medium	medium	low
x7	medium	high	low	low	low	high	high	medium	medium	high

5 Application

In this section we illustrate a possible application of the level-dependent Choquet integral to the decision problem with interacting parameters if weights are not provided by a decision maker in the query. Assume the following simple example of selection of n appropriate cars for a one person with respect to the set of criteria.

Example 7. Let $A = \{a_1, a_2, \ldots, a_{10}\}$ be the set of 10 cars (Hatchback, gas) we would like to choose from. We have a set of 7 criteria $X = \{x_1, x_2, \ldots, x_7\}$, where x_1 is the average gas mileage, x_2 is the price (in thousands EUR), x_3 is the power, x_4 is the safety of passengers, x_5 are the additional features, x_6 is the statistic of stolen cars, x_7 are the additional costs. It is easy to see that x_1, x_2, x_3 are quantitative criteria, x_4, x_5, x_6, x_7 are ordinal qualitative criteria. Moreover, it is obvious that utility functions for x_1, x_2 are decreasing, a utility function for x_3 is increasing, for x_4, x_5 the maximum of a linguistic scale is the optimal value, for x_6, x_7 the minimum of a linguistic scale is the optimal value. To make the model as simple as possible, assume that utility functions are piecewise linear and linguistic scales have three levels, *low, medium, good*, forming the fuzzy partition (in the sense of Ruspini) of the unit interval. Let our query Q be represented by the following real numbers and linguistuc terms (6; 15; 50; high; medium; low, low). We need to select the best three cars. Let us have the following values of criteria for the set of objects:

We will deal with the following utility functions for quantitative criteria.

x_1 - mileage

$$u^d_{x_1,0}(x(a_i)) = \begin{cases} \frac{7-x(a_i)}{3} & \text{for } x(a_i) \in [4,7[, \\ 0 & \text{for } x(a_i) \geq 7, \end{cases}$$

x_2 - price (in thousands EUR)

$$u^d_{x_2,0}(x(a_i)) = \begin{cases} \frac{15-x(a_i)}{8} & \text{for } x(a_i) \in [7,15[, \\ 0 & \text{for } x(a_i) \geq 15, \end{cases}$$

x_3 - the power

$$u^i_{x_3,1}(x(a_i)) = \begin{cases} 0 & \text{for } x(a_i) \leq 50, \\ \frac{x-50}{20} & \text{for } x(a_i) \in]50,70], \\ 1 & \text{for } x(a_i) > 70. \end{cases}$$

We will use the following distance between two values in a selected scale:

$$d(z_1,z_2)=\begin{cases}0 & \text{if } z_1=z_2,\\ 0.5 & \text{if } z_1,z_2 \text{ are adjacent,}\\ 1 & \text{elsewhere.}\end{cases}$$

Using definitions 12-15 we obtain fuzzy preferences with respect to the given criteria and query listed in Tables 5-11.

Table 5 Pair-wise preferences of objects given for criterion x_1

x_1	a_1	a_2	a_3	a_4	a_5	a_6	a_7	a_8	a_9	a_{10}
a_1	0.00	0.00	0.00	0.00	0.00	0.10	0.17	0.27	0.00	0.00
a_2	0.17	0.00	0.00	0.00	0.00	1.00	1.00	1.00	0.00	0.00
a_3	0.17	0.00	0.00	0.00	0.00	1.00	1.00	1.00	0.00	0.00
a_4	0.27	0.10	0.10	0.00	0.00	1.00	1.00	1.00	0.00	0.00
a_5	0.50	0.33	0.33	0.23	0.00	1.00	1.00	1.00	0.00	0.00
a_6	0.00	0.00	0.00	0.00	0.00	0.00	0.07	0.17	0.00	0.00
a_7	0.00	0.00	0.00	0.00	0.00	0.00	0.00	0.10	0.00	0.00
a_8	0.00	0.00	0.00	0.00	0.00	0.00	0.00	0.00	0.00	0.00
a_9	0.83	0.67	0.67	0.57	0.33	0.93	1.00	1.00	0.00	0.00
a_{10}	0.83	0.67	0.67	0.57	0.33	0.93	1.00	1.00	0.00	0.00

Table 6 Pair-wise preferences of objects given for criterion x_2

x_2	a_1	a_2	a_3	a_4	a_5	a_6	a_7	a_8	a_9	a_{10}
a_1	0.00	0.00	0.00	0.35	0.00	1.00	1.00	0.00	0.00	0.00
a_2	0.19	0.00	0.11	0.54	0.00	1.00	1.00	0.00	0.00	0.00
a_3	0.08	0.00	0.00	0.43	0.00	1.00	1.00	0.00	0.00	0.00
a_4	0.00	0.00	0.00	0.00	0.00	1.00	1.00	0.00	0.00	0.00
a_5	0.39	0.20	0.31	0.74	0.00	1.00	1.00	0.02	0.01	0.06
a_6	0.00	0.00	0.00	0.00	0.00	0.00	0.81	0.00	0.00	0.00
a_7	0.00	0.00	0.00	0.00	0.00	0.00	0.00	0.00	0.00	0.00
a_8	0.36	0.18	0.29	0.71	0.00	1.00	1.00	0.00	0.00	0.04
a_9	0.38	0.19	0.30	0.73	0.00	1.00	1.00	0.01	0.00	0.05
a_{10}	0.33	0.14	0.25	0.68	0.00	1.00	1.00	0.00	0.00	0.00

Table 7 Pair-wise preferences of objects given for criterion x_3

x_3	a_1	a_2	a_3	a_4	a_5	a_6	a_7	a_8	a_9	a_{10}
a_1	0.00	0.10	0.00	0.00	0.95	0.00	0.00	1.00	0.20	1.00
a_2	0.00	0.00	0.00	0.00	0.85	0.00	0.00	0.90	0.10	1.00
a_3	0.90	1.00	0.00	0.55	1.00	0.00	0.00	1.00	1.00	1.00
a_4	0.35	0.45	0.00	0.00	1.00	0.00	0.00	1.00	0.55	1.00
a_5	0.00	0.00	0.00	0.00	0.00	0.00	0.00	0.05	0.00	0.65
a_6	1.00	1.00	0.85	1.00	1.00	0.00	0.00	1.00	1.00	1.00
a_7	1.00	1.00	1.00	1.00	1.00	0.25	0.00	1.00	1.00	1.00
a_8	0.00	0.00	0.00	0.00	0.00	0.00	0.00	0.00	0.00	0.60
a_9	0.00	0.00	0.00	0.00	0.75	0.00	0.00	0.80	0.00	1.00
a_{10}	0.00	0.00	0.00	0.00	0.00	0.00	0.00	0.00	0.00	0.00

Table 8 Pair-wise preferences of objects given for criterion x_4

x_4	a_1	a_2	a_3	a_4	a_5	a_6	a_7	a_8	a_9	a_{10}
a_1	0.00	0.00	0.50	0.00	0.00	0.00	0.00	0.50	0.50	0.50
a_2	0.50	0.00	1.00	0.50	0.50	0.50	0.00	1.00	1.00	1.00
a_3	0.00	0.00	0.00	0.00	0.00	0.00	0.00	0.00	0.00	0.00
a_4	0.00	0.00	0.50	0.00	0.00	0.00	0.00	0.50	0.50	0.50
a_5	0.00	0.00	0.50	0.00	0.00	0.00	0.00	0.50	0.50	0.50
a_6	0.00	0.00	0.50	0.00	0.00	0.00	0.00	0.50	0.50	0.50
a_7	0.50	0.00	1.00	0.50	0.50	0.50	0.00	1.00	1.00	1.00
a_8	0.00	0.00	0.00	0.00	0.00	0.00	0.00	0.00	0.00	0.00
a_9	0.00	0.00	0.00	0.00	0.00	0.00	0.00	0.00	0.00	0.00
a_{10}	0.00	0.00	0.00	0.00	0.00	0.00	0.00	0.00	0.00	0.00

Table 9 Pair-wise preferences of objects given for criterion x_5

x_5	a_1	a_2	a_3	a_4	a_5	a_6	a_7	a_8	a_9	a_{10}
a_1	0.00	0.00	0.00	0.00	0.00	0.00	0.00	0.50	0.50	0.50
a_2	0.50	0.00	0.50	0.00	0.00	0.00	0.00	1.00	1.00	1.00
a_3	0.00	0.00	0.00	0.00	0.00	0.00	0.00	0.50	0.50	0.50
a_4	0.50	0.00	0.50	0.00	0.00	0.00	0.00	1.00	1.00	1.00
a_5	0.50	0.00	0.50	0.00	0.00	0.00	0.00	1.00	1.00	1.00
a_6	0.50	0.00	0.50	0.00	0.00	0.00	0.00	1.00	1.00	1.00
a_7	0.50	0.00	0.50	0.00	0.00	0.00	0.00	1.00	1.00	1.00
a_8	0.00	0.00	0.00	0.00	0.00	0.00	0.00	0.00	0.00	0.00
a_9	0.00	0.00	0.00	0.00	0.00	0.00	0.00	0.00	0.00	0.00
a_{10}	0.00	0.00	0.00	0.00	0.00	0.00	0.00	0.00	0.00	0.00

Table 10 Pair-wise preferences of objects given for criterion x_6

x_6	a_1	a_2	a_3	a_4	a_5	a_6	a_7	a_8	a_9	a_{10}
a_1	0.00	0.00	0.00	0.00	0.50	0.00	0.00	0.00	0.00	0.00
a_2	0.00	0.00	0.00	0.00	0.50	0.00	0.00	0.00	0.00	0.00
a_3	0.00	0.00	0.00	0.00	0.50	0.00	0.00	0.00	0.00	0.00
a_4	0.00	0.00	0.00	0.00	0.50	0.00	0.00	0.00	0.00	0.00
a_5	0.00	0.00	0.00	0.00	0.00	0.00	0.00	0.00	0.00	0.00
a_6	0.50	0.50	0.50	0.50	1.00	0.00	0.00	0.50	0.50	0.00
a_7	0.50	0.50	0.50	0.50	1.00	0.00	0.00	0.50	0.50	0.00
a_8	0.00	0.00	0.00	0.00	0.50	0.00	0.00	0.00	0.00	0.00
a_9	0.00	0.00	0.00	0.00	0.50	0.00	0.00	0.00	0.00	0.00
a_{10}	0.50	0.50	0.50	0.50	1.00	0.00	0.00	0.50	0.50	0.00

Table 11 Pair-wise preferences of objects given for criterion x_7

x_7	a_1	a_2	a_3	a_4	a_5	a_6	a_7	a_8	a_9	a_{10}
a_1	0.00	0.50	0.00	0.00	0.00	0.50	0.50	0.00	0.00	0.50
a_2	0.00	0.00	0.00	0.00	0.00	0.00	0.00	0.00	0.00	0.00
a_3	0.50	1.00	0.00	0.00	0.00	1.00	1.00	0.50	0.50	1.00
a_4	0.50	1.00	0.00	0.00	0.00	1.00	1.00	0.50	0.50	1.00
a_5	0.50	1.00	0.00	0.00	0.00	1.00	1.00	0.50	0.50	1.00
a_6	0.00	0.00	0.00	0.00	0.00	0.00	0.00	0.00	0.00	0.00
a_7	0.00	0.00	0.00	0.00	0.00	0.00	0.00	0.00	0.00	0.00
a_8	0.00	0.50	0.00	0.00	0.00	0.50	0.50	0.00	0.00	0.50
a_9	0.00	0.50	0.00	0.00	0.00	0.50	0.50	0.00	0.00	0.50
a_{10}	0.00	0.00	0.00	0.00	0.00	0.00	0.00	0.00	0.00	0.00

To compute a similarity between preference matrices for criteria x_i, x_j, for all $i, j \in \{1, 2, \ldots, 7\}, i \neq j$, we can use for example the following formula:

$$Sim(x_i, x_j) = 1 - \frac{\sum_{(pk)} \left| FP_{x_i}(a_p, a_k) - FP_{x_j}(a_p, a_k) \right|}{90},$$

where $p, k \in \{1, 2, \ldots, 10\}$.

Then we obtain similarities for pairs of criteria (ordered descending) given in Table 12.

Table 12 Similarities counted for couples of criteria

$Sim(x_4,x_5)$	$Sim(x_1,x_2)$	$Sim(x_4,x_6)$	$Sim(x_2,x_7)$	$Sim(x_3,x_5)$	$Sim(x_1,x_7)$	$Sim(x_3,x_4)$
0.87	0.79	0.72	0.70	0.70	0.69	0.67
$Sim(x_5,x_6)$	$Sim(x_3,x_6)$	$Sim(x_5,x_7)$	$Sim(x_1,x_6)$	$Sim(x_4,x_7)$	$Sim(x_2,x_6)$	$Sim(x_6,x_7)$
0.67	0.66	0.65	0.61	0.61	0.58	0.58
$Sim(x_1,x_4)$	$Sim(x_1,x_5)$	$Sim(x_2,x_4)$	$Sim(x_3,x_7)$	$Sim(x_2,x_5)$	$Sim(x_1,x_3)$	$Sim(x_2,x_3)$
0.56	0.56	0.54	0.53	0.49	0.40	0.33

If we analyze Table 12, we can identify, e.g., the levels for the level-dependent Choquet integral and corresponding groups of criteria, given in Table 13.

Table 13 Levels of level-dependent Choquet integral and corresponding groups

0.87	x_4, x_5
0.79	x_1, x_2
0.69	x_1, x_2, x_7
0.66	x_4, x_5, x_6, x_3
0.33	all criteria

If we assume that criteria in groups are of the same importance, we obtain the additive weights listed in Table 14.

Table 14 Weights of criteria at corresponding levels

	x_1	x_2	x_3	x_4	x_5	x_6	x_7
$\alpha \leq 0.33$	1/7	1/7	1/7	1/7	1/7	1/7	1/7
$0.33 \leq \alpha \leq 0.66$	1/6	1/6	1/8	1/8	1/8	1/8	1/6
$0.66 \leq \alpha \leq 0.69$	1/12	1/12	1/4	1/12	1/12	1/4	1/12
$0.69 \leq \alpha \leq 0.79$	1/10	1/10	1/5	1/10	1/10	1/5	1/5
$0.79 \leq \alpha \leq 0.87$	1/6	1/6	1/6	1/12	1/12	1/6	1/6
$0.87 \leq \alpha \leq 1$	1//7	1/7	1/7	1/7	1/7	1/7	1/7

From Table 14 we can define the level dependent capacity μ^G and compute strict preferences. We illustrate it by means of the level-dependent Choquet integral for pairs $(a_1, a_2), (a_1, a_3)\, (a_2, a_1)\, (a_2, a_3)\, (a_3, a_1)\, (a_3, a_2)$ (see Table 15).

Table 15 Strict preferences between objects a_1, a_2, a_3

FP	a_1	a_2	a_3
x_1	0	0.1	0.09
x_2	0.18	0	0.21
x_3	0.25	0.3	0

Using the constructed strict fuzzy preference we obtain the following ordering: $a_3 > a_2 > a_1$. If we compute the strict fuzzy preference for all 90 pairs of objects, we can derive an analogy for all objects.

6 Conclusions

In this paper we have discussed the possibility to apply the modified (level-dependent) Choquet integral to a monopersonal multicriterial decision-making problem. We have used the level-dependent Choquet integral to construct a simple decision-making algorithm. The proposed algorithm produces an outranking of objects taking into account an interaction between criteria. It is close to the ELECTRE method (see [8]). Finally, our approach is illustrated with choosing an appropriate car. In a real situation, this approach may be applied to choosing a set of objects from on-line available data set according to a query given by the client. We presented here mainly the mathematical foundation of the proposed algorithm. Its detailed comparison with ELECTRE methods, limitations of applicability and careful verification of proposed algorithm from the decision-making point of view is goal of our future research. Moreover, in our future work we intend to study further properties of the proposed algorithm, especially an appropriate construction of weights, and the possibility to deal with multipersonal decision-making problems.

Acknowledgements. The work of Martin Kalina has been supported by Science and Technology Assistance Agency under the Contract No. APVV-0375-06, and by the VEGA grant agency, Grant Number 1/4026/07. The work of Dana Hliněná has been supported by Project 1ET100300517 of the Program "Information Society" and by Project MSM0021630529 of the Ministry of Education. The work of Pavol Král' has been supported by the VEGA grant agency, Grant Number 1/0539/08.

References

1. De Baets, B., Mesiar, R.: $\mathcal{T}$-partitions. Fuzzy Sets and Systems 97, 211–223 (1998)
2. De Baets, B., Van de Walle, B., Kerre, E.: A plea for the use of Łukasiewicz triplets in the definition of fuzzy preference structures (II). The identity case. Fuzzy Sets and Systems 99, 303–310 (1998)
3. Bodjanova, S.: T_L- and S_L-evaluators. In: Proceedings of AGOP 2007, Ghent, pp. 165–172. Academia Press (2007)
4. Bodjanova, S., Kalina, M.: T-evaluators and S-evaluators. Fuzzy Sets and Systems 160, 1965–1983 (2009)
5. Burns, K.J., Demaree, H.A.: A chance to learn: On matching probabilities to optimize utilities. Information Sciences (to appear)
6. Choquet, G.: Theory of capacities. Ann. Inst. Fourier 5, 131–295 (1953-1954)
7. Cox, E.: Fuzzy Modeling and Genetic Algorithms for Data Mining and Exploration. Morgan Kaufmann Publishers, San Francisco (2005)
8. Figueira, J.R., Greco, S., Roy, B.: ELECTRE methods with interaction between criteria: An extension of the concordance index. European Journal of Operational Research 199, 478–495 (2008)
9. Fodor, J.C., Roubens, M.: Fuzzy Preference Modelling and Multicriteria Decision Support. Kluwer Academic Publishers, Dordrecht (1994)
10. Greco, S., Giove, S., Matarazzo, B.: The Choquet integral with respect to a level dependent capacity. Submited to FSS
11. Havranová, Z., Kalina, M.: Fuzzy preference relations and Łukasiewicz filters. In: Proceedings of the 5th EUSFLAT conference. New dimensions in Fuzzy logic and related Technologies, vol. II, pp. 337–341. University of Ostrava, Ostrava (2007)
12. Havranová, Z., Kalina, M.: Fuzzy Integrals as T_L- and S_L-Evaluators. In: Proceedings of AGOP 2007, Ghent, pp. 173–178. Academia Press (2007)
13. Havranová, Z., Kalina, M.: T-filters and T-ideals. Acta Univ. Mathaei Belii ser. Mathematics 14, 19–27 (2007)
14. Hliněná, D., Kalina, M., Král', P.: Choquet integral and its modifications. In: Magdalena, L., Ojeda-Aciego, M., Verdegay, J.L. (eds.) Proceedings of IPMU 2008, Torremolinos, Málaga, pp. 86–93 (2008)
15. Hliněná, D., Kalina, M., Král', P.: Choquet integral with respect to Łukasiewicz filters, and its modifications. Information Sciences 179, 2912–2922 (2009)
16. Hliněná, D., Král', P.: Similarity of fuzzy preference structures based on metrics. In: Proceedings of the 5th EUSFLAT conference. New dimensions in Fuzzy logic and related Technologies, vol. II, pp. 431–436 (2007)
17. Hliněná, D., Král', P.: Comparing fuzzy preference relations using fuzzy set operations. In: Proceedings of AGOP 2007, Ghent, pp. 135–140. Academia Press (2007)
18. Kalina, M.: Łukasiewicz filters and similarities. In: Proceedings of AGOP 2005, Lugano, pp. 5–57 (2005)
19. Klement, E.P., Mesiar, R., Pap, E.: Triangular norms. Kluwer, Dordrecht (2000)

20. Kojadinovic, I.: An axiomatic approach to the measurement of the amount of interaction among criteria or players. Fuzzy Sets and Systems 152, 417–435 (2005)
21. Menger, K.: Probabilistic theories of relations. Proc. Natl. Acad. Sci. USA 37(3), 178–180 (1951)
22. Mesiar, R.: Possibility measures, integration and fuzzy possibility measures. Fuzzy Sets and Systems 92, 191–196 (1997)
23. Murofushi, T., Soneda, S.: Techniques for reading fuzzy measures iii: Interaction index. In: Proc. 9th Fuzzy System Symposium, Sapporo, Japan, pp. 693–696 (1993)
24. Shilkret, N.: Maxitive measures and integration. Indag. Math. 33, 109–116 (1971)
25. Šipoš, J.: Integral with respect to a premeasure. Math. Slovaca 29, 141–145 (1979)
26. Sugeno, M.: Theory of fuzzy integrals and applications. PhD. thesis, Tokyo Inst. of Tech. (1974)
27. Vitali, G.: Sulla definizione di integrale delle funzioni di una variabile. Ann. Mat. Pura Appl. 2, 111–121 (1925)
28. Van de Walle, B., De Baets, B., Kerre, E.: A plea for the use of Łukasiewicz triplets in the definition of fuzzy preference structures (I). General argumentation. Fuzzy Sets and Systems 97, 349–359 (1998)
29. Chen, Y.-L., Wu, C.-C., Tang, K.: Building a cost-constrained decision tree with multiple condition attributes. Information Sciences (to appear)

Majority Rules Generated by Mixture Operators

Bonifacio Llamazares and Ricardo Alberto Marques Pereira

Abstract. Aggregation operators are a fundamental tool in multicriteria decision making procedures. Due to the huge variety of aggregation operators existing in the literature, one of the most important issues in this field is the choice of the best-suited operators in each aggregation process. Given that some aggregation operators can be seen as extensions of majority rules to the field of gradual preferences, we can choose the aggregation operators according to the class of majority rule that we want to obtain when individuals do not grade their preferences. Thus, in this paper we consider mixture operators to aggregate individual preferences and we characterize those that allow us to extend some majority rules, such as simple, Pareto, and absolute special majorities, to the field of gradual preferences.

1 Introduction

Aggregation operators are a fundamental tool in multicriteria decision making procedures. For this reason, they have received a great deal of attention in the literature (see, for instance, Marichal [6], Calvo, Mayor and Mesiar [1] and Xu and Da [15]).

An interesting kind of non-monotonic aggregation is obtained when mixture operators are used. Mixture operators are weighted averaging operators in which the weights depend on the given attribute satisfaction values through appropriate weighting functions. Mixture operators have been introduced in Marques Pereira and Pasi [8] and Marques Pereira [7], and further investigated in Ribeiro and Marques Pereira [11,12], Marques Pereira and Ribeiro [9], Mesiar and Špirková [10] and Špirková [13, 14].

Bonifacio Llamazares
Department of Applied Economics, Avda. Valle Esgueva 6, E-47010 Valladolid, Spain
e-mail: boni@eco.uva.es

Ricardo Alberto Marques Pereira
Department of Computer and Management Sciences, Via Inama 5, TN 38100 Trento, Italy
e-mail: ricalb.marper@unitn.it

B. Bouchon-Meunier et al. (Eds.) Found. of Reas. under Uncert., STUDFUZZ 249, pp. 189–201.
springerlink.com

Mixture operators can be used to aggregate individual preferences into a collective preference. Given that some aggregation operators can be seen as extensions of majority rules to the field of gradual preferences, the aim of this paper is to determine the mixture operators that correspond to extensions of some particular classes of majority rules.

We consider individual preferences expressed as pairwise comparisons between alternatives, with preference intensity values in the [0,1] interval. In this way each pairwise comparison is associated with a graded preference profile. Once an aggregation operator is chosen, each graded preference profile produces a collective preference intensity value in the unit interval. On the basis of this value and through a kind of strong α-cut, where $\alpha \in [0.5, 1)$, we can decide if an alternative is chosen or if both alternatives are collectively indifferent. When individuals do not grade their preferences (that is, when they are represented through the values 0, 0.5, and 1), the previous procedure allows us to obtain a majority rule. Hence, once α is fixed, it is possible to know which class of majority rule is present in the aggregation process according to the actual operator that is used.

We note that this procedure has already been used to characterize several classes of aggregation functions that extend some well-known majority rules. Thus, García-Lapresta and Llamazares [3] generalized two classes of majorities based on difference of votes by using quasi-arithmetic means and window OWA operators as aggregation functions. Likewise, Llamazares [4, 5] has characterized the OWA operators that generalize simple, Pareto, and absolute special majorities.

In this paper we characterize the mixture operators that allow us to extend simple, Pareto, and absolute special majorities to the field of gradual preferences.

The organization of the paper is as follows. In Section 2 we introduce the model used for extending a majority rule through an aggregation function. In Section 3 we show some characterizations of simple, Pareto, and absolute special majorities. In Section 4 we recall the construction of mixture operators and we determine those that satisfy the self-duality property. Section 5 presents the main results of the paper. We conclude in Section 6.

2 A Model for Extending Majority Rules

We consider m voters, with $m \geq 3$, and two alternatives x and y. Voters represent their preferences between x and y through variables r_i. If the individuals grade their preferences, then $r_i \in [0, 1]$ denotes the *intensity* with which voter i prefers x to y. We also suppose that $1 - r_i$ is the intensity with which voter i prefers y to x. If the individuals do not grade their preferences, then $r_i \in \{0, 0.5, 1\}$ represents that voter i prefers x to y ($r_i = 1$), prefers y to x ($r_i = 0$), or is indifferent between both alternatives ($r_i = 0.5$). The justification of this three-valued representation can be found in García-Lapresta and Llamazares [2].

A *profile of preferences* is a vector $\mathbf{r} = (r_1, \ldots, r_m)$ that describes voters' preferences between alternative x and alternative y. Clearly, $\mathbf{1} - \mathbf{r} = (1 - r_1, \ldots, 1 - r_m)$ shows voters' preferences between y and x. For each profile of preferences, the collective preference will be obtained by means of an aggregation function.

Definition 1. *An aggregation function is a mapping* $F : [0,1]^m \longrightarrow [0,1]$. *A discrete aggregation function (DAF) is a mapping* $H : \{0,0.5,1\}^m \longrightarrow \{0,0.5,1\}$.

The interpretation of the collective preference is consistent with the foregoing interpretation for individual preferences. Thus, if F is an aggregation function, then $F(\mathbf{r})$ is the intensity with which x is collectively preferred to y. Analogously, if H is a DAF, then $H(\mathbf{r})$ shows us if an alternative is collectively preferred to the other ($H(\mathbf{r}) \in \{0,1\}$), or instead if the two alternatives are collectively indifferent ($H(\mathbf{r}) = 0.5$).

Next we present some well-known properties of aggregation functions: symmetry, monotonicity, self-duality and idempotency. Symmetry means that the collective intensity of preference depends only on the set of individual intensities of preference, but not on which individuals have those preference intensities. Monotonicity means that the collective intensity of preference does not decrease if no individual intensity of preference decreases. Self-duality means that if everyone reverses their preference intensities between x and y, then the collective intensity of preference is also reversed. Finally, idempotency means that the collective intensity of preference coincides with the individual preference intensities when these are all the same.

Given $r \in [0,1]$, $\mathbf{r},\mathbf{s} \in [0,1]^m$ and σ a permutation on $\{1,\dots,m\}$, we will use the following notation: $\mathbf{r}_\sigma = (r_{\sigma(1)},\dots,r_{\sigma(m)})$; $\mathbf{1} = (1,\dots,1)$; $r \cdot \mathbf{1} = (r,\dots,r)$; and $\mathbf{r} \geq \mathbf{s}$ will denote $r_i \geq s_i$ for all $i \in \{1,\dots,m\}$.

Definition 2. *Let F be an aggregation function.*

1. *F is symmetric if for every profile $\mathbf{r} \in [0,1]^m$ and for every permutation σ of $\{1,\dots,m\}$ the following holds*
$$F(\mathbf{r}_\sigma) = F(\mathbf{r}).$$
2. *F is monotonic if for all pair of profiles $\mathbf{r},\mathbf{s} \in [0,1]^m$ the following holds*
$$\mathbf{r} \geq \mathbf{s} \Rightarrow F(\mathbf{r}) \geq F(\mathbf{s}).$$
3. *F is self-dual if for every profile $\mathbf{r} \in [0,1]^m$ the following holds*
$$F(\mathbf{1}-\mathbf{r}) = 1 - F(\mathbf{r}).$$
4. *F is idempotent if for every $r \in [0,1]$ the following holds*
$$F(r \cdot \mathbf{1}) = r.$$

All the previous properties are also valid for DAFs. Next we show some consequences of the previous properties. The cardinal of a set will be denoted by #.

Remark 1. If H is a symmetric DAF, then $H(\mathbf{r})$ depends only on the number of 1, 0.5, and 0. Given a discrete profile $\mathbf{r}$, if we consider

$$m_1 = \#\{i \mid r_i = 1\}, \quad m_2 = \#\{i \mid r_i = 0.5\}, \quad m_3 = \#\{i \mid r_i = 0\},$$

then $m_1 + m_2 + m_3 = m$.

Definition 3. *Let H be a symmetric DAF and*

$$\mathscr{M} = \{(m_1,m_2,m_3) \in \{0,\dots,m\}^3 \mid m_1+m_2+m_3 = m\}.$$

We say that H is represented by the function $\mathscr{H} : \mathscr{M} \longrightarrow \{0,0.5,1\}$*, defined by*

$$\mathscr{H}(m_1,m_2,m_3) = H(1,\overset{(m_1)}{\dots},1,0.5,\overset{(m_2)}{\dots},0.5,0,\overset{(m_3)}{\dots},0).$$

Definition 4. *The binary relation* $\succeq$ *on* $\mathscr{M}$ *is defined by*

$$(m_1,m_2,m_3) \succeq (n_1,n_2,n_3) \Leftrightarrow \begin{cases} m_1 \geq n_1, \\ m_1+m_2 \geq n_1+n_2. \end{cases}$$

We note that $\succeq$ is a partial order on $\mathscr{M}$ (reflexive, antisymmetric, and transitive binary relation).

Remark 2. If H is a symmetric DAF represented by $\mathscr{H}$, then it is monotonic if and only if $\mathscr{H}(m_1,m_2,m_3) \geq \mathscr{H}(n_1,n_2,n_3)$ for all $(m_1,m_2,m_3),(n_1,n_2,n_3) \in \mathscr{M}$ such that $(m_1,m_2,m_3) \succeq (n_1,n_2,n_3)$.

Remark 3. If H is a symmetric DAF represented by $\mathscr{H}$, then it is self-dual if and only if $\mathscr{H}(m_3,m_2,m_1) = 1-\mathscr{H}(m_1,m_2,m_3)$ for all $(m_1,m_2,m_3) \in \mathscr{M}$. In this case, H is characterized by the set $\mathscr{H}^{-1}(1)$, since

$$\mathscr{H}^{-1}(0) = \{(m_1,m_2,m_3) \in \mathscr{M} \mid \mathscr{H}(m_3,m_2,m_1) = 1\},$$
$$\mathscr{H}^{-1}(0.5) = \mathscr{M} \setminus \left(\mathscr{H}^{-1}(1) \cup \mathscr{H}^{-1}(0)\right).$$

When a DAF is self-dual, both alternatives have an egalitarian treatment. Therefore, if the DAF is also symmetric and the number of voters who prefer x to y coincides with the number of voters who prefer y to x, then x and y are collectively indifferent.

Remark 4. If H is a symmetric and self-dual DAF represented by $\mathscr{H}$, then for all $(m_1,m_2,m_3) \in \mathscr{M}$ such that $m_1 = m_3$ we have $\mathscr{H}(m_1,m_2,m_3) = 0.5$.

By Remark 3, it is possible to define a symmetric and self-dual DAF H by means of the elements $(m_1,m_2,m_3) \in \mathscr{M}$ where the mapping that represents H takes the value 1. Based on this, we now show some DAFs widely used in real decisions.

Definition 5

1. *The simple majority,* H_S*, is the symmetric and self-dual DAF defined by*

$$\mathscr{H}_S(m_1,m_2,m_3) = 1 \Leftrightarrow m_1 > m_3.$$

2. *The absolute majority,* H_A*, is the symmetric and self-dual DAF defined by*

$$\mathscr{H}_A(m_1,m_2,m_3) = 1 \Leftrightarrow m_1 > \frac{m}{2}.$$

3. *The Pareto majority, H_P, is the symmetric and self-dual DAF defined by*

$$\mathscr{H}_P(m_1,m_2,m_3) = 1 \Leftrightarrow m_1 > 0 \text{ and } m_3 = 0.$$

4. *The unanimous majority, H_U, is the symmetric and self-dual DAF defined by*

$$\mathscr{H}_U(m_1,m_2,m_3) = 1 \Leftrightarrow m_1 = m.$$

5. *Given $\beta \in [0.5,1)$, the absolute special majority Q_β is the symmetric and self-dual DAF defined by*

$$\mathscr{H}_\beta(m_1,m_2,m_3) = 1 \Leftrightarrow m_1 > \beta m.$$

It should be noted that absolute and unanimous majorities are specific cases of absolute special majorities. Moreover, given $\beta,\beta' \in [0.5,1)$, the absolute special majorities Q_β and $Q_{\beta'}$ are the same if and only if $[\beta m] = [\beta' m]$, where $[a]$ indicates the integer part of the number a.

Given an aggregation function, we can generate different DAFs by means of a parameter $\alpha \in [0.5,1)$. The procedure employed is based on strong α-cuts. Moreover, it is easy to check that the DAFs obtained are symmetric, monotonic, and self-dual when the original aggregation function satisfies these properties.

Definition 6. *Let F be an aggregation function and $\alpha \in [0.5,1)$. Then the α-DAF associated with F is the DAF F_α defined by*

$$F_\alpha(\mathbf{r}) = \begin{cases} 1, & \text{if } F(\mathbf{r}) > \alpha, \\ 0.5, & \text{if } 1-\alpha \le F(\mathbf{r}) \le \alpha, \\ 0, & \text{if } F(\mathbf{r}) < 1-\alpha. \end{cases}$$

Remark 5. Given an aggregation function F and $\alpha \in [0.5,1)$, the following statements hold:

1. If F is symmetric, then F_α is also symmetric.
2. If F is monotonic, then F_α is also monotonic.
3. If F is self-dual, then F_α is also self-dual.

Similar to the case of symmetric DAFs, when F is a symmetric aggregation function, the restriction $F|_{\{0,0.5,1\}^m}$ can be represented by $\mathscr{F} : \mathscr{M} \longrightarrow [0,1]$, where

$$\mathscr{F}(m_1,m_2,m_3) = F(1,\overset{(m_1)}{\ldots},1,0.5,\overset{(m_2)}{\ldots},0.5,0,\overset{(m_3)}{\ldots},0).$$

Now we show the relationship between $\mathscr{F}$ and the family of mappings $\mathscr{F}_\alpha$ that represent the α-DAFs associated with F.

Remark 6. Let F be a symmetric aggregation function and $\alpha \in [0.5,1)$. Then F_α and $F|_{\{0,0.5,1\}^m}$ can be represented by the mappings $\mathscr{F}_\alpha$ and $\mathscr{F}$, respectively. The following relationship between these mappings exists:

$$\mathscr{F}_\alpha(m_1,m_2,m_3) = \begin{cases} 1, & \text{if } \mathscr{F}(m_1,m_2,m_3) > \alpha, \\ 0.5, & \text{if } 1-\alpha \le \mathscr{F}(m_1,m_2,m_3) \le \alpha, \\ 0, & \text{if } \mathscr{F}(m_1,m_2,m_3) < 1-\alpha. \end{cases}$$

3 Characterization of Simple, Pareto, and Absolute Special Majorities

In order to generalize simple, Pareto and absolute special majorities by means of mixture operators, we show in this section some characterizations of these majority rules. The proofs of these results can be found in Llamazares [4, 5].

Simple majority is characterized through the elements $(m_1,m_2,m_3) \in \mathscr{M}$ such that $m_1 = m_3 + 1$.

Proposition 1. *Let H be a symmetric, monotonic, and self-dual DAF represented by $\mathscr{H}$. Then the following statements are equivalent:*

1. $H = H_S$.
2. $\mathscr{H}(m_3+1, m-(2m_3+1), m_3) = 1$ *for all* $m_3 \in \{0,\ldots,[\frac{m-1}{2}]\}$.

Pareto and absolute special majorities are both characterized through two elements of $\mathscr{M}$. The first one corresponds to the minimum support that alternative x needs to be selected. The second one corresponds to the maximum support that alternative x can obtain without being selected.

Proposition 2. *Let H be a symmetric, monotonic, and self-dual DAF represented by $\mathscr{H}$. Then the following statements are equivalent:*

1. $H = H_P$.
2. $\mathscr{H}(1,m-1,0) = 1$ *and* $\mathscr{H}(m-1,0,1) < 1$.

Proposition 3. *Let H be a symmetric, monotonic, and self-dual DAF represented by $\mathscr{H}$ and $\beta \in [0.5,1)$. Then the following statements are equivalent:*

1. $H = Q_\beta$.
2. $\mathscr{H}([\beta m]+1, 0, m-[\beta m]-1) = 1$ *and* $\mathscr{H}([\beta m], m-[\beta m], 0) < 1$.

4 Mixture Operators

Mixture operators have been introduced in Marques Pereira and Pasi [8] and Marques Pereira [7] as weighted averaging operators in which the usual constant weights are replaced by appropriate weighting functions depending on the attribute satisfaction values. The usual weighted averaging operators remain a particular case.

Definition 7. *Let $\varphi : [0,1] \longrightarrow (0,\infty)$ be a continuous function. The mixture operator $W^\varphi : [0,1]^m \longrightarrow [0,1]$ generated by φ is the aggregation function defined by*

$$W^{\varphi}(\mathbf{r}) = \frac{\sum_{i=1}^{m} \varphi(r_i) r_i}{\sum_{j=1}^{m} \varphi(r_j)}.$$

The mixture operator $W^{\varphi}(\mathbf{r})$ can be written as a weighted average of the various r_i,

$$W^{\varphi}(\mathbf{r}) = \sum_{i=1}^{m} w_i(\mathbf{r}) r_i,$$

where the classical constant weights w_i are replaced by the weighting functions

$$w_i(\mathbf{r}) = \frac{\varphi(r_i)}{\sum_{j=1}^{m} \varphi(r_j)}.$$

Mixture operators are symmetric and idempotent aggregation functions. The monotonicity of mixture operators has been studied by Marques Pereira and Pasi [8], Marques Pereira [7], Ribeiro and Marques Pereira [11, 12], Marques Pereira and Ribeiro [9], Mesiar and Špirková [10] and Špirková [13, 14].

With regard to the self-duality property, it is easy to check that the mixture operator W^{φ} is self-dual if and only if for every $\mathbf{r} \in [0,1]^m$ the following holds

$$\sum_{i=1}^{m} \Big(w_i(\mathbf{r}) - w_i(\mathbf{1} - \mathbf{r}) \Big) r_i = 0. \tag{1}$$

From this relationship, it is possible to obtain a characterization of self-dual mixture operators based on the fulfillment of a similar property by the function φ.

Proposition 4. *Let W^{φ} be the mixture operator generated by φ. W^{φ} is self-dual if and only if $\varphi(r) = \varphi(1-r)$ for all $r \in [0,1]$.*

Proof. Suppose that $W^{\varphi}(\mathbf{r})$ is self-dual. Given $r \in (0,1]$, consider $\mathbf{r_0} = (r, 0, \dots, 0)$. Since

$$w_1(\mathbf{r_0}) = \frac{\varphi(r)}{\varphi(r) + (m-1)\varphi(0)}, \quad w_1(\mathbf{1} - \mathbf{r_0}) = \frac{\varphi(1-r)}{\varphi(1-r) + (m-1)\varphi(1)},$$

by (1), we have

$$\frac{\varphi(r)}{\varphi(r) + (m-1)\varphi(0)} = \frac{\varphi(1-r)}{\varphi(1-r) + (m-1)\varphi(1)}.$$

Therefore $\varphi(r)\varphi(1) = \varphi(1-r)\varphi(0)$. Particularly, if $r = 1/2$ then $\varphi(1) = \varphi(0)$ and, consequently, $\varphi(r) = \varphi(1-r)$ for all $r \in [0,1]$.

On the other hand, if $\varphi(r) = \varphi(1-r)$ for all $r \in [0,1]$, then $w_i(\mathbf{r}) = w_i(\mathbf{1} - \mathbf{r})$ for all $\mathbf{r} \in [0,1]^m$ and for all $i \in \{1, \dots, m\}$. Therefore, condition (1) is satisfied. □

Remark 7. If W^{φ} is the self-dual mixture operator generated by φ and $\gamma = \dfrac{\varphi(0.5)}{\varphi(1)}$, then the mapping $\mathscr{W}^{\varphi}$ that represents $W^{\varphi}|_{\{0,0.5,1\}^m}$ takes the following values:

$$\begin{aligned}\mathscr{W}^{\varphi}(m_1,m_2,m_3) &= W^{\varphi}(1, \overset{(m_1)}{\ldots}, 1, 0.5, \overset{(m_2)}{\ldots}, 0.5, 0, \overset{(m_3)}{\ldots}, 0)\\ &= \frac{2m_1\varphi(1)+m_2\varphi(0.5)}{2((m_1+m_3)\varphi(1)+m_2\varphi(0.5))} = \frac{2m_1+m_2\gamma}{2(m_1+m_3+m_2\gamma)}.\end{aligned}$$

5 Majority Rules Obtained through Mixture Operators

In this section we establish the main results of the paper. Simple, Pareto, and absolute special majorities are generated through α-DAFs associated with self-dual mixture operators. In this way, the outcomes of this section allow us to extend these majority rules to the framework of gradual preferences by means of mixture operators.

First of all, we give a necessary and sufficient condition in order to obtain the simple majority through the α-DAFs associated with self-dual mixture operators.

Theorem 1. *Let W^{φ} be a self-dual mixture operator and $\gamma = \dfrac{\varphi(0.5)}{\varphi(1)}$. The following statements hold:*

1. If $\gamma \geq 1$:

$$W_{\alpha}^{\varphi} = H_S \Leftrightarrow \alpha < \frac{1}{2} + \frac{1}{2(1+(m-1)\gamma)}.$$

2. If $\gamma < 1$:

a. If m is odd:

$$W_{\alpha}^{\varphi} = H_S \Leftrightarrow \alpha < \frac{1}{2} + \frac{1}{2m}.$$

b. If m is even:

$$W_{\alpha}^{\varphi} = H_S \Leftrightarrow \alpha < \frac{1}{2} + \frac{1}{2(m+\gamma-1)}.$$

Proof. Let $\mathscr{W}^{\varphi}$ be the mapping that represents $W^{\varphi}|_{\{0,0.5,1\}^m}$. By Proposition 1 and Remark 6 we have that the condition $W_{\alpha}^{\varphi} = H_S$ is equivalent to $\mathscr{W}^{\varphi}(m_3+1, m-(2m_3+1), m_3) > \alpha$ for all $m_3 \in \{0,\ldots,[\frac{m-1}{2}]\}$. By Remark 7 we have

$$\begin{aligned}\mathscr{W}^{\varphi}(m_3+1, m-(2m_3+1), m_3) &= \frac{2(m_3+1)+(m-(2m_3+1))\gamma}{2(2m_3+1+(m-(2m_3+1))\gamma)}\\ &= \frac{1}{2} + \frac{1}{2(2m_3+1+(m-(2m_3+1))\gamma)}\\ &= \frac{1}{2} + \frac{1}{2(m\gamma+(2m_3+1)(1-\gamma))}\end{aligned}$$

for all $m_3 \in \{0,\dots,[\frac{m-1}{2}]\}$. Therefore,

$$W_\alpha^\varphi = H_S \Leftrightarrow \alpha < \min_{m_3\in\{0,\dots,[\frac{m-1}{2}]\}} \left\{\frac{1}{2} + \frac{1}{2(m\gamma + (2m_3+1)(1-\gamma))}\right\}.$$

We distinguish two cases:

1. If $\gamma \geq 1$, then the minimum is reached when $m_3 = 0$. Therefore,

$$W_\alpha^\varphi = H_S \Leftrightarrow \alpha < \frac{1}{2} + \frac{1}{2(1+(m-1)\gamma)}.$$

2. If $\gamma < 1$, then the minimum is reached when $m_3 = [\frac{m-1}{2}]$. We distinguish two cases:

   a. If m is odd, then $m_3 = \frac{m-1}{2}$. Therefore,

$$W_\alpha^\varphi = H_S \Leftrightarrow \alpha < \frac{1}{2} + \frac{1}{2m}.$$

   b. If m is even, then $m_3 = \frac{m}{2} - 1$. Therefore,

$$W_\alpha^\varphi = H_S \Leftrightarrow \alpha < \frac{1}{2} + \frac{1}{2(m+\gamma-1)}.$$ □

From the previous result it is possible to obtain the values of α for which the simple majority can be generated through the α-DAFs associated with self-dual mixture operators.

Corollary 1

1. *If m is odd, then there exists a self-dual mixture operator W^φ such that $W_\alpha^\varphi = H_S$ if and only if $\alpha < \frac{1}{2} + \frac{1}{2m}$.*
2. *If m is even, then there exists a self-dual mixture operator W^φ such that $W_\alpha^\varphi = H_S$ if and only if $\alpha < \frac{1}{2} + \frac{1}{2(m-1)}$.*

Proof

1. Suppose first that there exists a self-dual mixture operator W^φ such that $W_\alpha^\varphi = H_S$. By (1) and (2a) of Theorem 1, and given that

$$\max_{\gamma\in[1,\infty)} \left\{\frac{1}{2} + \frac{1}{2(1+(m-1)\gamma)}\right\} = \frac{1}{2} + \frac{1}{2m},$$

   we have $\alpha < \frac{1}{2} + \frac{1}{2m}$.

   For the converse, it is sufficient to consider a self-dual mixture operator W^φ such that $\gamma \leq 1$, i.e., $\varphi(0.5) \leq \varphi(1)$.

2. Suppose first that there exists a self-dual mixture operator W^φ such that $W^\varphi_\alpha = H_S$. By (1) and (2b) of Theorem 1, and given that

$$\max_{\gamma\in[1,\infty)} \left\{ \frac{1}{2} + \frac{1}{2(1+(m-1)\gamma)} \right\} = \frac{1}{2} + \frac{1}{2m},$$

$$\sup_{\gamma\in(0,1)} \left\{ \frac{1}{2} + \frac{1}{2(m+\gamma-1)} \right\} = \frac{1}{2} + \frac{1}{2(m-1)}$$

and

$$\max \left\{ \frac{1}{2} + \frac{1}{2m}, \frac{1}{2} + \frac{1}{2(m-1)} \right\} = \frac{1}{2} + \frac{1}{2(m-1)},$$

we have $\alpha < \dfrac{1}{2} + \dfrac{1}{2(m-1)}$.

For the converse, given $\alpha < \dfrac{1}{2} + \dfrac{1}{2(m-1)}$, it is sufficient to consider a self-dual mixture operator W^φ such that $\gamma < \dfrac{1}{2\alpha - 1} - (m-1)$, i.e., $\varphi(0.5) < \varphi(1)\left(\dfrac{1}{2\alpha-1} - (m-1)\right)$. □

In the following theorem we characterize the self-dual mixture operators whose associated α-DAFs correspond with the Pareto majority.

Theorem 2. *Let W^φ be a self-dual mixture operator and $\gamma = \dfrac{\varphi(0.5)}{\varphi(1)}$. The following statement holds:*

$$W^\varphi_\alpha = H_P \Leftrightarrow \begin{cases} \gamma < \dfrac{2}{(m-1)(m-2)} \quad \text{and} \\ 1 - \dfrac{1}{m} \le \alpha < \dfrac{1}{2} + \dfrac{1}{2(1+(m-1)\gamma)}. \end{cases}$$

Proof. Let $\mathscr{W}^\varphi$ be the mapping that represents $W^\varphi|_{\{0,0.5,1\}^m}$. By Proposition 2 and Remark 6 we have that the condition $W^\varphi_\alpha = H_P$ is equivalent to $\mathscr{W}^\varphi(m-1,0,1) \le \alpha$ and $\mathscr{W}^\varphi(1,m-1,0) > \alpha$. Since by Remark 7 we have

$$\mathscr{W}^\varphi(m-1,0,1) = \frac{2(m-1)}{2m}, \quad \mathscr{W}^\varphi(1,m-1,0) = \frac{2+(m-1)\gamma}{2(1+(m-1)\gamma)};$$

then

$$W^\varphi_\alpha = H_P \Leftrightarrow 1 - \frac{1}{m} \le \alpha < \frac{1}{2} + \frac{1}{2(1+(m-1)\gamma)}.$$

To demonstrate the thesis of the theorem it is sufficient to take into account that

$$\begin{aligned}1-\frac{1}{m}<\frac{1}{2}+\frac{1}{2(1+(m-1)\gamma)} &\Leftrightarrow \frac{m-2}{2m}<\frac{1}{2(1+(m-1)\gamma)}\\ &\Leftrightarrow 1+(m-1)\gamma<\frac{m}{m-2}\\ &\Leftrightarrow \gamma<\frac{2}{(m-1)(m-2)}.\end{aligned}$$

□

In the next theorem we give a necessary and sufficient condition in order to obtain the absolute special majorities through the α-DAFs associated with self-dual mixture operators.

Theorem 3. *Let W^{φ} be a self-dual mixture operator and $\gamma=\dfrac{\varphi(0.5)}{\varphi(1)}$. The following statement holds:*

$$W_{\alpha}^{\varphi}=Q_{\beta} \Leftrightarrow \begin{cases}\gamma>2[\beta m]\dfrac{m-[\beta m]-1}{(m-[\beta m])(2[\beta m]+2-m)} \text{ and}\\ \dfrac{1}{2}+\dfrac{1}{2\left(1+\dfrac{m-[\beta m]}{[\beta m]}\gamma\right)}\leq\alpha<\dfrac{[\beta m]+1}{m}.\end{cases}$$

Proof. Let $\mathscr{W}^{\varphi}$ be the mapping that represents $W^{\varphi}|_{\{0,0.5,1\}^m}$. By Proposition 3 and Remark 6 we have that the condition $W_{\alpha}^{\varphi}=Q_{\beta}$ is equivalent to $\mathscr{W}^{\varphi}([\beta m],m-[\beta m],0)\leq\alpha$ and $\mathscr{W}^{\varphi}([\beta m]+1,0,m-[\beta m]-1)>\alpha$. By Remark 7 we have

$$\mathscr{W}^{\varphi}([\beta m],m-[\beta m],0)=\frac{2[\beta m]+(m-[\beta m])\gamma}{2([\beta m]+(m-[\beta m])\gamma)},$$

$$\mathscr{W}^{\varphi}([\beta m]+1,0,m-[\beta m]-1)=\frac{2([\beta m]+1)}{2m}.$$

Since

$$\begin{aligned}\frac{2[\beta m]+(m-[\beta m])\gamma}{2([\beta m]+(m-[\beta m])\gamma)}&=\frac{1}{2}+\frac{[\beta m]}{2([\beta m]+(m-[\beta m])\gamma)}\\ &=\frac{1}{2}+\frac{1}{2\left(1+\dfrac{m-[\beta m]}{[\beta m]}\gamma\right)},\end{aligned}$$

then

$$W_{\alpha}^{\varphi}=Q_{\beta} \Leftrightarrow \frac{1}{2}+\frac{1}{2\left(1+\dfrac{m-[\beta m]}{[\beta m]}\gamma\right)}\leq\alpha<\frac{[\beta m]+1}{m}.$$

To demonstrate the thesis of the theorem it is sufficient to take into account that

$$\begin{aligned}\frac{1}{2}+\frac{1}{2\left(1+\frac{m-[\beta m]}{[\beta m]}\gamma\right)} < \frac{[\beta m]+1}{m} &\Leftrightarrow \\ &\Leftrightarrow \frac{[\beta m]}{2([\beta m]+(m-[\beta m])\gamma)} < \frac{2[\beta m]+2-m}{2m} \\ &\Leftrightarrow [\beta m]+(m-[\beta m])\gamma > \frac{m[\beta m]}{2[\beta m]+2-m} \\ &\Leftrightarrow \gamma > 2[\beta m]\frac{m-[\beta m]-1}{(m-[\beta m])(2[\beta m]+2-m)}.\end{aligned}$$

□

As particular cases of this theorem it is straightforward to give necessary and sufficient conditions to obtain absolute and unanimous majorities through α-DAFs associated with self-dual mixture operators.

Corollary 2. *Let W^{φ} be a self-dual mixture operator and $\gamma = \frac{\varphi(0.5)}{\varphi(1)}$. The following statements hold:*

1. a. *If m is odd:*

$$W^{\varphi}_{\alpha} = H_A \Leftrightarrow \begin{cases} \gamma > \dfrac{(m-1)^2}{m+1} \quad \text{and} \\ \dfrac{1}{2}+\dfrac{1}{2\left(1+\frac{m+1}{m-1}\gamma\right)} \leq \alpha < \dfrac{1}{2}+\dfrac{1}{2m}. \end{cases}$$

   b. *If m is even:*

$$W^{\varphi}_{\alpha} = H_A \Leftrightarrow \gamma > \frac{m}{2}-1 \text{ and } \frac{1}{2}+\frac{1}{2(1+\gamma)} \leq \alpha < \frac{1}{2}+\frac{1}{m}.$$

2. $W^{\varphi}_{\alpha} = H_U \Leftrightarrow \alpha \geq \frac{1}{2}+\frac{m-1}{2(m+\gamma-1)}$.

Proof

1. It is sufficient to take into account the following:
   a. If m is odd: $W^{\varphi}_{\alpha} = H_A \Leftrightarrow [\beta m] = \frac{m-1}{2}$.
   b. If m is even: $W^{\varphi}_{\alpha} = H_A \Leftrightarrow [\beta m] = \frac{m}{2}$.
2. It is sufficient to take into account that $W^{\varphi}_{\alpha} = H_U \Leftrightarrow [\beta m] = m-1$. □

6 Concluding Remarks

In this paper we have investigated under which conditions the α-DAF associated with a self-dual mixture operator corresponds to one of the three classical majority rules used in social choice theory: simple majority H_S, Pareto majority H_P, and absolute special majority Q_β. Interestingly, the necessary and sufficient conditions presented in Theorems 1, 2 and 3 depend only on the dimension m and on the ratio $\gamma = \varphi(0.5)/\varphi(1)$, whose difference from the unit value encodes a global measure of how much the self-dual mixture operator considered differs from the plain averaging operator, with constant weights $1/m$.

Acknowledgements. This work is partially financed by the Junta de Castilla y León (Project VA002B08), the Spanish Ministry of Education and Science (Project SEJ2006-04267) and ERDF.

References

1. Calvo, T., Mayor, G., Mesiar, R. (eds.): Aggregation Operators: New Trends and Applications. Physica-Verlag, Heidelberg (2002)
2. García Lapresta, J.L., Llamazares, B.: Aggregation of fuzzy preferences: Some rules of the mean. Social Choice and Welfare 17, 673–690 (2000)
3. García Lapresta, J.L., Llamazares, B.: Majority decisions based on difference of votes. Journal of Mathematical Economics 35, 463–481 (2001)
4. Llamazares, B.: Simple and absolute special majorities generated by OWA operators. European Journal of Operational Research 158, 707–720 (2004)
5. Llamazares, B.: Choosing OWA operator weights in the field of Social Choice. Information Sciences 177, 4745–4756 (2007)
6. Marichal, J.L.: Aggregation Operators for Multicriteria Decision Aid. MA Thesis, Liège University, Liège (1998)
7. Marques Pereira, R.A.: The orness of mixture operators: The exponencial case. In: Proceedings of the 8th International Conference on Information Processing and Management of Uncertainty in Knowledge-based Systems (IPMU 2000), Madrid, Spain, pp. 974–978 (2000)
8. Marques Pereira, R.A., Pasi, G.: On non-monotonic aggregation: Mixture operators. In: Proceedings of the 4th Meeting of the EURO Working Group on Fuzzy Sets (EUROFUSE 1999) and 2nd International Conference on Soft and Intelligent Computing (SIC 1999), Budapest, Hungary, pp. 513–517 (1999)
9. Marques Pereira, R.A., Ribeiro, R.A.: Aggregation with generalized mixture operators using weighting functions. Fuzzy Sets and Systems 137, 43–58 (2003)
10. Mesiar, R., Špirková, J.: Weighted means and weighting functions. Kybernetika 42, 151–160 (2006)
11. Ribeiro, R.A., Marques Pereira, R.A.: Weights as functions of attribute satisfaction values. In: Proceedings of Workshop on Preference Modelling and Applications (EUROFUSE), Granada, Spain, pp. 131–137 (2001)
12. Ribeiro, R.A., Marques Pereira, R.A.: Generalized mixture operators using weighting functions: A comparative study with WA and OWA. European Journal of Operational Research 145, 329–342 (2003)
13. Špirková, J.: Mixture and quasi-mixture operators. In: Proceedings of the 11th International Conference on Information Processing and Management of Uncertainty in Knowledge-based Systems (IPMU 2006), Paris, France, pp. 603–608 (2006)
14. Špirková, J.: Monotonicity of mixture and quasi-mixture operators. In: Proceedings of the 13th Zittau Fuzzy Colloquium, Zittau, Germany, pp. 212–219 (2006)
15. Xu, Z.S., Da, Q.L.: An overview of operators for aggregating information. International Journal of Intelligent Systems 18, 953–969 (2003)

Belief Function Correction Mechanisms

David Mercier, Thierry Denœux, and Marie-Hélène Masson

Abstract. Different operations can be used in the theory of belief functions to correct the information provided by a source, given metaknowledge about that source. Examples of such operations are discounting, de-discounting, extended discounting and contextual discounting. In this article, the links between these operations are explored. New interpretations of these schemes, as well as two families of belief function correction mechanisms are introduced and justified. The first family generalizes previous non-contextual discounting operations, whereas the second generalizes the contextual discounting.

Keywords: Dempster-Shafer theory, Belief functions, discounting operations, disjunctive and conjunctive canonical decompositions.

1 Introduction

Introduced by Dempster [1] and Shafer [13], belief functions constitute one of the main frameworks for reasoning with imperfect information.

When receiving a piece of information represented by a belief function, some metaknowledge regarding the quality or reliability of the source that provides the information, can be available. To correct the information according to this metaknowledge, different tools can be used:

David Mercier
Univ. Lille Nord de France, UArtois, EA 3926 LGI2A, France
e-mail: david.mercier@univ-artois.fr

Thierry Denœux
Université de Technologie de Compiègne, UMR CNRS 6599 Heudiasyc, France
e-mail: tdenoeux@hds.utc.fr

Marie-Hélène Masson
Université de Picardie Jules Verne, UMR CNRS 6599 Heudiasyc, France
e-mail: mmasson@hds.utc.fr

B. Bouchon-Meunier et al. (Eds.) Found. of Reas. under Uncert., STUDFUZZ 249, pp. 203–222.
springerlink.com

- The *discounting operation*, introduced by Shafer in his seminal book [13], allows one to weaken the information provided by the source;
- The *de-discounting operation*, introduced by Denœux and Smets [2], has the effect of strengthening the information;
- The *extended discounting operation*, introduced by Zhu and Basir [20], makes it possible to weaken, strengthen or contradict the information;
- The *contextual discounting operation*, a refining of the discounting operation, introduced by Mercier et al. [10], weakens the information by taking into account more detailed knowledge regarding the reliability of the source in different contexts, i.e., conditionally on different hypotheses regarding the answer to the question of interest.

In this article, the links between these operations are explored. Belief function correction mechanisms encompassing these schemes are introduced and justified.

First, discounting, de-discounting, and extended discounting are shown to be particular cases of a parameterized family of transformations [9]. This family includes all possible transformations, expressed by a belief function, based on the different states in which the source can be when the information is supplied.

Secondly, another family of correction mechanisms based on the concepts of negation [4] and canonical decompositions [3, 12, 16] of a belief function is explored. This family is shown to generalize the contextual discounting operation.

Belief functions are used in different theories of uncertainty such as, for instance, models based on lower and upper probabilities including Dempster's model [1] and the related Hint model [7], random set theory [6], or the Transferable Belief Model developed by Smets [15, 19]. In the latter model, belief functions are interpreted as weighted opinions of an agent or a sensor. This model is adopted in this article.

This article is organized as follows. Background material on belief functions is recalled in Section 2. All the discounting operations are presented in Section 3. A new interpretation of non-contextual discounting as well as a parameterized family of correction mechanisms are introduced and justified in Section 4. Another family of correction mechanisms based on the disjunctive and conjunctive canonical decompositions of a belief function is presented in Section 5. In Section 6, an example of a correction mechanism introduced in this article is tested with real data in a postal address recognition application, in which decisions associated with confidence scores are combined. Finally, Section 7 concludes this paper.

2 Belief Functions: Basic Concepts

2.1 Representing Information

Let us consider an agent Ag in charge of making a decision regarding the answer to a given question Q of interest.

Let $\Omega = \{\omega_1, \ldots, \omega_K\}$, called the *frame of discernment*, be the finite set containing the possible answers to question Q.

The information held by agent Ag regarding the answer to question Q can be quantified by a *basic belief assignment (BBA) or mass function* m^{Ω}_{Ag}, defined as a function from 2^{Ω} to $[0,1]$, and verifying:

$$\sum_{A \subseteq \Omega} m^{\Omega}_{Ag}(A) = 1 . \quad (1)$$

Function m^{Ω}_{Ag} describes the state of knowledge of agent Ag regarding the answer to question Q belonging to Ω. By extension, it also represents an item of evidence that induces such a state of knowledge. The quantity $m^{\Omega}_{Ag}(A)$ is interpreted as the part of the unit mass allocated to the hypothesis: "the answer to question Q is in the subset A of Ω".

When there is no ambiguity, the full notation m^{Ω}_{Ag} will be simplified to m^{Ω}, or even m.

Definition 1. *The following definitions are considered.*

- *A subset A of Ω such that $m(A) > 0$ is called a* focal element *of m.*
- *A BBA m with only one focal element A is said to be* categorical *and is denoted m_A; we thus have $m_A(A) = 1$.*
- *Total ignorance is represented by the BBA m_{Ω}, called the* vacuous belief function.
- *A BBA m is said to be:*
  - dogmatic *if $m(\Omega) = 0$;*
  - non-dogmatic *if $m(\Omega) > 0$;*
  - normal *if $m(\emptyset) = 0$;*
  - subnormal *if $m(\emptyset) > 0$;*
  - simple *if m has no more than two focal sets, Ω being included.*

Finally, $\overline{m}$ denotes the *negation* of m [4], defined by $\overline{m}(A) = m(\overline{A})$, for all $A \subseteq \Omega$.

2.2 Combining Pieces of Information

Two BBAs m_1 and m_2 induced by distinct and reliable sources of information can be combined using the *conjunctive rule of combination (CRC)*, also referred to as the *unnormalized Dempster's rule of combination*, defined for all $A \subseteq \Omega$ by:

$$m_1 \textcircled{\cap} m_2(A) = \sum_{B \cap C = A} m_1(B) m_2(C) . \quad (2)$$

Alternatively, if we only know that at least one of the sources is reliable, BBAs m_1 and m_2 can be combined using the *disjunctive rule of combination (DRC)*, defined for all $A \subseteq \Omega$ by:

$$m_1 \textcircled{\cup} m_2(A) = \sum_{B \cup C = A} m_1(B) m_2(C) . \quad (3)$$

2.3 Marginalization and Vacuous Extension

A mass function defined on a product space $\Omega \times \Theta$ may be *marginalized* on Ω by transferring each mass $m^{\Omega\times\Theta}(B)$ for $B \subseteq \Omega \times \Theta$ to its projection on Ω:

$$m^{\Omega\times\Theta\downarrow\Omega}(A) = \sum_{\substack{B\subseteq\Omega\times\Theta,\\ \mathrm{Proj}(B\downarrow\Omega)=A}} m^{\Omega\times\Theta}(B), \quad (4)$$

for all $A \subseteq \Omega$ where $\mathrm{Proj}(B \downarrow \Omega)$ denotes the projection of B onto Ω.

Conversely, it is usually not possible to retrieve the original BBA $m^{\Omega\times\Theta}$ from its marginal $m^{\Omega\times\Theta\downarrow\Omega}$ on Ω. However, the *least committed*, or *least informative BBA* [14] such that its projection on Ω is $m^{\Omega\times\Theta\downarrow\Omega}$ may be computed. This defines the *vacuous extension* of m^{Ω} in the product space $\Omega \times \Theta$ [14], noted $m^{\Omega\uparrow\Omega\times\Theta}$, and given by:

$$m^{\Omega\uparrow\Omega\times\Theta}(B) = \begin{cases} m^{\Omega}(A) \text{ if } B = A \times \Theta, A \subseteq \Omega, \\ 0 \qquad\quad \text{otherwise.} \end{cases} \quad (5)$$

2.4 Conditioning and Ballooning Extension

Conditional beliefs represent knowledge that is valid provided that an hypothesis is satisfied. Let m be a mass function and $B \subseteq \Omega$ an hypothesis; the *conditional belief function* $m[B]$ is given by:

$$m[B] = m \cap\!\!\!\!\bigcirc\, m_B. \quad (6)$$

If $m^{\Omega\times\Theta}$ is defined on the product space $\Omega \times \Theta$, and θ is a subset of Θ, the conditional BBA $m^{\Omega}[\theta]$ is defined by combining $m^{\Omega\times\Theta}$ with $m_{\theta}^{\Theta\uparrow\Omega\times\Theta}$, and marginalizing the result on Ω:

$$m^{\Omega}[\theta] = \left(m^{\Omega\times\Theta} \cap\!\!\!\!\bigcirc\, m_{\theta}^{\Theta\uparrow\Omega\times\Theta}\right)^{\downarrow\Omega}. \quad (7)$$

Assume now that $m^{\Omega}[\theta]$ represents the agent's beliefs on Ω conditionally on θ, i.e., in a context where θ holds. There are usually many BBAs on $\Omega \times \Theta$, whose conditioning on θ yields $m^{\Omega}[\theta]$. Among these, the least committed one is defined for all $A \subseteq \Omega$ by:

$$m^{\Omega}[\theta]^{\Uparrow\Omega\times\Theta}(A \times \theta \cup \Omega \times \overline{\theta}) = m^{\Omega}[\theta](A). \quad (8)$$

This operation is referred to as the *deconditioning* or *ballooning extension* [14] of $m^{\Omega}[\theta]$ on $\Omega \times \Theta$.

3 Correction Mechanisms

3.1 Discounting

When receiving a piece of information represented by a mass function m, agent Ag may have some doubts regarding the reliability of the source that provided this information. Such metaknowledge can be taken into account using the discounting operation introduced by Shafer [13, page 252], and defined by:

$$^{\alpha}m = (1-\alpha)m + \alpha\, m_{\Omega}\ , \tag{9}$$

where $\alpha \in [0,1]$.

A discount rate α equal to 1, means that the source is not reliable and the piece of information it provides cannot be taken into account, so Ag's knowledge remains vacuous: $m^{\Omega}_{Ag} = {}^{1}m = m_{\Omega}$. On the contrary, a null discount rate indicates that the source is fully reliable and the piece of information is entirely accepted: $m^{\Omega}_{Ag} = {}^{0}m = m$. In practice, however, agent Ag usually does not know for sure whether the source is reliable or not, but has some degree of belief expressed by:

$$\begin{cases} m^{\mathscr{R}}_{Ag}(\{R\}) = 1-\alpha \\ m^{\mathscr{R}}_{Ag}(\mathscr{R}) \quad = \alpha, \end{cases} \tag{10}$$

where $\mathscr{R} = \{R, NR\}$, R and NR standing, respectively, for *"the source is reliable"* and *"the source is not reliable"*. This formalization yields expression (9), as demonstrated by Smets in [14, Section 5.7].

Let us consider a BBA m^{Ω}_{0} defined by

$$m^{\Omega}_{0}(A) = \begin{cases} \beta & \text{if } A = \emptyset \\ \alpha & \text{if } A = \Omega \\ 0 & \text{otherwise,} \end{cases} \tag{11}$$

with $\alpha \in [0,1]$ and $\beta = 1-\alpha$. The discounting operation (9) of a BBA m is equivalent to the disjunctive combination (3) of m with m^{Ω}_{0}. Indeed:

$$m \circledcirc m^{\Omega}_{0}(A) = m(A)m^{\Omega}_{0}(\emptyset) = \beta m(A) = {}^{\alpha}m(A),\ \forall A \subset \Omega\ ,$$

and

$$m \circledcirc m^{\Omega}_{0}(\Omega) = m(\Omega)m^{\Omega}_{0}(\emptyset) + m^{\Omega}_{0}(\Omega)\sum_{A\subseteq\Omega} m(A) = \beta m(\Omega) + \alpha = {}^{\alpha}m(\Omega).$$

3.2 De-discounting

In this process, agent Ag receives a piece of information ${}^{\alpha}m$ from a source S, different from m_{Ω} and discounted with a discount rate $\alpha < 1$.

If Ag knows the discount rate α, then it can recompute m by reversing the discounting operation (9):

$$m_{Ag} = m = \frac{{}^{\alpha}m - \alpha\, m_{\Omega}}{1-\alpha}\ . \tag{12}$$

This procedure is called *de-discounting* by Denœux and Smets in [2].

If the agent receives a mass function m discounted with an unknown discount rate α, it can imagine all possible values in the range $[0, m(\Omega)]$. Indeed, as shown in [2], $m(\Omega)$ is the largest value for α such that the de-discounting operation (12) yields a BBA. De-discounting m with this maximal value is called *maximal de-discounting*. The result is the *totally reinforced belief function*, noted ${}^{tr}m$ and defined as follows:

$$ {}^{tr}m(A) = \begin{cases} \frac{m(A)}{1-m(\Omega)} & \forall A \subset \Omega, \\ 0 & \text{otherwise.} \end{cases} \quad (13) $$

The mass function ${}^{tr}m$ is thus obtained from m by redistributing the mass $m(\Omega)$ among the strict subsets of Ω.

3.3 Extended Discounting Scheme

In [20], Zhu and Basir proposed to extend the discounting process in order to *strengthen, discount or contradict* belief functions. The extended discounting scheme is composed of two transformations.

The first transformation, allowing us to strengthen or weaken a source of information, is introduced by retaining the discounting equation (9), while allowing the discount rate α to be in the range $\left[\frac{-m(\Omega)}{1-m(\Omega)}, 1\right]$.

- If $\alpha \in [0,1]$, this transformation is the discounting operation.
- If $\alpha \in [\frac{-m(\Omega)}{1-m(\Omega)}, 0]$, this transformation is equivalent to the de-discounting operation (12) with the reparameterization $\alpha = \frac{-\alpha'}{1-\alpha'}$ with $\alpha' \in [0, m(\Omega)]$. Indeed,

$$ {}^{\alpha}m = \left(1 - \frac{-\alpha'}{1-\alpha'}\right) m + \frac{-\alpha'}{1-\alpha'} m_{\Omega} = \frac{m - \alpha' m_{\Omega}}{1-\alpha'} . \quad (14) $$

The second transformation, allowing us to contradict a non-vacuous and normal belief function m, is defined by the following equation:

$$ \begin{cases} {}^{\alpha}m(\overline{A}) = (\alpha - 1) m(A) & \text{if } A \subset \Omega, \\ {}^{\alpha}m(\Omega) = (\alpha - 1) m(\Omega) + 2 - \alpha & \text{otherwise,} \end{cases} \quad (15) $$

where $\alpha \in \left[1, 1 + \frac{1}{1-m(\Omega)}\right]$.

- If $\alpha = 1$, ${}^{\alpha}m = m_{\Omega}$.
- If $\alpha = 1 + \frac{1}{1-m(\Omega)}$, ${}^{\alpha}m = \overline{{}^{tr}m}$, where $\overline{m}$ denotes the negation of m [4], defined by $\overline{m}(A) = m(\overline{A})$, $\forall A \subseteq \Omega$. In other words, after being totally reinforced, each basic belief mass $m(A)$ is transferred to its complement. The BBA m is then fully contradicted.

This scheme has been successfully applied in medical imaging [20]. However, it suffers from a lack of formal justification. Indeed, the number $(1 - \alpha)$ can no longer be interpreted as a degree of belief as it can take values greater than 1 and smaller than 0.

3.4 Contextual Discounting Based on a Coarsening

Contextual discounting was introduced in [10]. It makes it possible to take into account the fact that the reliability of the source of information can be expected to depend on the true answer of the question of interest.

For instance, in medical diagnosis, depending on his/her specialty, experience or training, a physician may be more or less competent to diagnose some types of diseases. Likewise, in target recognition, a sensor may be more capable of recognizing some types of targets while being less effective for other types.

Let $\Theta = \{\theta_1, \ldots, \theta_L\}$ be a coarsening [13, chapter 6] of Ω, in other words $\theta_1, \ldots, \theta_L$ form a partition of Ω.

Unlike (10), in the contextual model, agent Ag is assumed to hold beliefs on the reliability of the source of information conditionally on each θ_ℓ, $\ell \in \{1, \ldots, L\}$:

$$\begin{cases} m_{Ag}^{\mathscr{R}}[\theta_\ell](\{R\}) = 1 - \alpha_\ell = \beta_\ell \\ m_{Ag}^{\mathscr{R}}[\theta_\ell](\mathscr{R}) \quad = \alpha_\ell \,. \end{cases} \tag{16}$$

For all $\ell \in \{1, \ldots, L\}$, $\beta_\ell + \alpha_\ell = 1$, and β_ℓ represents the degree of belief that the source is reliable knowing that the true answer of the question of interest belongs to θ_ℓ.

In the same way as in the discounting operation (9), agent Ag considers that the source can be in two states: reliable or not reliable [10, 14]:

- If the source is reliable (state R), the information m_S^Ω it provides becomes Ag's knowledge. Formally, $m_{Ag}^\Omega[\{R\}] = m_S^\Omega$.
- If the source is not reliable (state NR), the information m_S^Ω it provides is discarded, and Ag remains in a state of ignorance: $m_{Ag}^\Omega[\{NR\}] = m_\Omega$.

The knowledge held by agent Ag, based on the information m_S^Ω from a source S as well as metaknowledge $m_{Ag}^{\mathscr{R}}$ concerning the reliability of the source can then be computed by:

- Deconditioning the L BBAs $m_{Ag}^{\mathscr{R}}[\theta_\ell]$ on the product space $\Omega \times \mathscr{R}$ using (8);
- Deconditioning $m_{Ag}^\Omega[\{R\}]$ on the same product space $\Omega \times \mathscr{R}$ using (8) as well;
- Combining them using the CRC (2);
- Marginalizing the result on Ω using (4).

Formally:

$$m_{Ag}^\Omega[m_S^\Omega, m_{Ag}^{\mathscr{R}}] = \left(\textcircled{\cap}_{\ell=1}^{L} m_{Ag}^{\mathscr{R}}[\theta_\ell]^{\Uparrow \Omega \times \mathscr{R}} \textcircled{\cap} m_{Ag}^\Omega[\{R\}]^{\Uparrow \Omega \times \mathscr{R}} \right)^{\downarrow \Omega} . \tag{17}$$

As shown in [10], the resulting BBA m_{Ag}^Ω, only depends on m_S and on the vector $\alpha = (\alpha_1, \ldots, \alpha_L)$ of discount rates. It is then denoted by ${}^{\alpha}_{\Theta}m$.

Proposition 1 ([10, Proposition 8]). *The contextual discounting* ${}^{\alpha}_{\Theta}m$ *on a coarsening* Θ *of a BBA* m *is equal to the disjunctive combination of* m *with a BBA* m_0^Ω *such that:*

$$m_0^{\Omega} = m_1^{\Omega} \cup\!\!\!\bigcirc\, m_2^{\Omega} \cup\!\!\!\bigcirc \ldots \cup\!\!\!\bigcirc\, m_L^{\Omega}\,, \tag{18}$$

where each m_ℓ^Ω, $\ell \in \{1,\ldots,L\}$, is defined by:

$$m_\ell^\Omega(A) = \begin{cases} \beta_\ell & \text{if } A = \emptyset \\ \alpha_\ell & \text{if } A = \theta_\ell \\ 0 & \text{otherwise.} \end{cases} \tag{19}$$

Remark 1. Two special cases of this discounting operation can be considered.

- If $\Theta = \{\Omega\}$ denotes the trivial partition of Ω in one class, combining m with m_0 defined by (11) is equivalent to combining m with m_0 defined by (18), so this contextual discounting operation is identical to the classical discounting operation.
- If $\Theta = \Omega$, the finest partition of Ω, this discounting is simply called contextual discounting and denoted ${}^{\alpha}m$. It is defined by the disjunctive combination of m with the BBA $m_1^\Omega \cup\!\!\!\bigcirc\, m_2^\Omega \cup\!\!\!\bigcirc \ldots \cup\!\!\!\bigcirc\, m_K^\Omega$, where each m_k^Ω, $k \in \{1,\ldots,K\}$ is defined by $m_k^\Omega(\emptyset) = \beta_k$ and $m_k^\Omega(\{\omega_k\}) = \alpha_k$.

In the following section, a new parameterized family of transformations encompassing all the non-contextual schemes presented in this section, is introduced and justified.

4 A Parameterized Family of Correction Mechanisms

In this section, the hypotheses concerning the states in which agent Ag considers that the source can be, are extended in the following way.

Let us assume that the source can be in N states R_i, $i \in \{1,\ldots,N\}$, whose interpretations are given by transformations m_i of m: if the source is in the state R_i then $m_{Ag}^\Omega = m_i$.

$$m_{Ag}^\Omega[\{R_i\}] = m_i, \quad \forall i \in \{1,\ldots,N\}\,. \tag{20}$$

Let $\mathscr{R} = \{R_1,\ldots,R_N\}$, and let us suppose that, for all $i \in \{1,\ldots,N\}$:

$$m_{Ag}^{\mathscr{R}}(\{R_i\}) = \nu_i, \text{ with } \sum_{i=1}^{N} \nu_i = 1\,. \tag{21}$$

The knowledge held by agent Ag, based on the information m_S^Ω from a source S and on metaknowledge $m_{Ag}^{\mathscr{R}}$ regarding the different states in which the source can be, can then be computed by:

- Deconditioning the N BBAs $m_{Ag}^\Omega[\{R_i\}]$ on the product space $\Omega \times \mathscr{R}$ using (8);
- Vacuously extending $m_{Ag}^{\mathscr{R}}$ on the same product space $\Omega \times \mathscr{R}$ using (5);
- Combining all BBAs using the CRC (2);
- Marginalizing the result on Ω using (4).

Formally:

$$m_{Ag}^{\Omega}[m_S^{\Omega}, m_{Ag}^{\mathscr{R}}] = \left(\textcircled{\cap}_{i=1}^{N} m_{Ag}^{\Omega}[\{R_i\}]^{\Uparrow \Omega\times\mathscr{R}} \textcircled{\cap} m_{Ag}^{\mathscr{R}\uparrow\Omega\times\mathscr{R}} \right)^{\downarrow\Omega} . \tag{22}$$

Proposition 2. *The BBA m_{Ag}^{Ω} defined by (22) only depends on m_i and ν_i, $i \in \{1,\ldots,N\}$. The result is noted ${}^{\nu}m$, ν denoting the vector of ν_i, and verifies:*

$$m_{Ag}^{\Omega} = {}^{\nu}m = \sum_{i=1}^{N} \nu_i \, m_i . \tag{23}$$

Proof. For all $i \in \{1,\ldots,N\}$ and $A \subseteq \Omega$:

- from (5) and (21), the vacuous extension of $m_{Ag}^{\mathscr{R}}$ is given by:

$$m_{Ag}^{\mathscr{R}\uparrow\Omega\times\mathscr{R}}(\Omega \times \{R_i\}) = \nu_i ; \tag{24}$$

- from (8) and (20), the deconditioning of $m_{Ag}^{\Omega}[\{R_i\}]$ verifies:

$$m_{Ag}^{\Omega}[\{R_i\}]^{\Uparrow\Omega\times\mathscr{R}}(A \times \{R_i\} \cup \Omega \times \overline{\{R_i\}}) = m_i(A) . \tag{25}$$

However, $\forall i \in \{1,\ldots,N\}$ and $\forall A_i \subseteq \Omega$:

$$\cap_{i=1}^{N}(A_i \times \{R_i\} \cup \Omega \times \overline{\{R_i\}}) = \cup_{i=1}^{N} A_i \times \{R_i\} , \tag{26}$$

and, $\forall j \in \{1,\ldots,N\}$:

$$(\cup_{i=1}^{N} A_i \times \{R_i\}) \cap \Omega \times \{R_j\} = A_j \times \{R_j\} . \tag{27}$$

Therefore, the conjunctive combination of $m_{Ag}^{\Omega}[\{R_i\}]^{\Uparrow\Omega\times\mathscr{R}}$, $i \in \{1,\ldots,N\}$, with $m_{Ag}^{\mathscr{R}\uparrow\Omega\times\mathscr{R}}$, denoted $\textcircled{\cap} m_{Ag}^{\Omega\times\mathscr{R}}$, has N focal elements such that:

$$\textcircled{\cap} m_{Ag}^{\Omega\times\mathscr{R}}(A_j \times \{R_j\}) = \nu_j \, m_j(A_j) \underbrace{\prod_{i\neq j} \sum_{A\subseteq\Omega} m_i(A)}_{=1}, \quad \forall j \in \{1,\ldots,N\} , \tag{28}$$

or, equivalently, $\forall A \subseteq \Omega$ and $\forall i \in \{1,\ldots,N\}$:

$$\textcircled{\cap} m_{Ag}^{\Omega\times\mathscr{R}}(A \times \{R_i\}) = \nu_i \, m_i(A). \tag{29}$$

Then, after projecting onto Ω:

$$m_{Ag}^{\Omega}(A) = \sum_{i=1}^{N} \nu_i \, m_i(A) \quad \forall A \subseteq \Omega, \tag{30}$$

which completes the proof. □

Proposition 3. *Discounting, de-discounting and extended discounting operations are particular cases of correction mechanisms expressed by (23):*

- *Discounting corresponds to the case of two states R_1 and R_2 such that $m_1 = m_\Omega$ and $m_2 = m$ (as already exposed in Section 3.1).*
- *De-discounting corresponds to the case of two states such that $m_1 = m$ and $m_2 = {}^{tr}m$, which means a first state where the information provided by the source is accepted, and a second one where this information is totally reinforced.*
- *The first transformation of the extended discounting operation, discounting equation (9) with $\alpha \in [-m(\Omega)/(1-m(\Omega)), 1]$, is obtained in the particular case of two states such that $m_1 = m_\Omega$ and $m_2 = {}^{tr}m$.*
- *The second transformation of the extended discounting operation (15) is retrieved by considering two states: a first one where the source is fully contradicted ($m_1 = \overline{{}^{tr}m}$) [18], and a second one where the information provided by the source is rejected ($m_2 = m_\Omega$).*

Proof. By considering two states such that $m_1 = m$ and $m_2 = {}^{tr}m$, ${}^{\nu}m = \nu_1 m + \nu_2 {}^{tr}m$ is a reparameterization of the de-discounting operation (12) with $\nu_1 = \frac{m(\Omega)-\alpha}{(1-\alpha)m(\Omega)}$, $\alpha \in [0, m(\Omega)]$. Indeed:

$$\begin{aligned}
{}^{\nu}m(A) &= \frac{m(\Omega)-\alpha}{(1-\alpha)m(\Omega)} m(A) + \left(1 - \frac{m(\Omega)-\alpha}{(1-\alpha)m(\Omega)}\right) \frac{m(A)}{1-m(\Omega)} \\
&= \frac{m(\Omega)-\alpha}{(1-\alpha)m(\Omega)} m(A) + \frac{(1-\alpha)m(\Omega) - m(\Omega) + \alpha}{(1-\alpha)m(\Omega)} \frac{m(A)}{1-m(\Omega)} \\
&= \frac{m(\Omega)-\alpha}{(1-\alpha)m(\Omega)} m(A) + \frac{\alpha\,(1-m(\Omega))}{(1-\alpha)m(\Omega)} \frac{m(A)}{1-m(\Omega)} \\
&= \frac{m(A)}{1-\alpha} \quad \forall A \subset \Omega,
\end{aligned}$$

and

$${}^{\nu}m(\Omega) = \frac{m(\Omega)-\alpha}{(1-\alpha)m(\Omega)} m(\Omega) = \frac{m(\Omega)-\alpha}{1-\alpha}.$$

The first transformation of the extended discounting operation, Equation (9) with $\alpha \in [-m(\Omega)/(1-m(\Omega)), 1]$, concerning the discounting or reinforcement of the source, is a reparameterization of (23) in the particular case of two states such that $m_1 = m_\Omega$ and $m_2 = {}^{tr}m$ with $\nu_1 = (1-\alpha)m(\Omega) + \alpha$. Indeed:

$$\begin{aligned}
{}^{\nu}m(A) &= (1-(1-\alpha)m(\Omega)-\alpha)\frac{m(A)}{1-m(\Omega)} = (1-\alpha)\, m(A), \quad \forall A \subset \Omega, \\
{}^{\nu}m(\Omega) &= (1-\alpha)m(\Omega) + \alpha.
\end{aligned}$$

Finally, the second transformation of the extended discounting operation (15), allowing one to contradict a source, is also a reparameterization of (23) by considering two states such that $m_1 = \overline{{}^{tr}m}$ and $m_2 = m_\Omega$, and setting $\nu_1 = (\alpha-1)(1-m(\Omega))$ with $\alpha \in [1, 1 + \frac{1}{1-m(\Omega)}]$:

$$
\begin{aligned}
{}^{\nu}m(\overline{A}) &= (\alpha-1)(1-m(\Omega))\frac{m(A)}{1-m(\Omega)} = (\alpha-1)\,m(A), \quad \forall A \subset \Omega, \\
{}^{\nu}m(\Omega) &= 1-(\alpha-1)(1-m(\Omega)) = 1-\alpha+\alpha\,m(\Omega)+1-m(\Omega) \\
&= (\alpha-1)m(\Omega)+2-\alpha. \qquad \square
\end{aligned}
$$

Models of correction mechanisms corresponding to discounting, de-discounting and extended discount operations are summarized in Table 1.

Table 1 Models corresponding to the correction mechanisms presented in Section 3.

Interpretations	Operation
$m_1 = m_\Omega$ $m_2 = m$	discounting
$m_1 = m$ $m_2 = {}^{tr}m$	de-discounting
$m_1 = m_\Omega$ $m_2 = {}^{tr}m$	extended disc. (1)
$m_1 = \overline{{}^{tr}m}$ $m_2 = m_\Omega$	extended disc. (2)

Remark 2. The first transformation of the extended discounting operation is a discounting of ${}^{tr}m$, while the second transformation is a discounting of $\overline{{}^{tr}m}$.

Remark 3. De-discounting operation is a particular reinforcement process. A more informative reinforcement than ${}^{tr}m$ can be chosen, for instance, the "pignistic BBA" defined, $\forall \omega \in \Omega$, by:

$$
{}^{bet}m(\{\omega\}) = \sum_{\{A\subseteq\Omega,\omega\in A\}} \frac{m(A)}{(1-m(\emptyset))|A|} . \tag{31}
$$

Thus, another reinforcement process is given by:

$$
{}^{\nu}m = \nu_1 m + (1-\nu_1)\,{}^{bet}m . \tag{32}
$$

Remark 4. By choosing $m_{Ag}^{\mathscr{R}}$ as follows:

$$
\begin{cases} m_{Ag}^{\mathscr{R}}(\{R_i\}) = \nu_i & \forall i \in \{1,\dots,N\}, \\ m_{Ag}^{\mathscr{R}}(\mathscr{R}) = 1-\sum_{i=1}^{N}\nu_i, & \end{cases} \tag{33}
$$

with $\sum_{i=1}^{N}\nu_i \leq 1$, Equation (22) leads to:

$$
{}^{\nu}m = \sum_{i=1}^{N}\nu_i\, m_i + (1-\sum_{i=1}^{N}\nu_i)m_\Omega , \tag{34}
$$

which is similar to (23) if one considers a state such that $m_i = m_\Omega$.

5 Correction Mechanisms Based on Decompositions

The preceding section has introduced a general form of correction mechanisms encompassing, in particular, the discounting, de-discounting and extended discounting operations. As mentioned in Remark 1 in Section 3.4, the discounting operation can also be seen as a particular case of the contextual discounting. However, the contextual discounting does not belong to the family of correction mechanisms presented in the previous section. In this section, contextual discounting is shown to be a particular member of another family of correction mechanisms based on the disjunctive decomposition of a subnormal BBA introduced by Denœux in [3].

5.1 Canonical Conjunctive and Disjunctive Decompositions

In [16], extending the notion of separable BBA introduced by Shafer [13, chapter 4], Smets shows that each non-dogmatic BBA m can be uniquely decomposed into a conjunctive combination of *generalized simple BBAs (GSBBAs)*, denoted $A^{w(A)}$ with $A \subset \Omega$, and defined from 2^Ω to $\mathbb{R}$ by:

$$\begin{aligned} A^{w(A)} : \Omega &\mapsto w(A) \\ A &\mapsto 1 - w(A) \\ B &\mapsto 0\,,\ \forall B \in 2^\Omega \setminus \{A, \Omega\}\,, \end{aligned} \tag{35}$$

with $w(A) \in [0, \infty)$.

Every non-dogmatic BBA m can then be canonically decomposed into a conjunctive combination of GSBBAs:

$$m = \bigcirc\!\!\!\!\cap_{A \subset \Omega} A^{w(A)}\,. \tag{36}$$

In [3], Denœux introduces another decomposition: the canonical disjunctive decomposition of a subnormal BBA into *negative GSBBAs (NGSBBAs)*, denoted $A_{v(A)}$ with $A \supset \emptyset$, and defined from 2^Ω to $\mathbb{R}$ by:

$$\begin{aligned} A_{v(A)} : \emptyset &\mapsto v(A) \\ A &\mapsto 1 - v(A) \\ B &\mapsto 0\,,\ \forall B \in 2^\Omega \setminus \{\emptyset, A\}\,, \end{aligned} \tag{37}$$

with $v(A) \in [0, \infty)$.

Every subnormal BBA m can be canonically decomposed into a disjunctive combination of NGSBBAs:

$$m = \bigcirc\!\!\!\!\cup_{A \supset \emptyset} A_{v(A)}\,. \tag{38}$$

Indeed, as remarked in [3], the negation of m can also be conjunctively decomposed as soon as m is subnormal (in this case, $\overline{m}$ is non-dogmatic). Then:

$$\overline{m} = \bigcirc\!\!\!\!\cap_{A \subset \Omega} A^{\overline{w}(A)} \Rightarrow m = \overline{\bigcirc\!\!\!\!\cap_{A \subset \Omega} A^{\overline{w}(A)}} = \bigcirc\!\!\!\!\cup_{A \subset \Omega} \overline{A^{\overline{w}(A)}} = \bigcirc\!\!\!\!\cup_{A \supset \emptyset} A_{\overline{w}(\overline{A})}\,. \tag{39}$$

The relation between functions v and w is then $v(A) = \overline{w}(\overline{A})$ for all $A \supset \emptyset$.

5.2 A Correction Mechanism Based on the Disjunctive Decomposition

According to the previous definitions (35) and (37), BBAs m_ℓ, $\ell \in \{1,\ldots,L\}$, defined in (19) by $m_\ell(\emptyset) = \beta_\ell$ and $m_\ell(\theta_\ell) = \alpha_\ell$, can be denoted $\theta_{\ell_{\beta_\ell}}$ or θ_{β_ℓ} in a simple way.

From (18) and (38), the contextual discounting on a coarsening $\Theta = \{\theta_1,\ldots,\theta_L\}$ of Ω of a subnormal BBA m is thus defined by:

$$^{\alpha}_{\Theta}m = m \textcircled{\cup} \theta_{\beta_1} \textcircled{\cup} \ldots \textcircled{\cup} \theta_{\beta_L} = \textcircled{\cup}_{A \supset \emptyset} A_{\nu(A)} \textcircled{\cup} \theta_{\beta_1} \textcircled{\cup} \ldots \textcircled{\cup} \theta_{\beta_L} .$$

In particular, as $A_{\nu_1(A)} \textcircled{\cup} A_{\nu_2(A)} = A_{\nu_1\nu_2(A)}$ for all non empty subet A of Ω:

- The classical discounting of a subnormal BBA $m = \textcircled{\cup}_{A \supset \emptyset} A_{\nu(A)}$ is defined by:

$$^{\alpha}m = \Omega_{\beta\nu(\Omega)} \textcircled{\cup}_{\Omega \supset A \supset \emptyset} A_{\nu(A)} ; \tag{40}$$

- The contextual discounting (Remark 1) of a subnormal BBA $m = \textcircled{\cup}_{A \supset \emptyset} A_{\nu(A)}$ is defined by:

$$^{\alpha}m = \textcircled{\cup}_{\omega_k \in \Omega} \{\omega_k\}_{\beta_k \nu(\{\omega_k\})} \textcircled{\cup}_{A \subset \Omega, |A|>1} A_{\nu(A)} . \tag{41}$$

These contextual discounting operations are then particular cases of a more general correction mechanism defined by:

$$^{\alpha\cup}m = \textcircled{\cup}_{A \supset \emptyset} A_{\beta_A \nu(A)}, \tag{42}$$

where $\beta_A \in [0,1]$ for all $A \supset \emptyset$ and α is the vector $\{\alpha_A\}_{A \supset \emptyset}$.

In [10], the interpretation of each β_A has been given only in the case where the union of the subsets A forms a partition of Ω, β_A being interpreted as the degree of belief held by the agent regarding the fact that the source is reliable, knowing that the value searched belongs to A.

Instead of considering (16), let us now suppose that agent Ag holds beliefs regarding the reliability of the source, conditionally on each subset A of Ω:

$$\begin{cases} m^{\mathscr{R}}_{Ag}[A](\{R\}) = 1 - \alpha_A = \beta_A \\ m^{\mathscr{R}}_{Ag}[A](\mathscr{R}) \quad = \alpha_A , \end{cases} \tag{43}$$

where $\alpha_A \in [0,1]$.

In the same way as in Section 3.4, the knowledge held by agent Ag, based on the information m^{Ω}_S from a source and on metaknowledge $m^{\mathscr{R}}_{Ag}$ (43) regarding the reliability of this source, can be computed as follows:

$$m^{\Omega}_{Ag}[m^{\Omega}_S, m^{\mathscr{R}}_{Ag}] = \left(\textcircled{\cap}_{A \subseteq \Omega} m^{\mathscr{R}}_{Ag}[A]^{\Uparrow \Omega \times \mathscr{R}} \textcircled{\cap} m^{\Omega}_{Ag}[\{R\}]^{\Uparrow \Omega \times \mathscr{R}} \right)^{\downarrow \Omega} . \tag{44}$$

Proposition 4. *The BBA m^{Ω}_{Ag} resulting from (44) only depends on m^{Ω}_S and the vector $\alpha = \{\alpha_A\}_{A \subseteq \Omega}$. The result is denoted $^{\alpha}_{2^\Omega}m$ and is equal to the disjunctive combination of m^{Ω}_S with a BBA m^{Ω}_0 defined by:*

$$m_0^{\Omega}(C) = \prod_{\cup A = C} \alpha_A \prod_{\cup B = \overline{C}} \beta_B, \quad \forall C \subseteq \Omega. \tag{45}$$

Proof. For each $A \subseteq \Omega$, the deconditioning of $m_{Ag}^{\mathscr{R}}[A]$ on $\Omega \times \mathscr{R}$ is given by:

$$m_{Ag}^{\mathscr{R}}[A]^{\Uparrow \Omega \times \mathscr{R}}(A \times \{R\} \cup \overline{A} \times \mathscr{R}) = \beta_A, \tag{46}$$

$$m_{Ag}^{\mathscr{R}}[A]^{\Uparrow \Omega \times \mathscr{R}}(\Omega \times \mathscr{R}) = \alpha_A. \tag{47}$$

With $A \neq B$:

$$(A \times \{R\} \cup \overline{A} \times \mathscr{R}) \cap (B \times \{R\} \cup \overline{B} \times \mathscr{R}) = (A \cup B) \times \{R\} \cup \overline{(A \cup B)} \times \mathscr{R} .$$

Then:

$$\circledcap_{A \subseteq \Omega} m_{Ag}^{\mathscr{R}}[A]^{\Uparrow \Omega \times \mathscr{R}}(C \times \{R\} \cup \overline{C} \times \mathscr{R}) = \prod_{\cup D = \overline{C}} \alpha_D \prod_{\cup E = C} \beta_E, \quad \forall C \subseteq \Omega , \tag{48}$$

or, by exchanging the roles of C and $\overline{C}$:

$$\circledcap_{A \subseteq \Omega} m_{Ag}^{\mathscr{R}}[A]^{\Uparrow \Omega \times \mathscr{R}}(\overline{C} \times \{R\} \cup C \times \mathscr{R}) = \prod_{\cup D = C} \alpha_D \prod_{\cup E = \overline{C}} \beta_E, \quad \forall C \subseteq \Omega . \tag{49}$$

It remains to combine conjunctively $m_{Ag}^{\Omega}[\{R\}]^{\Uparrow \Omega \times \mathscr{R}}$ and $\circledcap_{A \subseteq \Omega} m_{Ag}^{\mathscr{R}}[A]^{\Uparrow \Omega \times \mathscr{R}}$ which have focal sets of the form $B \times \{R\} \cup \Omega \times \{NR\}$ and $\overline{C} \times \{R\} \cup C \times \mathscr{R}$, respectively, with $B, C \subseteq \Omega$. The intersection of two such focal sets is:

$$(\overline{C} \times \{R\} \cup C \times \mathscr{R}) \cap (B \times \{R\} \cup \Omega \times \{NR\}) = B \times \{R\} \cup C \times \{NR\} ,$$

and it can be obtained only for a particular choice of B and C. Then:

$$\circledcap_{A \subseteq \Omega} m_{Ag}^{\mathscr{R}}[A]^{\Uparrow \Omega \times \mathscr{R}} \circledcap m_{Ag}^{\Omega}[\{R\}]^{\Uparrow \Omega \times \mathscr{R}}(B \times \{R\} \cup C \times \{NR\}) =$$

$$= \left[\prod_{\cup D = C} \alpha_D \prod_{\cup E = \overline{C}} \beta_E \right] m_S^{\Omega}(B).$$

Finally, the marginalization of this BBA on Ω is given by:

$$^{\alpha}m(A) = \sum_{B \cup C = A} \left[\prod_{\cup D = C} \alpha_D \prod_{\cup E = \overline{C}} \beta_E \right] m_S^{\Omega}(B), \quad \forall A \subseteq \Omega, \tag{50}$$

□

Let us note that the above proof has many similarities with proofs presented in [10, Sections A.1 and A.3].

As in the case of contextual discounting operations considered in Section 3.4, the BBA m_0^{Ω} defined in Proposition 4 admits a simple decomposition described in the following proposition.

Proposition 5. *The BBA m_0^Ω defined in Proposition 4 can be rewritten as:*

$$m_0^\Omega = \textcircled{\cup}_{A \supset \emptyset} A_{\beta_A} \, . \tag{51}$$

Proof. Directly from (45) and the definition of the DRC (3). □

From (51), the contextual discounting ${}^{\alpha}_{2^\Omega}m$ of a subnormal BBA $m = \textcircled{\cup}_{A \supset \emptyset} A_{v(A)}$ is defined by:

$${}^{\alpha}_{2^\Omega}m = \textcircled{\cup}_{A \supset \emptyset} A_{v(A)} \textcircled{\cup}_{A \supset \emptyset} A_{\beta_A} = \textcircled{\cup}_{A \supset \emptyset} A_{\beta_A v(A)} = {}^{\alpha\cup} m \, . \tag{52}$$

This contextual discounting is thus equivalent to the correction mechanism introduced in this section. Each coefficient β_A of this correction mechanism can then be interpreted as the degree of belief held by the agent Ag regarding the fact that the source is reliable knowing that the true answer to the question Q of interest belongs to A.

5.3 A Correction Mechanism Based on the Conjunctive Decomposition

In a similar way, a correction mechanism for a non-dogmatic BBA m can be defined, from the conjunctive decomposition of m, by:

$${}^{\alpha\cap}m = \textcircled{\cap}_{A \subset \Omega} A^{\beta_A w(A)} \, ; \tag{53}$$

where $\forall A \subset \Omega, \beta_A \in [0,1]$, and α is the vector $\{\alpha_A\}_{A \subset \Omega}$.

Correction mechanisms ${}^{\alpha\cap}m$ (42) and ${}^{\alpha\cup}m$ (53) are related in the following way. Let us consider a subnormal BBA m, $\overline{m}$ is then non-dogmatic:

$${}^{\alpha\cap}\overline{m} = \textcircled{\cap}_{A \subset \Omega} A^{\beta_A \overline{w}(A)} \, . \tag{54}$$

Then:

$$\begin{aligned}
\overline{{}^{\alpha\cap}\overline{m}} &= \overline{\textcircled{\cap}_{A \subset \Omega} A^{\beta_A \overline{w}(A)}} \\
&= \textcircled{\cup}_{A \subset \Omega} \overline{A^{\beta_A \overline{w}(A)}} \\
&= \textcircled{\cup}_{A \supset \emptyset} A_{\beta_A \overline{w}(\overline{A})} \\
&= \textcircled{\cup}_{A \supset \emptyset} A_{\beta_A v(A)} \\
&= {}^{\alpha\cup}m
\end{aligned} \tag{55}$$

These two correction mechanisms can thus be seen as belonging to a general family of correction mechanisms.

6 Application Example

In this section, an application example in the domain of postal address recognition illustrates the potential benefits of using a particular correction mechanism of the form (23).

In this application, three postal address readers (PARs) are available, each one providing pieces of information regarding the address lying on the image of a mail. These pieces of knowledge are represented by belief functions on a frame of discernment gathering all postal addresses. Belief functions can then be combined in order to make a decision. This fusion scheme is represented in Fig. 1. Details of this application can be found in [11].

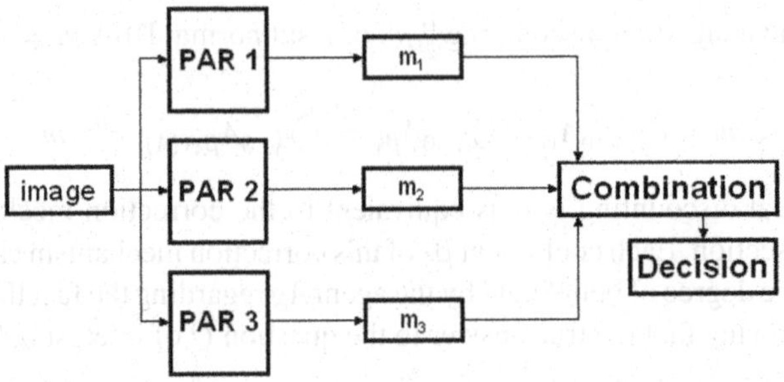

Fig. 1 Fusion scheme with three PARs in the belief function framework.

An extension of this scheme is obtained by considering the fact that PAR 1 and PAR 2 each output an address and a confidence score regarding the town part of the address.

To visualize the real information provided by these confidence scores, scores of correct and incorrect towns output by PAR 1 for a set of postal addresses are shown in Fig. 2.

It can be observed that the greater the score is, the more important is the proportion of addresses with a correct town. Hence, this score carries useful information regarding the reliability of the town information in the output address. Similar observations were made with PAR 2. Therefore, BBAs m_1 and m_2 representing the information provided by PAR 1 and PAR 2, should be corrected according to these scores. An idea consists in reinforcing the information provided by a PAR when the score is high, and, conversely, discounting it when the score is too low. For that purpose, we defined four thresholds T_1, T_2, T_3, and T_4 illustrated in Fig. 2, such that information provided by the PAR is:

- totally discounted, if the score is lower than T_1;
- discounted according to the score, if the score belongs to $[T_1, T_2]$;
- kept unchanged, if the score belongs to $[T_2, T_3]$;
- reinforced according to the score, if the score belongs to $[T_3, T_4]$;
- at last, totally reinforced, if the score is greater than T_4.

Formally, this adjustment can be realized, for both PARs 1 and 2 (Fig. 3), by using the correction mechanism defined by:

$$^{v}m = v_1\, m_{\Omega} + v_2\, m + v_3\, {}^{tr}m\,, \tag{56}$$

where parameters v_i are set as illustrated in Fig. 4.

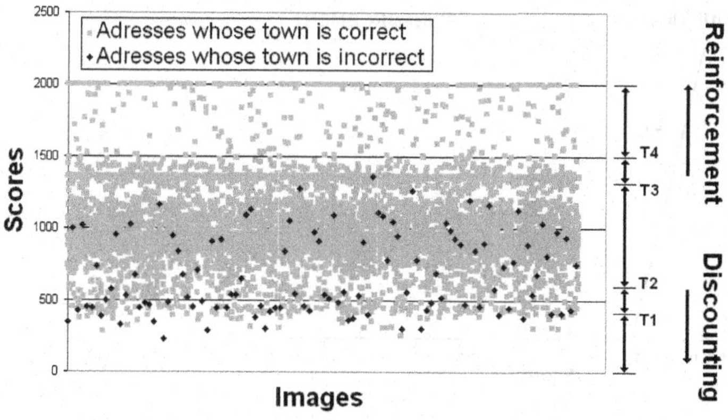

Fig. 2 Confidence scores and addresses provided by PAR 1 regarding images of a learning set. A dark rhomb corresponds to an address whose town is incorrect. A clear square is associated with an address whose town is correct.

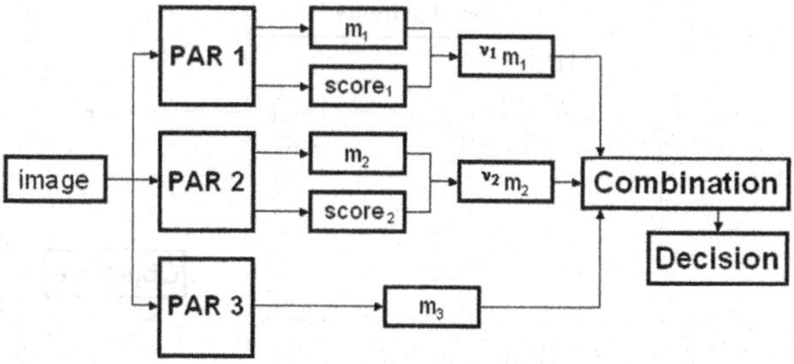

Fig. 3 An extended model adjusting BBAs provided by *PAR* 1 and *PAR* 2 according to supplied scores.

Performances of this combination, on a test set of mails, are reported in Fig. 5. To preserve the confidentiality of PARs performances, reference values were used when representing performance rates. Correct recognition rates, represented on the x-axis, are expressed relatively to a reference correct recognition rate, denoted by R. Error rates, represented on the y-axis, are expressed relatively to a reference error rate, denoted by E. The rate R has a value greater than 80%. The rate E has a value smaller than 0.1%.

As different PARs are available in this application, we can expect the combination to yield the greatest possible recognition rates while keeping error rate at an acceptable level. In this article, the maximal tolerated error rate is chosen equal to the least PARs error rates.

This extended model allows us to obtain a combination point denoted by C_+, which is associated with an acceptable error rate and a higher recognition rate than the previous combination point C, obtained with the model illustrated in Fig. 1. The

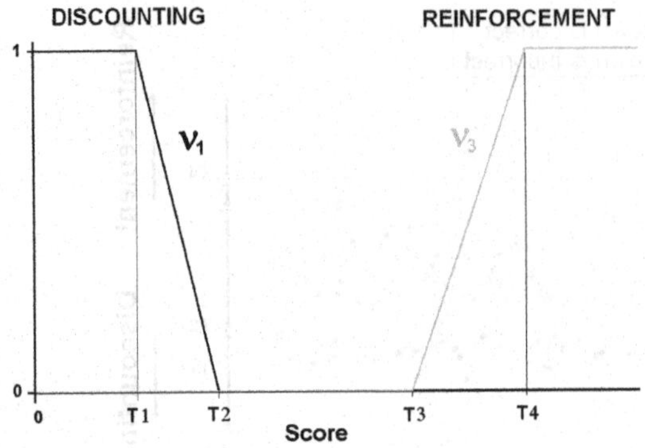

Fig. 4 Correction parameters as function of the scores ($v_1 + v_2 + v_3 = 1$).

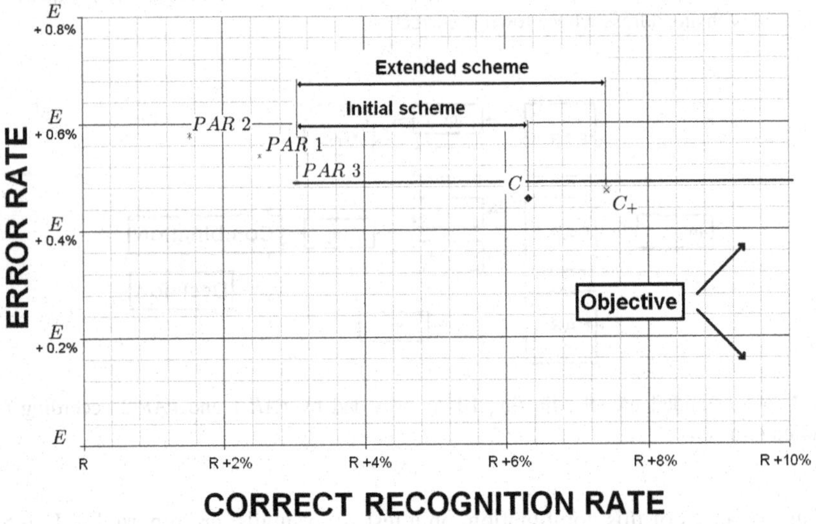

Fig. 5 PARs and combination performances regarding towns written on mails.

individual performances of PARs are further improved using the extended model based on a correction mechanism.

In [8], another example of improvement in the same application can be found using contextual discounting.

7 Conclusion

In this article, two families of belief function correction mechanisms have been introduced and justified.

The first family of correction mechanisms highlights the links between the discounting, the de-discounting, and the extended discounting, and generalizes these three operations. Different transformations, expressed by belief functions, can be associated to different states in which the source can be: reliable, not reliable, too cautious, lying, etc.

The second family, based on the concepts of negation of a BBA and disjunctive and conjunctive decompositions of a BBA, generalizes the contextual discounting operation.

An application example has illustrated a practical interest of the first family. It introduces a way to combine scores with decisions to improve the recognition performances.

Future works will aim at exploring more deeply the second family of correction mechanisms and testing it on real data.

It would also be interesting to automatically learn the coefficients of the correction mechanisms from data, as done in the "expert tuning" method for the classical or the contextual discounting operations [5, 10].

Acknowledgements. The authors would like to thank the anonymous reviewer for both the careful review as well as the useful comments. The first author's works have been financed by the French region Nord-Pas de Calais under the project CISIT (Campus International pour la Sécurité et l'Intermodalité des Transports).

References

1. Dempster, A.: Upper and Lower Probabilities Induced by Multivalued Mapping. Annals of Mathematical Statistics AMS-38, 325–339 (1967)
2. Denœux, T., Smets, P.: Classification using Belief Functions: the Relationship between the Case-Based and Model-Based Approaches. IEEE Transactions on Systems, Man and Cybernetics, Part B 36(6), 1395–1406 (2006)
3. Denœux, T.: Conjunctive and Disjunctive Combination of Belief Functions Induced by Non Distinct Bodies of Evidence. Artificial Intelligence 172, 234–264 (2008)
4. Dubois, D., Prade, H.: A set-theoretic view of belief functions: logical operations and approximations by fuzzy sets. International Journal of General Systems 12, 193–226 (1986)
5. Elouedi, Z., Mellouli, K., Smets, P.: Assessing sensor reliability for multisensor data fusion with the transferable belief model. IEEE Transactions on Systems, Man and Cybernetics B 34, 782–787 (2004)
6. Goodman, I.R., Mahler, R.P., Nguyen, H.T.: Mathematics of Data Fusion. Kluwer Academic Publishers, Norwell (1997)
7. Kohlas, J., Monney, P.-A.: A Mathematical Theory of Hints. An Approach to the Dempster-Shafer Theory of Evidence. Lecture Notes in Economics and Mathematical Systems, vol. 425. Springer, Berlin (1995)
8. Mercier, D.: Fusion d'informations pour la reconnaissance automatique d'adresses postales dans le cadre de la théorie des fonctions de croyance, PhD Thesis, Université de Technologie de Compiègne (December 2006)

9. Mercier, D., Denœux, T., Masson, M.-H.: A parameterized family of belief functions correction mechanisms. In: Magdalena, L., Ojeda-Aciego, M., Verdegay, J.L. (eds.) Proceedings of IPMU 2008, Torremolinos, Malaga, June 22-27, pp. 306–313 (2008)
10. Mercier, D., Quost, B., Denœux, T.: Refined modeling of sensor reliability in the belief function framework using contextual discounting. Information Fusion 9, 246–258 (2008)
11. Mercier, D., Cron, G., Denœux, T., Masson, M.-H.: Decision fusion for postal address recognition using belief functions. Expert Systems with Applications, part 1 36(3), 5643–5653 (2009)
12. Pichon, F.: Belief functions: canonical decompositions and combination rules, PhD Thesis, Université de Technologie de Compiègne (March 2009)
13. Shafer, G.: A mathematical theory of evidence. Princeton University Press, Princeton (1976)
14. Smets, Ph.: Belief functions: the disjunctive rule of combination and the generalized Bayesian theorem. International Journal of Approximate Reasoning 9, 1–35 (1993)
15. Smets, Ph.: What is Dempster-Shafer's model? In: Yager, R.R., Fedrizzi, M., Kacprzyk, J. (eds.) Advances in the Dempster-Shafer theory of evidence, pp. 5–34. Wiley, New-York (1994)
16. Smets, Ph.: The canonical decomposition of a weighted belief. In: Proceedings of the 14th International Joint Conference on Artificial Intelligence (IJCAI 1995), pp. 1896–1901. Morgan Kaufman, San Mateo (1995)
17. Smets, Ph.: The Transferable Belief Model for quantified belief representation. In: Gabbay, D.M., Smets, P. (eds.) Handbook of Defeasible reasoning and uncertainty management systems, vol. 1, pp. 267–301. Kluwer Academic Publishers, Dordrecht (1998)
18. Smets, Ph.: Managing Deceitful Reports with the Transferable Belief Model. In: Proceedings of the 8th International Conference On Information Fusion, FUSION 2005, Philadelphia, USA, July 25-29, paper C8-3 (2005)
19. Smets, Ph., Kennes, R.: The Transferable Belief Model. Artificial Intelligence 66, 191–243 (1994)
20. Zhu, H., Basir, O.: Extended discounting scheme for evidential reasoning as applied to MS lesion detection. In: Svensson, P., Schubert, J. (eds.) Proceedings of the 7th International Conference on Information Fusion, FUSION 2004, June 2004, pp. 280–287 (2004)

A Comparison of Conditional Coherence Concepts for Finite Spaces

Enrique Miranda

Abstract. We compare the different notions of conditional coherence within the behavioural theory of imprecise probabilities when all the referential spaces are finite. We show that the difference between weak and strong coherence comes from conditioning on sets of (lower, and in some cases upper) probability zero. Next, we characterise the range of coherent extensions, proving that the greatest coherent extensions can always be calculated using the notion of regular extension. Finally, we investigate which consistency conditions are preserved by convex combinations point-wise limits, and whether it is possible to update a coherent lower prevision while maintaining 2-monotonicity.

1 Introduction

This paper focuses on the theory of coherent lower previsions, established mainly by Walley [16]. This theory is based on the work on coherent previsions by de Finetti [7, 8], and considers some of its distinct features: the *behavioural* approach to probability as a fair price, or supremum acceptable betting rate; the focus on *gambles* (bounded random variables) instead of events; and the use of a consistency criteria for the acceptable betting rates, which is called *coherence*. At the same time, it accounts for imprecision by allowing us to be undecided between buying and selling gambles for some prices, producing then what are called *lower* and *upper* previsions.

The resulting theory, which we summarise in Section 2, not only generalises de Finetti's work to the imprecise case, but also includes as particular cases most of the other models of non-additive measures considered in the literature, such as 2- and n-monotone lower probabilities [1], sets of probability measures [9], or Choquet

Enrique Miranda
University of Oviedo, Dep. of Statistics and Operations Research. C-Calvo Sotelo, s/n 33007 Oviedo (Spain)
e-mail: mirandaenrique@uniovi.es

B. Bouchon-Meunier et al. (Eds.) Found. of Reas. under Uncert., STUDFUZZ 249, pp. 223–245.
springerlink.com

integrals [6]. It has, however, a number of drawbacks: one of them are the difficulties that arise when we want to update our information, producing then the so-called conditional lower previsions. In that case, it becomes necessary, when we have infinite spaces, to assume a principle called *conglomerability* in order to maintain the consistency with the initial assessments. This principle is also one of the points of disagreement between Walley's and de Finetti's work. But even if we restrict ourselves to finite spaces, as we shall do in this paper, there is not a unique way to guarantee the consistency of the conditional assessments, and also the verification of the different possibilities is not straightforward.

Our goal in this paper is to compare and to provide more manageable expressions for the notions of *weak* and *strong* coherence of a number of conditional lower previsions. This is the subject of our work in Section 3. In Section 4, we investigate the different ways of updating a coherent lower prevision under the light of new information, establishing the most conservative and the more informative ways of doing so.

The results we obtain in these two sections allow us to investigate a bit further in Section 5 the properties of weak and strong coherence. Specifically, we investigate whether the class of weakly coherent (resp. coherent) models is closed under convex combinations and point-wise limits, and if the property of 2-monotonicity is preserved when we update in a coherent way. Finally, we give some concluding remarks in Section 6.

As we said before, we restrict ourselves here to finite spaces. This simplifies a bit the mathematical developments we make in the paper, and allows for some properties that do not hold on infinite spaces. The finite case is also interesting for a number of applications, for instance in the context of credal networks [5]. Some comments on the infinite case can be found in [11].

2 Coherence Notions on Finite Spaces

Let us give a short introduction to the concepts and results from the behavioural theory of imprecise probabilities that we shall use in the rest of the paper. We refer to [16] for an in-depth study of these and other properties, and to [10] for a brief survey.

Given a possibility space Ω, a *gamble* is a bounded real-valued function on Ω. This function represents a random reward $f(\omega)$, which depends on the a priori unknown value ω in Ω. We shall denote by $\mathscr{L}(\Omega)$ the set of all gambles on Ω. A *lower prevision* $\underline{P}$ is a real functional defined on some set of gambles $\mathscr{K} \subseteq \mathscr{L}(\Omega)$. It is used to represent a subject's supremum acceptable buying prices for these gambles, in the sense that for any $\varepsilon > 0$ and any f in $\mathscr{K}$ the subject is disposed to accept the uncertain reward $f - \underline{P}(f) + \varepsilon$. Given a lower prevision $\underline{P}$, we will denote by $\overline{P}$ its conjugate *upper prevision*, given by $\overline{P}(f) = -\underline{P}(-f)$ for any gamble f. $\overline{P}(f)$ represents the infimum acceptable selling price for the gamble f for our subject.

We can also consider the supremum buying prices for a gamble, *conditional* on a subset of Ω. Given such a set B and a gamble f on Ω, the lower prevision $\underline{P}(f|B)$

represents the subject's supremum acceptable buying price for the gamble f, provided he later comes to know that the unknown value ω belongs to B, and nothing else. If we consider a partition $\mathscr{B}$ of Ω (for instance a set of categories), then we shall represent by $\underline{P}(f|\mathscr{B})$ the gamble on Ω that takes the value $\underline{P}(f|B)$ if and only if $\omega \in B$. The functional $\underline{P}(\cdot|\mathscr{B})$ that maps any gamble f on its domain into the gamble $\underline{P}(f|\mathscr{B})$ is called a *conditional lower prevision.*

Let us now re-formulate the above concepts in terms of random variables, which are the focus of our attention in this paper. Consider random variables $X_1,\dots,X_n$, taking values in respective *finite* sets $\mathscr{X}_1,\dots,\mathscr{X}_n$. For any subset $J \subseteq \{1,\dots,n\}$ we shall denote by X_J the (new) random variable

$$X_J := (X_j)_{j\in J},$$

which takes values in the product space

$$\mathscr{X}_J := \times_{j\in J}\mathscr{X}_j.$$

We shall also use the notation $\mathscr{X}^n$ for $\mathscr{X}_{\{1,\dots,n\}}$. In the current formulation made by random variables, $\mathscr{X}^n$ is just the possibility space Ω.

Definition 1. *Let J be a subset of $\{1,\dots,n\}$, and let $\pi_J : \mathscr{X}^n \to \mathscr{X}_J$ be the* projection operator, *i.e., the operator that drops the elements of a vector in $\mathscr{X}^n$ that do not correspond to indexes in J. A gamble f on $\mathscr{X}^n$ is called* $\mathscr{X}_J$-measurable *when for any $x,y \in \mathscr{X}^n$, $\pi_J(x) = \pi_J(y)$ implies that $f(x) = f(y)$.*

There is a one-to-one correspondence between the gambles on $\mathscr{X}^n$ that are $\mathscr{X}_J$-measurable and the gambles on $\mathscr{X}_J$. We shall denote by $\mathscr{K}_J$ the set of $\mathscr{X}_J$-measurable gambles.

Consider two disjoint subsets O,I of $\{1,\dots,n\}$. $\underline{P}(X_O|X_I)$ represents a subject's behavioural dispositions about the gambles that depend on the outcome of the variables $\{X_k, k \in O\}$, after coming to know the outcome of the variables $\{X_k, k \in I\}$. As such, it is defined on the set of gambles that depend on the variables in $O \cup I$ only, i.e., on the set $\mathscr{K}_{O\cup I}$ of the $\mathscr{X}_{O\cup I}$-measurable gambles on $\mathscr{X}^n$. Given such a gamble f and $x \in \mathscr{X}_I$, $\underline{P}(f|X_I = x)$ represents a subject's supremum acceptable buying price for the gamble f, if he later came to know that the variable X_I took the value x (and nothing else). Under the notation we gave above for lower previsions conditional on events and partitions, this would be $\underline{P}(f|B)$, where $B := \pi_I^{-1}(x)$. When there is no possible confusion about the variables involved in the lower prevision, we shall use the notation $\underline{P}(f|x)$ for $\underline{P}(f|X_I = x)$. The sets $\{\pi_I^{-1}(x) : x \in \mathscr{X}_I\}$ form a partition of $\mathscr{X}^n$. Hence, we can define the gamble $\underline{P}(f|X_I)$, which takes the value $\underline{P}(f|x)$ on $x \in \mathscr{X}_I$. This is a conditional lower prevision.

These assessments can be made for any disjoint subsets O,I of $\{1,\dots,n\}$, and therefore it is not uncommon to model a subject's beliefs using a finite number of different conditional previsions. We should verify then that all the assessments modelled by these conditional previsions are coherent with each other. The first requirement we make is that for any disjoint $O,I \subseteq \{1,\dots,n\}$, the conditional lower prevision $\underline{P}(X_O|X_I)$ defined on $\mathscr{K}_{O\cup I}$ should be *separately coherent*. In the context of this paper, where the domain is a linear set of gambles, separate coherence holds if and only if the following conditions are satisfied for any $x \in \mathscr{X}_I, f, g \in \mathscr{K}_{O \cup I}$, and $\lambda > 0$:

$$\underline{P}(f|x) \geq \min_{\omega \in \pi_I^{-1}(x)} f(\omega). \tag{SC1}$$

$$\underline{P}(\lambda f|x) = \lambda \underline{P}(f|x). \tag{SC2}$$

$$\underline{P}(f+g|x) \geq \underline{P}(f|x) + \underline{P}(g|x). \tag{SC3}$$

It is also useful for this paper to consider the particular case where $I = \emptyset$, that is, when we have (unconditional) information about the variables X_O. We have then an *(unconditional) lower prevision* $\underline{P}(X_O)$ on the set $\mathscr{K}_O$ of $\mathscr{X}_O$-measurable gambles. Separate coherence is called then simply *coherence*, and it holds if and only if the following three conditions hold for any $f, g \in \mathscr{K}_O$, and $\lambda > 0$:

$$\underline{P}(f) \geq \min f. \tag{C1}$$

$$\underline{P}(\lambda f) = \lambda \underline{P}(f). \tag{C2}$$

$$\underline{P}(f+g) \geq \underline{P}(f) + \underline{P}(g). \tag{C3}$$

Separate coherence is a notion of internal consistency for the assessments represented by a lower prevision, which means that the acceptable buying prices do not lead to a loss, no matter the outcome, and that the supremum acceptable buying price cannot be raised using the other assessments.

In general, separate coherence is not enough to guarantee the consistency of the lower previsions: conditional lower previsions can be conditional on the values of many different variables, and still we should verify that the assessments they provide are consistent not only separately, but also with each other. Formally, we are going to consider what we shall call *collections* of conditional lower previsions.

Definition 2. *Let* $\{\underline{P}_1(X_{O_1}|X_{I_1}), \ldots, \underline{P}_m(X_{O_m}|X_{I_m})\}$ *be conditional lower previsions with respective domains* $\mathscr{K}^1, \ldots, \mathscr{K}^m \subseteq \mathscr{L}(\mathscr{X}^n)$, *where* $\mathscr{K}^j$ *is the set of* $\mathscr{X}_{O_j \cup I_j}$-*measurable gambles,*[1] *for* $j = 1, \ldots, m$. *This is called a* collection *on* X^n *when for each* $j_1 \neq j_2$ *in* $\{1, \ldots, m\}$, *either* $O_{j_1} \neq O_{j_2}$ *or* $I_{j_1} \neq I_{j_2}$.

This means that we do not have two different conditional lower previsions giving information about the same set of variables X_O, conditional on the same set of variables X_I. Indeed, since all these conditional lower previsions represent the behavioural dispositions of the same subject, it does not make sense to consider twice the same $\underline{P}(X_O|X_I)$.

Given a collection $\underline{P}_1(X_{O_1}|X_{I_1}), \ldots, \underline{P}_m(X_{O_m}|X_{I_m})$ of conditional lower previsions, there are different ways in which we can guarantee their consistency[2]. The first one is called avoiding partial loss.

[1] We use $\mathscr{K}^j$ instead of $\mathscr{K}_{O_j \cup I_j}$ in order to alleviate the notation when no confusion is possible about the variables involved.

[2] We give the particular definitions of these notions for finite spaces. See [12, 16] for the general definitions of these notions on infinite spaces and non-linear domains.

The $\mathscr{X}_I$*-support* $S(f)$ of a gamble f in $\mathscr{K}_{O\cup I}$ is given by

$$S(f) := \{\pi_I^{-1}(x) : x \in \mathscr{X}_I, f\mathbb{I}_{\pi_I^{-1}(x)} \neq 0\}, \tag{1}$$

i.e., it is the set of conditioning events for which the restriction of f is not identically zero. Here, and in the rest of the paper, we use the notation $\mathbb{I}_A$ to denote the *indicator function* of a set A, i.e., the gamble whose value is 1 on the elements of A and 0 elsewhere. We shall also use the notations

$$G(f|x) = \mathbb{I}_x(f - \underline{P}(f|x)),\ G(f|X_I) = \sum_{x \in \mathscr{X}_I} G(f|x) = f - \underline{P}(f|X_I)$$

for any $f \in \mathscr{K}_{O\cup I}$ and any $x \in \mathscr{X}_I$.

Definition 3. *Let* $\underline{P}_1(X_{O_1}|X_{I_1}),\dots,\underline{P}_m(X_{O_m}|X_{I_m})$ *be separately coherent. We say that they* avoid partial loss *when for any* $f_j \in \mathscr{K}^j$, $j = 1,\dots,m$,

$$\max_{\omega \in \mathbb{S}(f_j)} \left[\sum_{j=1}^m G_j(f_j|X_{I_j})\right](\omega) \geq 0,$$

where $\mathbb{S}(f_j) := \{x \in \mathscr{X}^n : x \in B \text{ for some } B \in \cup_{j=1}^m S_j(f_j)\}$ *denotes the elements* $x \in \mathscr{X}^n$ *in the union of the supports.*

The idea behind this notion is that a combination of transactions that are acceptable for our subject should not make him lose utiles. It is based on the rationality requirement that a gamble $f \leq 0$ such that $f < 0$ on some set A should not be desirable.

Definition 4. *Let* $\underline{P}_1(X_{O_1}|X_{I_1}),\dots,\underline{P}_m(X_{O_m}|X_{I_m})$ *be separately coherent conditional lower previsions. We say that they are* weakly coherent *when for any* $f_j \in \mathscr{K}^j$, $j = 1,\dots,m$, $j_0 \in \{1,\dots,m\}$, $f_0 \in \mathscr{K}^{j_0}, x_0 \in \mathscr{X}_{I_{j_0}}$,

$$\max_{\omega \in \mathscr{X}^n} \left[\sum_{j=1}^m G_j(f_j|X_{I_j}) - G_{j_0}(f_0|x_0)\right](\omega) \geq 0.$$

With this condition we require that our subject should not be able to raise his supremum acceptable buying price $\underline{P}_{j_0}(f_0|x_0)$ for a gamble f_0 contingent on x_0 by taking into account other conditional assessments. However, under the behavioural interpretation, a number of weakly coherent conditional lower previsions can still present some forms of inconsistency with each other; see [16, Example 7.3.5] for an example and [16, Chapter 7] and [17] for some discussion. On the other hand, weak coherence neither implies or is implied by the notion of avoiding partial loss. Because of these two facts, we consider another notion which is stronger than both, and which is called (*joint* or *strong*) coherence:[3]

[3] The distinction between this and the unconditional notion of coherence mentioned above will always be clear from the context.

Definition 5. *Let $\underline{P}_1(X_{O_1}|X_{I_1}),\dots,\underline{P}_m(X_{O_m}|X_{I_m})$ be separately coherent conditional lower previsions. We say that they are* coherent *when for every $f_j \in \mathscr{K}^j$, $j = 1,\dots,m$, $j_0 \in \{1,\dots,m\}$, $f_0 \in \mathscr{K}^{j_0}, x_0 \in \mathscr{X}_{I_{j_0}}$,*

$$\left[\sum_{j=1}^{m} G_j(f_j|X_{I_j}) - G_{j_0}(f_0|x_0)\right](\omega) \geq 0 \tag{2}$$

for some $\omega \in \pi_{I_{j_0}}^{-1}(x_0) \cup \mathbb{S}(f_j)$.

Because we are dealing with finite spaces, this notion coincides with the one given by Williams in [18]. The coherence of a collection of conditional lower previsions implies their weak coherence; although the converse does not hold in general, it does in the particular case when we only have a conditional and an unconditional lower prevision.

It is important at this point to introduce a particular case of conditional lower previsions that will be of special interest for us: that of *conditional linear previsions*. We say that a conditional lower prevision $\underline{P}(X_O|X_I)$ on the set $\mathscr{K}_{O\cup I}$ is linear if and only if it is separately coherent and moreover $\underline{P}(f+g|x) = \underline{P}(f|x) + \underline{P}(g|x)$ for any $x \in \mathscr{X}_I$ and $f,g \in \mathscr{K}_{O\cup I}$. Conditional linear previsions correspond to the case where a subject's supremum acceptable buying price (lower prevision) coincides with his infimum acceptable selling price (upper prevision) for any gamble on the domain. When a separately coherent conditional lower prevision $\underline{P}(X_O|X_I)$ is linear we shall denote it by $P(X_O|X_I)$; in the unconditional case, we shall use the notation $P(X_O)$. In this paper, where all the referential spaces are finite, the model can be given a sensitivity analysis interpretation: a coherent lower prevision $\underline{P}(X_O)$ is always the lower envelope of a closed and convex set of linear previsions $P(X_O)$, the ones *dominating* it, in the sense that

$$P(f) \geq \underline{P}(f)\ \forall f \in \mathscr{K}_O.$$

Similarly, a separately coherent conditional lower prevision $\underline{P}(X_O|X_I)$ is the lower envelope of the closed and convex set of conditional linear previsions $P(X_O|X_I)$ that dominate it.

One interesting particular case is that where we are given only an unconditional lower prevision $\underline{P}$ on $\mathscr{L}(\mathscr{X}^n)$ and a conditional lower prevision $\underline{P}(X_O|X_I)$ on $\mathscr{K}_{O\cup I}$. Then weak and strong coherence are equivalent, and they both hold if and only if, for any $\mathscr{X}_{O\cup I}$-measurable f and any $x \in \mathscr{X}_I$,

$$\underline{P}(G(f|x)) = 0. \tag{GBR}$$

This is called the generalised Bayes rule (GBR). When $\underline{P}(x) > 0$, GBR can be used to determine the value $\underline{P}(f|x)$: it is then the *unique* value for which $\underline{P}(G(f|x)) = \underline{P}(\mathbb{I}_x(f - \underline{P}(f|x))) = 0$ holds.

If P and $P(X_O|X_I)$ are linear, they are coherent if and only if for any $\mathscr{X}_{O\cup I}$-measurable f, $P(f) = P(P(f|X_I))$. This is equivalent to requiring that $P(f|x) = \frac{P(f\mathbb{I}_x)}{P(x)}$ for all $f \in \mathscr{K}_{O\cup I}$ and all $x \in \mathscr{X}_I$ with $P(x) > 0$.

3 Relationships between Weak and Strong Coherence

Let us study in more detail the notions of avoiding partial loss, weak coherence and strong coherence. We start by recalling a recent characterisation of weak coherence:

Theorem 1. *[13, Theorem 1] $\underline{P}_1(X_{O_1}|X_{I_1}),\dots,\underline{P}_m(X_{O_m}|X_{I_m})$ are weakly coherent if and only if there is a lower prevision $\underline{P}$ on $\mathscr{L}(\mathscr{X}^n)$ that is pairwise coherent with each conditional lower prevision $\underline{P}_j(X_{O_j}|X_{I_j})$. In particular, given conditional linear previsions $P_j(X_{O_j}|X_{I_j})$ for $j=1,\dots,m$, they are weakly coherent if and only if there is a linear prevision P which is coherent with each $P_j(X_{O_j}|X_{I_j})$.*

This theorem shows one of the differences between weak and strong coherence: weak coherence is equivalent to the existence of a joint which is coherent with each of the assessments, considered separately; coherence on the other hand is equivalent to the existence of a joint which is coherent with all the assessments, taken together.

Weakly coherent conditional previsions can also be given a sensitivity analysis interpretation as lower envelopes of precise models; a similar result for coherence can be found in [16, Theorem 8.1.9].

Theorem 2. *Any weakly coherent $\underline{P}_1(X_{O_1}|X_{I_1}),\dots,\underline{P}_m(X_{O_m}|X_{I_m})$ are the lower envelope of a family of weakly coherent conditional linear previsions.*

Proof. From Theorem 1, there is a coherent lower prevision $\underline{P}$ on $\mathscr{X}^n$ which is coherent with $\underline{P}_j(X_{O_j}|X_{I_j})$ for $j=1,\dots,m$. Consider $j\in\{1,\dots,m\}$, $x\in\mathscr{X}_{I_j}$ and $f\in\mathscr{K}^j$. If $\underline{P}(x)>0$, then $\underline{P}_j(f|x)$ is uniquely determined by (GBR), and from [16, Section 6.4.2], it coincides with the lower envelope of $\{P(f|x): P\geq \underline{P}\}$. Hence, for any $\varepsilon>0$ there is some $P\geq\underline{P}$ such that $P(x)>0$ and $P(f|x)-\underline{P}_j(f|x)<\varepsilon$. Given this P, we can apply Lemma 2 further on[4] to define conditional previsions $P_j(X_{O_j}|x')$ for $x'\in\mathscr{X}_{I_j}, x'\neq x$ and $P_i(X_{O_i}|X_{I_i})$ for $i\neq j$ such that P and $P_k(X_{O_k}|X_{I_k})$ are coherent for $k=1,\dots,m$.

If $\underline{P}(x)=0$, we consider some $P\geq\underline{P}$ such that $P(x)=0$, and take $P(X_{O_j}|x)\in\mathscr{M}(\underline{P}_j(X_{O_j}|x))$ such that $P(f|x)=\underline{P}(f|x)$. For any other x' we can apply Lemma 2 to define conditional previsions $P_i(X_{O_i}|x')$ for $i\neq j, x'\in\mathscr{X}_{I_i}$ and for $i=j$, $x'\in\mathscr{X}_{I_i}, x'\neq x$ such that P and $P_k(X_{O_k}|X_{I_k})$ are coherent for $k=1,\dots,m$.

In any of the two cases, we obtain a family of conditional previsions $P_1(X_{O_1}|X_{I_1}),\dots,P_m(X_{O_m}|X_{I_m})$ which are weakly coherent (P is a compatible joint), dominate $\underline{P}_1(X_{O_1}|X_{I_1}),\dots,\underline{P}_m(X_{O_m}|X_{I_m})$ and s.t. $P_j(f|x)-\underline{P}_j(f|x)<\varepsilon$. This shows that $\underline{P}_1(X_{O_1}|X_{I_1}),\dots,\underline{P}_m(X_{O_m}|X_{I_m})$ are the lower envelope of a family of weakly coherent conditional previsions. ∎

This tells us that we could also establish our results assuming the existence of precise models $P_1(X_{O_1}|X_{I_1}),\dots,P_m(X_{O_m}|X_{I_m})$, for which our imprecise knowledge makes us consider a set of possible candidates

---

[4] We have put this lemma in Section 4.1 because we thought this helped to improve the clarity of the paper; it is easy to see that none of the results established prior to it (and in particular this one) are used in its proof, i.e., that there are no loops in our results.

$$\{P_1^\lambda(X_{O_1}|X_{I_1}),\ldots,P_m^\lambda(X_{O_m}|X_{I_m}) : \lambda \in \Lambda\}.$$

We see that the consistency requirements we make on this set of possible models (weak or strong coherence) also hold for the conditional lower previsions that summarise them by taking their lower envelopes.

From Theorem 1, weakly coherent conditional lower previsions always have a compatible joint $\underline{P}$. Our following result establishes the smallest such joint:

Theorem 3. *Let $\underline{P}_1(X_{O_1}|X_{I_1}),\ldots,\underline{P}_m(X_{O_m}|X_{I_m})$ be weakly coherent conditional lower previsions, and let $\underline{E}$ be given on $\mathcal{L}(\mathcal{X}^n)$ by*

$$\underline{E}(f) := \sup\{\alpha : \exists f_j \in \mathcal{K}^j, j=1,\ldots,m, s.t. \\ \max_{\omega\in\mathcal{X}^n}[\sum_{j=1}^m G(f_j|X_{I_j}) - (f-\alpha)](\omega) < 0\}. \quad (3)$$

$\underline{E}$ is the smallest coherent lower prevision which is coherent with each of the conditional lower previsions $\underline{P}_j(X_{O_j}|X_{I_j})$.

Proof. We prove in [13, Theorem 1] that $\underline{E}$ is a coherent lower prevision that is also coherent with $\underline{P}_j(X_{O_j}|X_{I_j})$ for $j=1,\ldots,m$. Let $\underline{F}$ be another coherent lower prevision with this property. Assume that there is some gamble f such that $\underline{F}(f) = \underline{E}(f) - \delta$ for some $\delta > 0$. It follows from the definition of $\underline{E}$ that there are $f_j \in \mathcal{K}^j$ for $j=1,\ldots,m$ such that

$$\max_{\omega\in\mathcal{X}^n}\left[\sum_{j=1}^m G(f_j|X_{I_j}) - (f - (\underline{F}(f) + \frac{\delta}{2}))\right](\omega) < 0,$$

whence

$$\max_{\omega\in\mathcal{X}^n}\left[\sum_{j=1}^m G(f_j|X_{I_j}) - (f - \underline{F}(f))\right](\omega) < -\frac{\delta}{2},$$

contradicting the weak coherence of $\underline{F}, \underline{P}_j(X_{O_j}|X_{I_j})$, $j=1,\ldots,m$. ∎

Using this result and Theorem 2, we can also give a sensitivity analysis interpretation to $\underline{E}$ in the precise case.

Corollary 1. *Let $P_1(X_{O_1}|X_{I_1}),\ldots,P_m(X_{O_m}|X_{I_m})$ be weakly coherent conditional linear previsions. The lower prevision $\underline{E}$ defined in* (3) *is the lower envelope of the set $\mathcal{M}$ of linear previsions which are coherent with each $P_j(X_{O_j}|X_{I_j}), j=1,\ldots,m$.*

Proof. From Theorem 3, $\underline{E}$ is the smallest coherent lower prevision such that $\underline{E}, P_1(X_{O_1}|X_{I_1}),\ldots,P_m(X_{O_m}|X_{I_m})$ are weakly coherent. From Theorem 2, the previsions $\underline{E}, P_1(X_{O_1}|X_{I_1}),\ldots,P_m(X_{O_m}|X_{I_m})$ are the lower envelope of a class of dominating weakly coherent linear previsions. But since our conditional previsions are all linear, this means that $\underline{E}$ is the lower envelope of a class $\mathcal{M}$ of linear previsions P which are weakly coherent with the conditional previsions $P_1(X_{O_1}|X_{I_1}),\ldots,P_m(X_{O_m}|X_{I_m})$.

Assume the existence of a linear prevision P which is weakly coherent with $P_1(X_{O_1}|X_{I_1}),\dots,P_m(X_{O_m}|X_{I_m})$ and such that $P(f) < \underline{E}(f)$ for some gamble f. If we define the coherent lower prevision $\underline{P}_1 := \min\{P,\underline{E}\}$, we would deduce that $\underline{P}_1, P_1(X_{O_1}|X_{I_1}),\dots,P_m(X_{O_m}|X_{I_m})$ are also weakly coherent, because weak coherence is closed under lower envelopes. This contradicts Theorem 3. Therefore, $\underline{E}$ is the lower envelope of the set of linear previsions which are coherent with $P_j(X_{O_j}|X_{I_j})$ for $j = 1,\dots,m$. ∎

Let us focus now on the relationship between weak and strong coherence and avoiding partial loss. We start by considering this problem in the precise case. In this case coherence is equivalent to avoiding partial loss, and is in general greater than weak coherence; see [16, Example 7.3.5] for an example of weakly coherent conditional previsions that incur a partial loss. We are going to show next that when a number of conditional previsions are weakly coherent but not coherent, this is due to the definition of the conditional previsions on some sets of probability zero.

Theorem 4. *Let $P_1(X_{O_1}|X_{I_1}),\dots,P_m(X_{O_m}|X_{I_m})$ be weakly coherent conditional linear previsions with respective domains $\mathcal{K}^1,\dots,\mathcal{K}^m$, and let $\overline{E}$ be the conjugate of the functional $\underline{E}$ defined in* (3). *They are coherent if and only if for all gambles $f_j \in \mathcal{K}^j$, $j = 1,\dots,m$ with $\overline{E}(\mathbb{S}(f_j)) = 0$, $\max_{\omega\in\mathbb{S}(f_j)} \sum_{j=1}^m [f_j - P_j(f_j|X_{I_j})](\omega) \geq 0$.*

Proof. Because we are dealing with conditional linear previsions, coherence is equivalent to avoiding partial loss. Hence, we must verify whether for any $f_j \in \mathcal{K}^j$, $j = 1,\dots,m$,

$$\max_{\omega\in\mathbb{S}(f_j)} \sum_{j=1}^m [f_j - P_j(f_j|X_{I_j})](\omega) \geq 0. \tag{4}$$

It is clear that if Equation (4) holds for any $f_j \in \mathcal{K}^j$, $j = 1,\dots,m$, it also holds for any gambles $f_1,\dots,f_m$ satisfying $\overline{E}(\mathbb{S}(f_j)) = 0$. Conversely, assume that this condition holds. If $P_1(X_{O_1}|X_{I_1}),\dots,P_m(X_{O_m}|X_{I_m})$ are not coherent, there must be $f_j \in \mathcal{K}^j$, $j = 1,\dots,m$, such that $\overline{E}(\mathbb{S}(f_j)) > 0$ and

$$\max_{\omega\in\mathbb{S}(f_j)} \sum_{j=1}^m [f_j - P_j(f_j|X_{I_j})](\omega) \leq -\delta < 0$$

for some $\delta > 0$. Applying Corollary 1, there is some linear prevision P which is coherent with $P_j(X_{O_j}|X_{I_j})$ for $j = 1,\dots,m$ and such that $P(\mathbb{S}(f_j)) > 0$.

Let us define $g := \sum_{j=1}^m [f_j - P_j(f_j|X_{I_j})]$. The coherence of $P, P_j(X_{O_j}|X_{I_j})$ for $j = 1,\dots,m$ implies that $P(f_j) = P(P_j(f_j|X_{I_j}))$ for $j = 1,\dots,m$, and the linearity of P implies then that

$$P(g) = \sum_{j=1}^m P(f_j - P_j(f_j|X_{I_j})) = 0.$$

But on the other hand we have that

$$P(g) = P(gI_{\mathbb{S}(f_j)}) \leq P(-\delta I_{\mathbb{S}(f_j)}) = -\delta P(\mathbb{S}(f_j)) < 0.$$

This is a contradiction. Therefore, it suffices to verify the coherence condition on those gambles whose union of supports have upper probability zero under the upper prevision $\overline{E}$ determined by Eq. (3). ■

Taking into account this theorem and the envelope result established in Theorem 2, we can characterise the difference between weak coherence and avoiding partial loss for conditional lower previsions:

Corollary 2. *Let* $\underline{P}_1(X_{O_1}|X_{I_1}),\ldots,\underline{P}_m(X_{O_m}|X_{I_m})$ *be weakly coherent conditional lower previsions. They avoid partial loss if and only if for all* $f_j \in \mathscr{K}^j$, $j=1,\ldots,m$ *with* $\overline{E}(\mathbb{S}(f_j))=0$, $\max_{\omega\in\mathbb{S}(f_j)}\sum_{j=1}^m[f_j-\underline{P}_j(f_j|X_{I_j})](\omega)\geq 0$, *where* $\overline{E}$ *is the conjugate of the functional defined in* (3).

Proof. $\underline{P}_1(X_{O_1}|X_{I_1}),\ldots,\underline{P}_m(X_{O_m}|X_{I_m})$ avoid partial loss if and only if for any $f_j \in \mathscr{K}^j$, $j=1,\ldots,m$,

$$\max_{\omega\in\mathbb{S}(f_j)}\sum_{j=1}^m[f_j-\underline{P}_j(f_j|X_{I_j})](\omega)\geq 0.$$

It is clear that if this condition holds it also holds in particular for gambles $f_1,\ldots,f_m$ with $\overline{E}(\mathbb{S}(f_j))=0$. Conversely, assume that this holds but that there are $f_1,\ldots,f_m$ such that $\overline{E}(\mathbb{S}(f_j))>0$ and

$$\max_{\omega\in\mathbb{S}(f_j)}\sum_{j=1}^m[f_j-\underline{P}_j(f_j|X_{I_j})](\omega)\leq -\delta<0.$$

Let us define $g:=\sum_{j=1}^m[f_j-\underline{P}_j(f_j|X_{I_j})]$. Since $\underline{E}$ and $\underline{P}_j(X_{O_j}|X_{I_j})$ are coherent for $j=1,\ldots,m$, we deduce that $\underline{E}(f_j-\underline{P}_j(f_j|X_{I_j}))\geq 0$ for $j=1,\ldots,m$. The super-additivity (C3) of the coherent lower prevision $\underline{E}$ implies then that

$$\underline{E}(g)=\underline{E}\left(\sum_{j=1}^m[f_j-\underline{P}_j(f_j|X_{I_j})]\right)\geq\sum_{j=1}^m\underline{E}\left(f_j-\underline{P}_j(f_j|X_{I_j})\right)\geq 0.$$

But on the other hand, we have that

$$\underline{E}(g)=\underline{E}(gI_{\mathbb{S}(f_j)})\leq\underline{E}(-\delta I_{\mathbb{S}(f_j)})=-\delta\overline{E}(\mathbb{S}(f_j))<0.$$

This is a contradiction. Therefore, it suffices to verify the avoiding partial loss condition on those gambles whose union of supports has upper probability zero under $\overline{E}$. ■

Hence, if a number of weakly coherent lower previsions incur sure loss, this incoherent behaviour is due to the definition of the conditional previsions on some sets of zero upper probability. It may be argued, specially since we are dealing with finite spaces, that we may modify the definition of these conditional lower previsions on these sets in order to avoid partial loss without further consequences, in the sense that this will not affect their weak coherence: they will still be weakly coherent with the same unconditional $\underline{P}$.

So let us consider a number of weakly coherent conditional lower previsions that avoid partial loss. Our next example shows that, unlike for precise previsions, this is not sufficient for coherence. Hence, Theorem 4 does not extend to the imprecise case. This is because the condition equivalent to avoiding partial loss in Corollary 2 is not in general equivalent to coherence, in the sense that the union of the supports of a number of gambles producing incoherence may have positive upper probability:

Example 1. Consider two random variables X_1, X_2 taking values in the finite space $\mathscr{X} := \{1,2,3\}$, and let us define conditional lower previsions $\underline{P}(X_2|X_1)$ and $\underline{P}(X_1|X_2)$ by

$$\begin{aligned}
&\underline{P}(f|X_1 = 1) = f(1,1)\\
&\underline{P}(f|X_1 = 2) = f(2,3)\\
&\underline{P}(f|X_1 = 3) = \min\{f(3,2), f(3,3)\}\\
&\underline{P}(f|X_2 = 1) = f(2,1)\\
&\underline{P}(f|X_2 = 2) = \min\{f(1,2), f(2,2), f(3,2)\}\\
&\underline{P}(f|X_2 = 3) = \min\{f(1,3), f(2,3), f(3,3)\},
\end{aligned}$$

for any gamble f in $\mathscr{L}(\mathscr{X}^2)$.

Let us consider the unconditional lower prevision $\underline{P}$ on $\mathscr{L}(\mathscr{X}^2)$ given by $\underline{P}(f) = \min\{f(3,2), f(3,3)\}$. Using Theorem 1, we can see that $\underline{P}, \underline{P}(X_1|X_2)$ and $\underline{P}(X_2|X_1)$ are weakly coherent.

To see that $\underline{P}(X_1|X_2)$ and $\underline{P}(X_2|X_1)$ avoid partial loss, we apply Corollary 2 and consider $f_1, f_2 \in \mathscr{L}(\mathscr{X}^2)$ such that $\overline{P}(\mathbb{S}(f_j)) = 0$. Let us prove that

$$\max_{\omega \in \mathbb{S}(f_j)} [G(f_1|X_2) + G(f_2|X_1)](\omega) \geq 0. \tag{5}$$

Assume $f_1 \neq 0 \neq f_2$; the other cases are similar (and easier). Since $\overline{P}(\mathbb{S}(f_j)) = 0$ for any coherent lower prevision that is weakly coherent with $\underline{P}(X_1|X_2)$ and $\underline{P}(X_2|X_1)$, neither $(3,2)$ nor $(3,3)$ belong to $\mathbb{S}(f_j)$, and consequently $f_1(x,2) = f_1(x,3) = 0$ for $x = 1,2,3$. If $(X_1 = 2) \in S_1(f_2)$, then $[G(f_1|X_2) + G(f_2|X_1)](2,3) = 0 + 0 = 0$, and therefore Equation (5) holds. If $(X_1 = 2) \notin S_1(f_2)$, then $[G(f_1|X_2) + G(f_2|X_1)](2,1) = 0 + 0 = 0$.

Let us prove finally that $\underline{P}(X_1|X_2), \underline{P}(X_2|X_1)$ are not coherent. Let $f_1 = -I_{\{(1,1),(3,1)\}}$, $f_2 = -I_{\{(1,2),(1,3),(2,1),(2,2)\}}$ and $f_3 = I_{\{(2,3),(3,3)\}}$, and let us show that

$$[G(f_1|X_2) + G(f_2|X_1) - G(f_3|X_2 = 3)](\omega) < 0$$

for all $\omega \in B := \pi_2^{-1}(3) \cup \mathbb{S}(f_j)$. In this case $S_2(f_1) = \{X_2 = 1\}$ and $S_1(f_2) = \{X_1 = 1, X_1 = 2\}$, whence $B = S_2(f_1) \cup S_1(f_2) \cup \{X_2 = 3\} = \mathscr{X}^2 \setminus \{(3,2)\}$. On the other hand, the gamble $g := G(f_1|X_2) + G(f_2|X_1) - G(f_3|X_2 = 3)$ satisfies $g(\omega) = -1$ for all $\omega \in B$. This shows that $\underline{P}(X_1|X_2), \underline{P}(X_2|X_1)$ are not coherent. However, $\overline{E}(B) = 1$ because $(3,3) \in B$. ♦

Hence, when a number of conditional lower previsions are weakly coherent but not coherent, the behaviour causing a contradiction can be caused by conditioning on sets of positive upper probability. It is interesting to look for conditions under which it suffices to check the weak coherence of a number of previsions to be able to deduce their coherence. One such condition was established, in a different context, in [13]. In the case of conditional linear previsions, Theorem 4 allows us to derive immediately the following result:

Lemma 1. *Consider weakly coherent $P_1(X_{O_1}|X_{I_1}),\dots,P_m(X_{O_m}|X_{I_m})$, and let P be a coherent prevision such that $P, P_j(X_{O_j}|X_{I_j})$ are coherent for $j=1,\dots,m$. If $P(x)>0$ for any $x\in\mathscr{X}_{I_j}$, $j=1,\dots,m$, then the conditional previsions $P_1(X_{O_1}|X_{I_1}),\dots,P_m(X_{O_m}|X_{I_m})$ are coherent.*

Proof. Since all the conditional previsions are linear, Theorem 4 tells us that it suffices to verify the avoiding partial loss condition on those gambles $f_1,\dots,f_m$ for which $\overline{E}(\mathbb{S}(f_j))=0$. But the hypotheses of the lemma imply that the gambles $f_1,\dots,f_m$ only satisfy $\overline{E}(\mathbb{S}(f_j))=0$ when $\mathbb{S}(f_j)=\emptyset$, which in turn holds if and only if they are all equal to 0, because $\overline{E}$ dominates P from Corollary 1. ■

From this result, we can easily derive a similar condition for conditional lower previsions.

Theorem 5. *Let $\underline{P}_1(X_{O_1}|X_{I_1}),\dots,\underline{P}_m(X_{O_m}|X_{I_m})$ be weakly coherent conditional lower previsions, and let $\underline{P}$ be a coherent lower prevision such that $\underline{P},\underline{P}_j(X_{O_j}|X_{I_j})$ are coherent for $j=1,\dots,m$. If $\underline{P}(x)>0$ for all $x\in\mathscr{X}_{I_j}$ and all $j=1,\dots,m$, then the conditional lower previsions $\underline{P}_1(X_{O_1}|X_{I_1}),\dots,\underline{P}_m(X_{O_m}|X_{I_m})$ are coherent.*

Proof. Consider $f_j\in\mathscr{K}^j$ for $j=1,\dots,m$, $j_0\in\{1,\dots,m\}$, $x_0\in\mathscr{X}_{I_{j_0}}$ and $f_0\in\mathscr{K}^{j_0}$. From Theorem 2, for any $\varepsilon>0$ there are weakly coherent conditional previsions $P_1(X_{O_1}|X_{I_1}),\dots,P_m(X_{O_m}|X_{I_m})$ such that $P_j(X_{O_j}|X_{I_j})\geq\underline{P}_j(X_{O_j}|X_{I_j})$ for $j=1,\dots,m$ and moreover $P_{j_0}(f_0|x_0)-\underline{P}_{j_0}(f_0|x_0)<\varepsilon$. As a consequence,

$$\left[\sum_{j=1}^{m}(f_j-\underline{P}_j(f_j|X_{I_j}))-\pi^{-1}_{I_{j_0}}(x_0)(f_0-\underline{P}_{j_0}(f_0|x_0))\right](\omega)$$
$$\geq\left[\sum_{j=1}^{m}(f_j-P_j(f_j|X_{I_j}))-\pi^{-1}_{I_{j_0}}(x_0)(f_0-P_{j_0}(f_0|x_0))\right](\omega)-\varepsilon.$$

for every $\omega\in\mathscr{X}^n$. We also deduce from the proof of Theorem 2 that there is a coherent prevision $P\geq\underline{P}$ such that P and $P_j(X_{O_j}|X_{I_j})$ are coherent for $j=1,\dots,m$. As a consequence, $P(x)>0$ for all $x\in\mathscr{X}_{I_j}$ and for all $j=1,\dots,m$, and applying Lemma 1 we deduce that $P_1(X_{O_1}|X_{I_1}),\dots,P_m(X_{O_m}|X_{I_m})$ are coherent. Therefore, there is some $\omega^*\in\pi^{-1}_{I_{j_0}}(x_0)\cup\mathbb{S}(f_j)$ such that

$$\left[\sum_{j=1}^{m}(f_j-P_j(f_j|X_{I_j}))-\pi^{-1}_{I_{j_0}}(x_0)(f_0-P_{j_0}(f_0|x_0))\right](\omega^*)\geq 0,$$

whence also

$$\left[\sum_{j=1}^{m}(f_j - \underline{P}_j(f_j|X_{I_j})) - \pi_{I_{j_0}}^{-1}(x_O)(f_O - \underline{P}_{j_0}(f_O|x_O))\right](\omega^*) \geq -\varepsilon.$$

Since we can do this for any $\varepsilon > 0$, the conditional lower previsions $\underline{P}_1(X_{O_1}|X_{I_1}),\ldots,\underline{P}_m(X_{O_m}|X_{I_m})$ are coherent. ■

The proof of this theorem shows that if a number of weakly coherent conditional lower previsions avoid partial loss but are not coherent, for any gambles $f_0,\ldots,f_m$ violating Eq. (2) it must be $\underline{E}(\pi_{I_{j_0}}^{-1}(x_O)\cup\mathbb{S}(f_j)) = 0$ (although, as Example 1 shows, it can be $\overline{E}(\pi_{I_{j_0}}^{-1}(x_O)\cup\mathbb{S}(f_j)) > 0$).

Let us recall that when all the conditioning events have positive lower probability, the conditional lower previsions are uniquely determined by the joint $\underline{P}$ and by (GBR). Hence, in that case $\underline{P}_1(X_{O_1}|X_{I_1}),\ldots,\underline{P}_m(X_{O_m}|X_{I_m})$ are the only conditional previsions which are coherent with $\underline{P}$.

4 Coherent Updating

Although our last result is interesting, it is fairly common in situations of imprecise information to be conditioning on events of lower probability zero and positive upper probability. In that case, there is an infinite number of conditional lower previsions which are coherent with the unconditional $\underline{P}$. In this section, we characterise them by determining the smallest and the greatest coherent extensions.

4.1 Updating with the Regular Extension

The first updating rule we consider is called the *regular extension*. Consider an unconditional lower prevision $\underline{P}$ and disjoint O,I in $\{1,\ldots,n\}$. The conditional lower prevision $\underline{R}(X_O|X_I)$ defined by regular extension is given, for any $f \in \mathcal{K}_{O\cup I}$ and any $x \in \mathcal{X}_I$, by

$$\underline{R}(f|x) := \inf\left\{\frac{P(f\mathbb{I}_x)}{P(x)} : P \geq \underline{P}, P(x) > 0\right\}.$$

For this definition to be applicable, we need that $\overline{P}(x) > 0$ for any $x \in \mathcal{X}_I$. The regular extension is the lower envelope of the updated linear previsions using Bayes's rule.

Lemma 2. *Let $\underline{P},\underline{P}(X_O|X_I)$ be coherent unconditional and conditional previsions, with $\mathcal{X}_I$ finite. Assume that $\overline{P}(x) > 0$ for all $x \in \mathcal{X}_I$, and define $\underline{R}(X_O|X_I)$ from $\underline{P}$ using regular extension.*

1. *$\underline{P},\underline{R}(X_O|X_I)$ are coherent.*
2. *$\underline{R}(X_O|X_I) \geq \underline{P}(X_O|X_I)$.*
3. *For any $P \geq \underline{P}$, there is some $P(X_O|X_I)$ which is coherent with P and dominates $\underline{P}(X_O|X_I)$.*

Proof. Since we are dealing with finite spaces, the coherence of $\underline{P}, \underline{R}(X_O|X_I)$ is equivalent to $\underline{P}(\mathbb{I}_{\pi_I^{-1}(x)}(f - \underline{R}(f|x))) = 0$ for all $x \in \mathscr{X}_I$, and this condition holds because of [16, Appendix (J3)].

For the second statement, consider some x in $\mathscr{X}_I$ and $f \in \mathscr{K}_{O\cup I}$. Assume ex-absurdo that $\underline{R}(f|x) < \underline{P}(f|x)$. It follows from the definition of the regular extension that there is some $P \geq \underline{P}$ such that $P(x) > 0$ and $P(f|x) < \underline{P}(f|x)$. Since $P(x) > 0$, it follows from the generalised Bayes rule that $P(f|x)$ is the *unique* value satisfying $0 = P(\mathbb{I}_{\pi_I^{-1}(x)}(f - P(f|x)))$. As a consequence, given $\underline{P}(f|x) > P(f|x)$, we have that $\mathbb{I}_{\pi_I^{-1}(x)}(f - P(f|x)) \geq \mathbb{I}_{\pi_I^{-1}(x)}(f - \underline{P}(f|x))$, whence

$$0 = P(\mathbb{I}_{\pi_I^{-1}(x)}(f - P(f|x))) \geq P(\mathbb{I}_{\pi_I^{-1}(x)}(f - \underline{P}(f|x))) \\ \geq \underline{P}(\mathbb{I}_{\pi_I^{-1}(x)}(f - \underline{P}(f|x))) = 0,$$

using the coherence of $\underline{P}, \underline{P}(X_O|X_I)$. But this implies that $P(\mathbb{I}_{\pi_I^{-1}(x)}(f - P(f|x))) = P(\mathbb{I}_{\pi_I^{-1}(x)}(f - \underline{P}(f|x))) = 0$, and then there are two different values of μ for which $P(\mathbb{I}_{\pi_I^{-1}(x)}(f - \mu)) = 0$. This is a contradiction.

Let us finally establish the third statement. Consider $P \geq \underline{P}$, and $x \in \mathscr{X}_I$. If $P(x) > 0$, then for all $f \in \mathscr{K}_{O\cup I}$ the value of $P(f|x)$ is uniquely determined by (GBR) and dominates the regular extension $\underline{R}(f|x)$. Hence, $P(f|x) \geq \underline{R}(f|x) \geq \underline{P}(f|x)$, where the last inequality follows from the second statement. Finally, if $P(x) = 0$, taking any element $P(X_O|x)$ of $\mathscr{M}(\underline{P}(X_O|x))$ we have that $P(\mathbb{I}_{\pi_I^{-1}(x)}(f - P(f|x))) = 0$ for all $f \in \mathscr{K}_{O\cup I}$. This completes the proof. ∎

From this lemma, we deduce that if we use regular extension to derive conditional lower previsions $\underline{R}_1(X_{O_1}|X_{I_1}), \ldots, \underline{R}_m(X_{O_m}|X_{I_m})$ from an unconditional $\underline{P}$, then $\underline{P}, \underline{R}_1(X_{O_1}|X_{I_1}), \ldots, \underline{R}_m(X_{O_m}|X_{I_m})$ are weakly coherent. Moreover, if we consider any other weakly coherent conditional lower previsions $\underline{P}_1(X_{O_1}|X_{I_1}), \ldots, \underline{P}_m(X_{O_m}|X_{I_m})$, it follows that $\underline{R}_j(X_{O_j}|X_{I_j}) \geq \underline{P}_j(X_{O_j}|X_{I_j})$ for $j = 1, \ldots, m$. Hence, the procedure of regular extension provides the greatest, or more informative, updated lower previsions that are weakly coherent with $\underline{P}$. In the following theorem we prove that they are also coherent.

Theorem 6. *Let $\underline{P}$ be a coherent lower prevision on $\mathscr{L}(\mathscr{X}^n)$, and consider disjoint O_j, I_j for $j = 1, \ldots, m$. Assume that $\overline{P}(x) > 0$ for all $x \in \mathscr{X}_{I_j}$, and let us define $\underline{R}_j(X_{O_j}|X_{I_j})$ using regular extension for $j = 1, \ldots, m$. Then $\underline{P}, \underline{R}_1(X_{O_1}|X_{I_1}), \ldots,$ $\underline{R}_m(X_{O_m}|X_{I_m})$ are coherent.*

Proof. Each of the conditional previsions defined by regular extension is coherent with $\underline{P}$ from Lemma 2, and therefore they are all weakly coherent. Consider $f_j \in \mathscr{K}^j$ for $j = 1, \ldots, m$, $j_0 \in \{1, \ldots, m\}, x_0 \in \mathscr{X}_{I_{j_0}}$, $f_0 \in \mathscr{K}^{j_0}$, and let us prove that

$$\left[\sum_{j=1}^{m}(f_j - \underline{R}_j(f_j|X_{I_j})) - \pi_{I_{j_0}}^{-1}(x_0)(f_0 - \underline{R}_{j_0}(f_0|x_0))\right](\omega) \geq 0$$

for some $\omega \in \pi_{I_{j_0}}^{-1}(x_0) \cup \mathbb{S}(f_j)$.

Assume ex-absurdo that the sum above is smaller than $-\delta$ for some $\delta > 0$ and for all $\omega \in \pi_{I_{j0}}^{-1}(x_O) \cup \mathbb{S}(f_j)$. It follows then from the definition of the regular extension that given $\frac{\delta}{2} > 0$ there is some $P \geq \underline{P}$ such that $P(x_O) > 0$ and $P_{j0}(f_O|x_O) - \underline{R}_{j0}(f_O|x_O) < \frac{\delta}{2}$. From Lemma 2, we can consider $P_1(X_{O_1}|X_{I_1}),\dots,P_m(X_{O_m}|X_{I_m})$ such that $P_j(X_{O_j}|X_{I_j})$ dominates $\underline{R}_j(X_{O_j}|X_{I_j})$ and is coherent with P for all j, and such that moreover $P_{j0}(f|x_O) - \underline{R}_{j0}(f|x_O) < \frac{\delta}{2}$. As a consequence,

$$\sum_{j=1}^{m}(f_j - \underline{R}_j(f_j|X_{I_j})) - \pi_{I_{j0}}^{-1}(x_O)(f_O - \underline{R}_{j0}(f_O|x_O))$$
$$\geq \sum_{j=1}^{m}(f_j - P_j(f_j|X_{I_j})) - \pi_{I_{j0}}^{-1}(x_O)(f_O - P_{j0}(f_O|x_O)) - \frac{\delta}{2},$$

and if we let $g := \sum_{j=1}^{m}(f_j - P_j(f_j|X_{I_j})) - \pi_{I_{j0}}^{-1}(x_O)(f_O - P_{j0}(f_O|x_O))$ then it follows from the coherence of P and $P_j(X_{O_j}|X_{I_j})$ for all j that $P(g) = 0$.

On the other hand, the above equations imply that $g(\omega) < -\frac{\delta}{2}$ for all $\omega \in \pi_{I_{j0}}^{-1}(x_O) \cup \mathbb{S}(f_j)$. The definition of the supports implies moreover that $g(\omega) = 0$ for all $\omega \notin \pi_{I_{j0}}^{-1}(x_O) \cup \mathbb{S}(f_j)$. Hence,

$$P(g) = P\left(gI_{\pi_{I_{j0}}^{-1}(x_O) \cup \mathbb{S}(f_j)}\right) < -\frac{\delta}{2}P\left(\pi_{I_{j0}}^{-1}(x_O) \cup \mathbb{S}(f_j)\right) < 0,$$

because $P(\pi_{I_{j0}}^{-1}(x_O) \cup \mathbb{S}(f_j)) \geq P(x_O) > 0$. This is a contradiction. Hence, there is some $\omega \in \pi_{I_{j0}}^{-1}(x_O) \cup \mathbb{S}(f_j)$ such that

$$\left[\sum_{j=1}^{m}(f_j - \underline{R}_j(f_j|X_{I_j})) - \pi_{I_{j0}}^{-1}(x_O)(f_O - \underline{R}_{j0}(f_O|x_O))\right](\omega) \geq 0,$$

and this implies that $\underline{R}_1(X_{O_1}|X_{I_1}),\dots,\underline{R}_m(X_{O_m}|X_{I_m})$ are coherent. ∎

When $\overline{P}(x) = 0$ for some $x \in \mathscr{X}_{I_j}$, $j = 1,\dots,m$, we cannot use regular extension to define $\underline{R}_j(X_{O_j}|x)$. It can be checked that in that case any separately coherent conditional lower prevision is weakly coherent with $\underline{P}$. However, we cannot guarantee the strong coherence:

Example 2. Let $\mathscr{X}_1 = \mathscr{X}_2 = \{1,2,3\}$, and $P(X_1), P(X_2|X_1)$ determined by $P(X_1 = 3) = 1$, and $P(X_2 = x|X_1 = x) = 1$ for $x = 1,2,3$. From [16, Theorem 6.7.2], $P(X_1), P(X_2|X_1)$ are coherent. However, if we define arbitrarily $P(X_1|X_2 = x)$ when $P(X_2 = x) = 0$ (that is, for $x = 1,2$), then $P(X_1|X_2)$ and $P(X_2|X_1)$ may not be coherent: make it for instance $P(X_1 = 1|X_2 = 2) = 1 = P(X_1 = 2|X_2 = 1) = P(X_1 = 3|X_2 = 3)$. Then [16, Example 7.3.5] shows that $P(X_1|X_2)$ and $P(X_2|X_1)$ are not coherent. ♦

4.2 Updating with the Natural Extension

Next, we introduce the notion of *natural extension*. Let $\underline{P}$ be a coherent lower prevision on $\mathcal{L}(\mathcal{X}^n)$. Consider disjoint subsets O_j, I_j of $\{1,\dots,n\}$ for $j = 1,\dots,m$. For each $j = 1,\dots,m$, the *natural extension* $\underline{E}_j(X_{O_j}|X_{I_j})$ is uniquely determined by (GBR) when $\underline{P}(x) > 0$ and is vacuous when $\underline{P}(x) = 0$, being then defined by $\underline{E}_j(f|x) = \min_{\omega \in \pi_{I_j}^{-1}(x)} f(\omega)$ for every $f \in \mathcal{K}^j$. Hence, in this respect the natural extensions can be calculated more easily than the regular extensions.

This notion of natural extension is a particular case of the notion of natural extension of conditional lower previsions $\underline{P}_1(X_{O_1}|X_{I_1}),\dots,\underline{P}_m(X_{O_m}|X_{I_m})$ which are defined on linear spaces $\mathcal{H}^1,\dots,\mathcal{H}^m$ and avoid partial loss. This notion is studied in great detail in [16, Section 8.1]. It is proven in [16, Theorem 8.1.9] that if we use natural extension to obtain conditional lower previsions $\underline{E}_1(X_{O_1}|X_{I_1}),\dots,\underline{E}_m(X_{O_m}|X_{I_m})$, then these are the smallest coherent conditional lower previsions to dominate $\underline{P}_1(X_{O_1}|X_{I_1}),\dots,\underline{P}_m(X_{O_m}|X_{I_m})$ on their domains.

Using this result with $\underline{P}_1(X_{O_1}|X_{I_1}),\dots,\underline{P}_m(X_{O_m}|X_{I_m})$ defined on the constant gambles only, it is immediate to establish the following theorem, whose proof is therefore omitted.

Theorem 7. *Let $\underline{P}$ be a coherent lower prevision on $\mathcal{L}(\mathcal{X}^n)$. Consider disjoint O_j, I_j for $j = 1,\dots,m$, and let us define $\underline{E}_j(X_{O_j}|X_j)$, $j = 1,\dots,m$ using natural extension. Then $\underline{P},\underline{E}_1(X_{O_1}|X_{I_1}),\dots,\underline{E}_m(X_{O_m}|X_{I_m})$ are coherent.*

The natural extension provides the smallest conditional lower previsions which are coherent together with $\underline{P}$. The previsions $\underline{E}_j(X_{O_j}|X_{I_j})$ are uniquely determined by (GBR) when $\underline{P}(x) > 0$ and are vacuous when $\underline{P}(x) = 0$, being then defined by $\underline{E}_j(f|x) = \min_{\omega \in \pi_{I_j}^{-1}(x)} f(\omega)$ for any $f \in \mathcal{K}^j$. Hence, in that respect the natural extensions can be calculated more easily than the regular extensions.

We showed before that the conditional previsions defined by regular extension were also the greatest conditional lower previsions that are weakly coherent with the unconditional lower prevision $\underline{P}$. Using Theorem 1 and the results in [16, Chapter 6], it is not difficult to show that the natural extensions are the smallest weakly coherent extensions:

Theorem 8. *Let $\underline{P}$ be coherent on $\mathcal{L}(\mathcal{X}^n)$, and define conditional lower previsions $\underline{E}_1(X_{O_1}|X_{I_1}),\dots,$ $\underline{E}_m(X_{O_m}|X_{I_m})$ using natural extension. Then $\underline{E}_1(X_{O_1}|X_{I_1}),\dots,\underline{E}_m(X_{O_m}|X_{I_m})$ are the smallest conditional lower previsions which are weakly coherent with $\underline{P}$.*

Proof. Since $\underline{P},\underline{E}_1(X_{O_1}|X_{I_1}),\dots,\underline{E}_m(X_{O_m}|X_{I_m})$ are coherent because of Theorem 7, they are also weakly coherent. Consider $j \in \{1,\dots,m\}$ and $\underline{P}_j(X_{O_j}|X_{I_j})$ which is coherent with $\underline{P}$. For any $x \in \mathcal{X}_{I_j}$, there are two possibilities: either $\underline{P}(x) > 0$, and then $\underline{P}_j(X_{O_j}|x)$ is uniquely determined by (GBR), whence $\underline{P}_j(X_{O_j}|x) = \underline{E}_j(X_{O_j}|x)$; or $\underline{P}(x) = 0$, and the separate coherence of $\underline{P}_j(X_{O_j}|x)$ implies that $\underline{P}_j(f|x) \geq \min_{\omega \in \pi_{I_j}^{-1}(x)} f(\omega) = \underline{E}_j(f|x)$ for any $f \in \mathcal{K}^j$.

Hence, for all $j=1,\dots,m$, any conditional lower prevision $\underline{P}_j(X_{O_j}|X_{I_j})$ which is coherent with $\underline{P}$ dominates the natural extension $\underline{E}_j(X_{O_j}|X_{I_j})$. Applying Theorem 1, $\underline{E}_1(X_{O_1}|X_{I_1}),\dots,\underline{E}_m(X_{O_m}|X_{I_m})$ are the smallest weakly coherent extensions. ∎

5 Additional Properties

Let us investigate a bit further the properties of weakly and strongly coherent models. Specifically, we are going to determine (i) if these models are closed under convex combinations and point-wise limits and (ii) if the property of 2-monotonicity is preserved by the coherent updating.

5.1 Convexity

We begin by studying the convexity of weakly coherent and coherent conditional lower previsions. In the unconditional case, it is proven in [16, Thm. 2.6.4] that a convex combination of coherent lower previsions produces again a coherent lower prevision. Let us investigate whether such a property holds in the conditional case. In this sense, it is easy to show that if we fix the subsets O,I of $\{1,\dots,n\}$ and consider a finite number of separately coherent conditional lower previsions $\underline{P}(X_O|X_I)$, their convex combination is also separately coherent.

We proved in Theorem 1 that weakly coherent conditional lower previsions are always pairwise coherent with some coherent lower prevision $\underline{P}$. Let us show that if we fix this $\underline{P}$, then the convex combination of the weakly coherent conditional lower previsions is again weakly coherent:

Theorem 9. *Let $\underline{P}$ be a coherent lower prevision on $\mathscr{L}(\mathscr{X}^n)$. Consider conditional lower previsions $\underline{P}_1^j(X_{O_1}|X_{I_1}),\dots,\underline{P}_m^j(X_{O_m}|X_{I_m})$ with respective domains $\mathscr{K}^1,\dots,\mathscr{K}^m$ and such that $\underline{P},\underline{P}_1^j(X_{O_1}|X_{I_1}),\dots,\underline{P}_m^j(X_{O_m}|X_{I_m})$ are weakly coherent, for $j=1,\dots,\ell$. Let $\alpha_1,\dots,\alpha_\ell\in[0,1]$ be real numbers such that $\alpha_1+\dots+\alpha_\ell=1$, and define, for $k=1,\dots,m$, $\underline{P}_k(X_{O_k}|X_{I_k})$ on $\mathscr{K}^k$ by*

$$\underline{P}_k(f|X_{I_k})=\sum_{i=1}^{\ell}\alpha_i\underline{P}_k^i(f|X_{I_k})$$

for every $f\in\mathscr{K}^k$. Then $\underline{P},\underline{P}_1(X_{O_1}|X_{I_1}),\dots,\underline{P}_m(X_{O_m}|X_{I_m})$ are weakly coherent.

Proof. From Theorem 1, for every $j=1,\dots,\ell$ and every $k=1,\dots,m$, $\underline{P},\underline{P}_j^k(X_{O_k}|X_{I_k})$ are coherent, and consequently they satisfy (GBR). Hence, for every $k=1,\dots,m$ and every $f\in\mathscr{K}^k$, Lemma 2 and Theorem 8 imply that

$$\underline{P}_k^{\ell}(f|X_{I_k})\in[\underline{E}_k(f|X_{I_k}),\underline{R}_k(f|X_{I_k})],$$

where $\underline{E}_k(X_{O_k}|X_{I_k}),\underline{R}_k(X_{O_k}|X_{I_k})$ are the natural and regular extensions derived from $\underline{P}$. As a consequence,

$$\underline{P}_k(f|X_{I_k}) \in [\underline{E}_k(f|X_{I_k}), \underline{R}_k(f|X_{I_k})],$$

whence it also satisfies (GBR) (and consequently it is coherent) with $\underline{P}$. Applying again Theorem 1, $\underline{P}, \underline{P}_1(X_{O_1}|X_{I_1}), \ldots, \underline{P}_m(X_{O_m}|X_{I_m})$ are weakly coherent. ■

This shows that the class of weakly coherent models is made of convex layers, associated to the different unconditional lower previsions. However, when the conditional lower previsions are not weakly coherent with the same unconditional lower prevision, weak coherence is not preserved by taking convex combinations. This is shown in the following example. Note that, because we are considering one conditional and one unconditional lower prevision, the result is also valid for coherence:

Example 3. Let $\mathscr{X}_1 = \mathscr{X}_2 = \{1,2\}$. Let $\underline{P}_1$ be the vacuous lower prevision on $\mathscr{X}_1 \times \mathscr{X}_2$, and $\underline{P}_1(X_1|X_2)$ be the vacuous conditional lower prevision. It follows from Theorem 7 that $\underline{P}_1, \underline{P}_1(X_1|X_2)$ are coherent. Consider on the other hand the linear prevision P_2 determined by a uniform probability distribution on $\mathscr{X}_1 \times \mathscr{X}_2$, and $P_2(X_1|X_2)$ the conditional prevision determined from P_2 by Bayes' rule. Then also $P_2, P_2(X_1|X_2)$ are coherent. Take now the convex combinations $\underline{P} = \frac{\underline{P}_1+P_2}{2}$, $\underline{P}(X_1|X_2) = \frac{\underline{P}_1(X_1|X_2)+P_2(X_1|X_2)}{2}$, and let us show that $\underline{P}, \underline{P}(X_1|X_2)$ do not satisfy (GBR) and therefore are not coherent.

Let $f \in \mathscr{L}(\mathscr{X}_1 \times \mathscr{X}_2)$. Then $\underline{P}(f|X_2 = 1) = \frac{f(1,1)+f(2,1)}{4} + \frac{\min\{f(1,1),f(2,1)\}}{2}$. Hence,

$$G(f|X_2 = 1)(1,1) = \begin{cases} \frac{f(1,1)-f(2,1)}{4} & \text{if } f(1,1) = \min\{f(1,1), f(2,1)\} \\ \frac{3(f(1,1)-f(2,1))}{4} & \text{if } f(2,1) = \min\{f(1,1), f(2,1)\}, \end{cases}$$

and similarly

$$G(f|X_2 = 1)(2,1) = \begin{cases} \frac{f(2,1)-f(1,1)}{4} & \text{if } f(2,1) = \min\{f(1,1), f(2,1)\} \\ \frac{3(f(2,1)-f(1,1))}{4} & \text{if } f(1,1) = \min\{f(1,1), f(2,1)\}. \end{cases}$$

As a consequence,

$$\begin{aligned} \underline{P}(G(f|X_2 = 1)) &= \frac{G(f|X_2 = 1)(1,1) + G(f|X_2 = 1)(2,1)}{8} \\ &\quad + \frac{\min\{G(f|X_2 = 1)(1,1), G(f|X_2 = 1)(2,1), 0\}}{2} \\ &= \frac{\max\{f(1,1), f(2,1)\} - \min\{f(1,1), f(2,1)\}}{16} \\ &\quad + \frac{\min\{f(1,1), f(2,1)\} - \max\{f(1,1), f(2,1)\}}{8} \neq 0 \end{aligned}$$

unless $f(1,1) = f(2,1)$. This shows that $\underline{P}, \underline{P}(X_1|X_2)$ do not satisfy (GBR) and therefore are not coherent. ♦

Next, we consider a number of coherent conditional lower previsions, and investigate if their convex combinations are also coherent. We show in the following example that this is not the case. Note that, since the example involves conditional linear previsions, the result also shows that the models that avoid partial loss are not closed under convex combinations.

Example 4. Let X_1, X_2, X_3 be binary random variables, and let us consider the conditional linear previsions $P_1(X_3|X_1), P_1(X_3|X_2)$ determined by

$$\begin{cases} P_1(X_3 = 1|X_1 = 1) = 0.5,\ P_1(X_3 = 1|X_1 = 0) = 0.3 \\ P_1(X_3 = 1|X_2 = 1) = 0.5,\ P_1(X_3 = 1|X_2 = 0) = 0.5. \end{cases}$$

Let us show that $P_1(X_3|X_1), P_1(X_3|X_2)$ are coherent. Since they are linear, coherence is equivalent to avoiding partial loss. Consider $f_1 \in \mathcal{K}_{1,3}, f_2 \in \mathcal{K}_{2,3}$, and let us show that

$$\max_{\omega \in \mathbb{S}(f_j)} [G_1(f_1|X_1) + G_1(f_2|X_2)](\omega) \geq 0. \quad (6)$$

Note first of all that they are weakly coherent: the linear prevision P_1 determined by a uniform probability distribution on $\{\omega \in \{0,1\}^3 : \pi_1(\omega) = 1\}$ is coherent with both $P_1(X_3|X_1), P_1(X_3|X_2)$. Indeed, it follows from the definition of $P_1(X_3|X_2)$ that any linear prevision P which is coherent with $P_1(X_3|X_2)$ must satisfy $P(X_3 = 1) = 0.5$, and as a consequence, if P is also coherent with $P_1(X_3|X_1)$ it must satisfy $P(X_1 = 1) = 1$. Applying Theorem 4, we can assume without loss of generality that $f_1 = 0$ on $\pi_1^{-1}(1)$.

Assume that $f_2 \neq 0$; otherwise, Eq. (6) follows from the separate coherence of $P_1(X_3|X_1)$. Let $x_2 \in \{0,1\}$ be such that $(X_2 = x_2)$ belongs to $S_2(f_2)$. Then it follows from the separate coherence of $P_2(X_3|X_2)$ that there is some $x_3 \in \{0,1\}$ such that $G_1(f_2|X_2)(x_1, x_2, x_3) \geq 0$ for every $x_1 \in \{0,1\}$, also taking into account that f_2 is $\mathcal{X}_{2,3}$-measurable. As a consequence,

$$[G_1(f_1|X_1) + G_1(f_2|X_2)](1, x_2, x_3) = 0 + G_1(f_2|X_2)(1, x_2, x_3) \geq 0,$$

taking into account that $X_1 = 1$ does not belong to $S_1(f_1)$. Since $(1, x_2, x_3) \in \mathbb{S}(f_j)$, we see that Eq. (6) holds and $P_1(X_3|X_1), P_1(X_3|X_2)$ are coherent.

Consider now the conditional linear previsions $P_2(X_3|X_1), P_2(X_3|X_2)$ determined by

$$\begin{cases} P_2(X_3 = 1|X_1 = 1) = 0.3,\ P_2(X_3 = 1|X_1 = 0) = 0.5 \\ P_2(X_3 = 1|X_2 = 1) = 0.5,\ P_2(X_3 = 1|X_2 = 0) = 0.5. \end{cases}$$

Reasoning as in the above case, we conclude that $P_2(X_3|X_1), P_2(X_3|X_2)$ are coherent (and in particular also weakly coherent).

Consider now $\alpha = 0.5$, and define

$$\begin{cases} P(X_3|X_1) = \alpha P_1(X_3|X_1) + (1-\alpha) P_2(X_3|X_1) \\ P(X_3|X_2) = \alpha P_1(X_3|X_2) + (1-\alpha) P_2(X_3|X_2). \end{cases}$$

These conditional previsions are determined by

$$\begin{cases} P(X_3 = 1|X_1 = 1) = 0.4, \; P(X_3 = 1|X_1 = 0) = 0.4 \\ P(X_3 = 1|X_2 = 1) = 0.5, \; P(X_3 = 1|X_2 = 0) = 0.5. \end{cases}$$

Let us show that $P(X_3|X_1), P(X_3|X_2)$ are not weakly coherent. From Theorem 1, if they were weakly coherent there would be a linear prevision P which would be weakly coherent with them. But the coherence of $P, P(X_3|X_1)$ implies that $P(X_3 = 1) = 0.4$, while it follows from the coherence of $P, P(X_3|X_2)$ that $P(X_3 = 1) = 0.5$. Hence, there is no such P and from Theorem 1 we conclude that $P(X_3|X_1), P(X_3|X_2)$ are not weakly coherent. ♦

5.2 Point-Wise Limits

We next investigate if weakly (resp., strongly) coherent models are closed under point-wise limits. This property holds in the unconditional case, as proven in [16, Thm. 2.6.5]. The following result shows that the same happens for conditional lower previsions:

Theorem 10. *[14, Lemma 4] Consider a sequence of conditional lower previsions* $\{\underline{P}_1^k(X_{O_1}|X_{I_1}), \ldots, \underline{P}_m^k(X_{O_m}|X_{I_m})\}_{k\in\mathbb{N}}$ *with domains* $\mathscr{H}^1, \ldots, \mathscr{H}^m$*. Assume their point-wise limits* $\underline{P}_1(X_{O_1}|X_{I_1}), \ldots, \underline{P}_m(X_{O_m}|X_{I_m})$ *exist.*

1. *If* $\underline{P}_1^k(X_{O_1}|X_{I_1}), \ldots, \underline{P}_m^k(X_{O_m}|X_{I_m})$ *are weakly coherent for all k, then so are* $\underline{P}_1(X_{O_1}|X_{I_1}), \ldots, \underline{P}_m(X_{O_m}|X_{I_m})$.
2. *If moreover* $\underline{P}_1^k(X_{O_1}|X_{I_1}), \ldots, \underline{P}_m^k(X_{O_m}|X_{I_m})$ *are coherent for all k, then so are* $\underline{P}_1(X_{O_1}|X_{I_1}), \ldots, \underline{P}_m(X_{O_m}|X_{I_m})$.

5.3 n-Monotonicity

One interesting particular case of coherent lower previsions are the so-called *n-monotone* lower previsions, which are those defined on a lattice of gambles and that satisfy the following condition:

$$\sum_{I\subseteq\{1,\ldots,p\}} (-1)^{|I|}\underline{P}\left(f \wedge \bigwedge_{i\in I} f_i\right) \geq 0$$

for every $p \leq n$ and gambles $f, f_1, \ldots, f_p$ on the domain, where $f_1 \wedge f_2$ denotes the point-wise minimum of f_1, f_2. These previsions were investigated in some detail in [15] and in [3, 4]. They have a number of interesting properties: for instance, they are characterised by their restrictions to events, through the Choquet integral [1, 6]; moreover, the property of 2-monotonicity is equivalent [4, Theorem 15] to comonotonic additivity, which is of interest in economics. A coherent lower prevision which is n-monotone for every natural number n is called *completely monotone*.

The notion of n-monotonicity can be easily extended to conditional lower previsions: given disjoint subsets O,I of $\{1,\dots,n\}$, we say that a conditional lower prevision $\underline{P}(X_O|X_I)$ with domain $\mathcal{K}_{O\cup I}$ is n-monotone when for every $x \in \mathcal{X}_I$, $\underline{P}(\cdot|x)$ is an n-monotone lower prevision on its domain.

In this section, we are going to study if, given an unconditional 2-monotone lower prevision, it is possible to define a conditional lower prevision which is coherent with it and still satisfies the property of 2-monotonicity. There are a few examples in the literature hinting that this is the case: since from [4, Theorem 11] a vacuous lower prevision is completely monotone, it follows from [16, Sect. 6.6.1] that the natural extension of the vacuous unconditional lower prevision is again vacuous (and therefore completely monotone).This, together with the results in Sect. 4.2, implies that the natural extension of a linear prevision (which is in particular completely monotone) is also completely monotone. That this is also the case for lower previsions is established in the following theorem:

Theorem 11. *[15, Theorem 7.2] Let $\underline{P}$ be a 2-monotone lower prevision on $\mathcal{L}(\mathcal{X}^n)$, and let $\underline{P}(X_O|X_I)$ be defined from $\underline{P}$ by natural extension. Then $\underline{P}(X_O|X_I)$ is also 2-monotone on events.*

Note, however, that $\underline{P}(X_O|X_I)$ need not be the Choquet integral with respect to its restriction to events, and as a consequence it is not 2-monotone in general.

It is an open problem at this stage whether this property generalises to n-monotonicity, for $n \geq 2$, and whether we can also define n-monotone conditional lower previsions by means of other procedures, such as regular extension.

6 Conclusions

In this paper we have studied the difference between the weak and strong coherence of a number of conditional lower previsions when all the referential spaces are finite. We have proven that one of the key points is the issue of conditioning on sets of (lower) probability zero. This problem has been considered in some detail by a number of authors (see for instance [16, Section 6.10] and [2, Chapter 12]).

On the other hand, we have also established the smallest (more conservative) and greatest (more informative) conditional lower previsions that we can derive from an unconditional lower prevision in a coherent way. Although weak and strong coherence are not equivalent when we want to derive more than one conditional lower prevision, we have proven that the smallest and greatest weakly coherent updated previsions coincide with the smallest and greatest coherent updated previsions, and are given by the natural and regular extensions, respectively. In this sense, it is interesting to remark some recent work [14], based on earlier results in [17], where it is established that the natural extension can be seen as a limit of conditional lower previsions defined using regular extension.

We have shown that the classes of weakly and strongly coherent models are closed by point-wise limits but not by convex combinations (although in the case of weak coherence we have convexity if we fix the compatible joint), and that we can use the natural extension not only to propagate coherence, but also the stronger notion of 2-monotonicity.

We would like to conclude remarking that most of the properties established in this paper do not extend to conditional previsions on infinite spaces. This is studied in detail in [11]. We also refer to this paper for a study of the coherent updating of possibility measures. On the other hand, an open problem derived from this paper would be to establish the results for previsions conditional on partitions, not necessarily related to product spaces. This approach is used in [16, Chapter 6] and [8, Chapter 4].

Acknowledgements. I would like to thank the financial support of the projects TIN2008-06796-C04-01 and MTM2007-61193.

References

1. Choquet, G.: Theory of capacities. Annales de l'Institut Fourier 5, 131–295 (1953-1954)
2. Coletti, G., Scozzafava, R.: Probabilistic logic in a coherent setting. Kluwer, New York (2002)
3. de Cooman, G., Troffaes, M., Miranda, E.: A unifying approach to integration of non-additive charges. Journal of Mathematical Analysis and Applications 340(2), 982–999 (2008)
4. de Cooman, G., Troffaes, M., Miranda, E.: n-monotone exact functionals. Journal of Mathematical Analysis and Applications 347(1), 133–146 (2008)
5. Cozman, F.G.: Credal networks. Artificial Intelligence 120(2), 199–233 (2000)
6. Denneberg, D.: Non-Additive Measure and Integral. Kluwer Academic, Dordrecht (1994)
7. de Finetti, B.: La prévision: ses lois logiques, ses sources subjectives. Annales de l'Institut Henri Poincaré 7, 1–68 (1937)
8. de Finetti, B.: Theory of probability. John Wiley and Sons, Chichester (1974-1975); English translation of Teoria delle Probabilità. Einaudi, Turin (1970)
9. Levi, I.: The enterprise of knowledge. MIT Press, Cambridge (1980)
10. Miranda, E.: A survey of the theory of coherent lower previsions. International Journal of Approximate Reasoning 48(2), 628–658 (2008)
11. Miranda, E.: Updating coherent previsions on finite spaces. Fuzzy Sets and Systems 160(9), 1286–1307 (2009)
12. Miranda, E., de Cooman, G.: Coherence and independence in non-linear spaces. Technical Report (2005), `http://bellman.ciencias.uniovi.es/~emiranda/`
13. Miranda, E., Zaffalon, M.: Coherence graphs. Artificial Intelligence 173(1), 104–144 (2009)
14. Miranda, E., Zaffalon, M.: Natural extension as a limit of regular extensions. In: Proceedings of ISIPTA 2009, Durham, England, July 2009, pp. 327–336 (2009)
15. Walley, P.: Coherent lower (and upper) probabilities. University of Warwick, Coventry. Statistics Research Report 22 (1981)

16. Walley, P.: Statistical reasoning with imprecise probabilities. Chapman and Hall, London (1991)
17. Walley, P., Pelessoni, R., Vicig, P.: Direct algorithms for checking consistecy and making inferences for conditional probability assessments. Journal of Statistical Planning and Inference 126(1), 119–151 (2004)
18. Williams, P.: Notes on conditional previsions. Technical report, University of Sussex (1975); Reprinted in a revised form in the International Journal of Approximate Reasoning 44(3), 366–383 (2007)

On Evidential Markov Chains

Hélène Soubaras

Abstract. Evidential Markov chains (EMCs) are a generalization of classical Markov chains to the Dempster-Shafer theory, replacing the involved states by sets of states. They have been proposed recently in the particular field of an image segmentation application, as hidden models. With the aim to propose them as a more general tool, this paper explores new theoretical aspects about the conditioning of belief functions and the comparison to classical Markov chains and HMMs will be discussed. New computation tools based on matrices are proposed. The potential application domains seem promising in the information-based decision-support systems and an example is given.

Keywords: Markov chains, belief functions, Dempster-Shafer theory, Hidden Markov Models, evidential networks.

1 Introduction

Markov chains [4] are well-known statistical models for memoryless systems. They are applied to a wide range of application domains, and they are a mathematically powerful tool [19, 9].

But the parameters they involve are precise probabilities, which will not be available in a family of decision-making problems where the data are imprecise or incomplete, or in systems whose behavior can be described only roughly. This is why the generalization of Markov chains to belief functions has recently been proposed in works around W. Pieczynski [3, 7]. This new model, called *Evidential Markov Chain* (EMC), was used as hidden model in a particular application of image segmentation. These works proposed an algorithm to solve the hidden model based on HMM approaches, and examined the computational complexity.

Hélène Soubaras
Thales R&T. Campus Polytechnique, 1. av. A. Fresnel - F91767 Palaiseau
e-mail: helene.soubaras@thalesgroup.com

B. Bouchon-Meunier et al. (Eds.) Found. of Reas. under Uncert., STUDFUZZ 249, pp. 247–264.
springerlink.com

The objective of this paper is to explore some theoretical aspects about EMCs, and to show their relevance to a wide panel of possible applications.

Basics of the Dempster-Shafer theory [13, 17] will first be reminded, then the Markov chains and the EMC will be defined. Aspects about conditioning will be discussed [15], and some possible applications will be proposed.

2 Basics of the Dempster-Shafer Theory

This section will remind the basics of the Dempster-Shafer theory and provide tools to understand them (probabilistic point of view and matrix) that will be useful in the sequel.

2.1 *Basic Belief Assignment*

One calls *frame of discernment* a set Ω of all possible hypotheses; Ω can be discrete or continuous.

A *mass function*, also called BBA (Basic Belief Assignment) [13], is a mapping m on the power set 2^Ω, which is the set of all subsets of Ω, to $[0;1]$ such that, if Ω is finite:

$$\sum_{A\subseteq\Omega} m(A) = 1$$

A subset $A \subset \Omega$ is called a *focal set* as soon as its mass is nonzero. If Ω is countable or continuous, the above expression is still valid if there is a finite number of focal sets.

If $m(\oslash) \neq 0$, some belief on an hypothesis that would be outside Ω. This is the *Open World Assumption (OWA)* [6]. Otherwise, the mass function is said normalized. m becomes a classical probability when the focal sets are disjoint singletons. $\mathscr{F} \subseteq 2^\Omega$ will denote the set of all focal sets.

2.2 *Induced Probability Space*

In this paragraph shows that the belief functions can be manipulated through a probability μ, as did Shafer [14].

The set $2^{\mathscr{F}}$ is then the set of all collections of focal sets. Note that $2^{\mathscr{F}} \subseteq 2^{2^\Omega}$, which is the set of all collections of subsets of Ω. The elements of $2^{\mathscr{F}}$ are then of the form:

$$A \in 2^{\mathscr{F}} \iff A = \{B_1, B_2...B_n\}$$

with $B_i \in \mathscr{F}\, \forall i$. So, $2^{\mathscr{F}}$ is a σ-algebra on $\mathscr{F}$. Let's define on $2^{\mathscr{F}}$ the following function:

$$\mu : 2^{\mathscr{F}} \to [0;1]$$

such that $\forall A \in 2^{\mathscr{F}}$,

$$\mu\left(A = \{B_1, B_2...B_n\}\right) = \sum_{i=1}^{n} m(B_i) \tag{1}$$

It is easy to see that μ is a measure, since $\mu(\oslash) = 0$ and μ is additive, i.e. $\mu(\bigcup_i A_i) = \sum_i \mu(A_i)$ as soon as the A_i are pairwise disjoint. Furthermore, $\mu(\mathscr{F}) = 1$. Thus, $(\mathscr{F}, 2^{\mathscr{F}}, \mu)$ is a probability space. In other words, focal sets can be seen as set-valued random variables. The probability $\mu(A = \{B_1, B_2 ... B_n\})$ corresponds to the fact that one of the focal sets B_i, $1 \leq i \leq n$ occurs (thus the truth is in one of these sets). At this stage one doesn't take into account the fact that the B_i are disjoint or not.

One can define two functions that provide collections in $2^{\mathscr{F}}$ for any given $A \subseteq \Omega$ (even if A is not in $\mathscr{F}$):

$$\overline{\mathscr{F}}(A) = \{B \in \mathscr{F} / B \cap A \neq \oslash\}$$

(these are the elements of $\mathscr{F}$ *hitting* the given subset A), and the dual collection, which is:

$$\underline{\mathscr{F}}(A) = \{B \in \mathscr{F} / B \subseteq A\}$$

Note that $\overline{\mathscr{F}}(A) = \underline{\mathscr{F}}^c(A^c)$.

$\underline{\mathscr{F}}(A)$ and $\overline{\mathscr{F}}(A)$ are called respectively the *inner* and the *outer restriction* of $\mathscr{F}$ with respect to A [21].

2.3 Belief Function, Plausibility and Commonality

For a given mass function m a *belief* function Bel, a *plausibility* function Pl and a *commonality* function q have been defined as follows [13] for all $A \subseteq \Omega$:

$$Bel(A) = \sum_{B \subseteq A, B \neq \oslash} m(B) \tag{2}$$

$$Pl(A) = \sum_{B \cap A \neq \oslash} m(B) \tag{3}$$

$$q(A) = \sum_{B \supseteq A} m(B) \tag{4}$$

They can also be written as:

$$Bel(A) = \mu(S \subseteq A, S \neq \oslash) = \mu(\underline{A})$$

(probability that the truth is *always* in A),

$$Pl(A) = \mu(S \cap A \neq \oslash) = \mu(\overline{A})$$

(probability that the truth is *possibly* in A), and

$$q(A) = \mu(A \subseteq S).$$

One can remark that Bel and Pl can be written using the inner and the outer extension of A in $\mathscr{F}* = \mathscr{F} \setminus \{\oslash\}$, which denotes the set of nonempty focal sets of the frame Ω. The belief function can be expressed as:

$$Bel(A) = \mu\left(\underline{\mathscr{F}*}(A)\right)$$

and the plausibility as:

$$Pl(A) = \mu(\overline{\mathscr{F}*(A)})$$

The two functions *Bel* and *Pl* are dual, related by:

$$Pl(A) = 1 - m(\oslash) - Bel(A^c) \tag{5}$$

where A^c denotes the complementary set of A in Ω.

Smets [17] introduced the *pignistic probability Bet* associated to m. When Ω is discrete, it is defined $\forall x \in \Omega$ by:

$$Bet(x) = \frac{1}{1 - m(\oslash)} \sum_{A/x \in A} \frac{m(A)}{|A|} \tag{6}$$

where $|A|$ is the cardinality of A, i.e. the number of elements of A. *Bel*, *Pl* and *Bet* are all equal to classical probabilities when the focal sets are disjoint singletons.

It is important to notice that the three functions (belief, plausibility and commonality) are not measures because they are not additive, but subadditive since:

$$Bel(A \cup B) \geq Bel(A) + Bel(B)$$

3 Matrix Tools

We consider a mass function on a finite discrete frame Ω. N_f is the number of focal sets. One will define the mass vector M by its coordinates:

$$M(j) = m(A_j) = m_j \tag{7}$$

for all focal set A_j, $1 \leq j \leq N_f$.

3.1 *Matrix Tools for Belief Functions*

It is known [5] that the relation between the mass function m and the belief function *Bel* is a bijection. For a given discrete space Ω containing N elements and any function *Bel* defined on a set of subsets $\mathscr{F} \subseteq 2^\Omega$, if *Bel* satisfies the three assumptions:

1. $Bel(\Omega) \leq 1$
2. *Bel* is completely monotone, i.e. if $A \subset B$ then $Bel(A) \leq Bel(B)$
3. *Bel* is subadditive

then a mass function m can be deduced thanks to the so-called *Möbius transform*:

$$m(A) = \sum_{B \subseteq A} (-1)^{|A \setminus B|} Bel(B) \tag{8}$$

If one denotes as M the column vector of the masses of all the subsets of Ω, its size will be 2^N. The column vector Bel containing all the values of the belief function on the nonempty subsets will also be of size 2^N, and it can be calculated from M thanks to a matrix product:

$$Bel = \mathbf{BfrM}.M$$

and the Möbius transform is then performed by the inverse matrix $\mathbf{BfrM}^{-1}$. **BfrM** is a $2^N \times 2^N$ generalization matrix G defined by Smets [16] as:

Definition 1. *A* generalization matrix *of a collection of subsets A_i is a stochastic matrix G satisfying $G(i,j) = 0$ if $A_j \not\subseteq A_i$.*

Smets [16] also defined, similarly:

Definition 2. *A* specialization matrix *is a stochastic matrix S satisfying $S(i,j) = 0$ if $A_i \not\subseteq A_j$.*

Those matrices are stochastic in Smets's general definition, but in this case, there nonzero elements are equal to 1.

In this paper we also propose a new matrix in order to compute the plausibility function. It will be called the gauge matrix:

Definition 3. *The* gauge *(pattern) matrix of a collection of subets A_i is defined by:*

$$G_a(i,j) = \begin{cases} 1 \text{ if} & A_i \cap A_j \neq \oslash \\ 0 \text{ otherwise} & \end{cases}$$

The 2^N-size column vector Pl of the plausibility function is then defined by:

$$Pl = G_a M$$

The commonality can also be computed through such a matrix product.

3.2 Markov Kernel Matrix

Definition 4. *Let be X and Y two discrete random variables. A* Markov kernel *is a matrix of the conditional probabilities $p(i|k)$ of the occurrence $Y = y_i$ given $X = x_k$ has occurred. (In Markov chains, as we shall see at paragraph 4.1, the state transition matrix is a Markov kernel whose random variables are two successive states of the system).*

Let Ω be a frame of discernment. One supposes there is a finite partition $\mathscr{H} = \{X_i \; / \; 1 \leq i \leq N_c\} \subseteq 2^\Omega$ on the frame Ω. The couple $(\Omega, \mathscr{H})$ is called a *propositional space*. Each subset X_i can be called a *class*. Let m be a mass function on Ω, with a finite set of focal sets $\mathscr{F} = \{A_k / 1 \leq k \leq N_f\}$. One would like to estimate in which class X the truth is for a given mass function.

Classes and focal sets can be viewed as random variables X and A., taking values in $\mathcal{H}$ and $\mathcal{F}$ respectively. Each focal set A_k can occur with a probability m_k.

The assumption that will be made now is that there exists a fixed Markov kernel K defined by $K(i,k) = p(i|k)$ such that

$$p_i = \sum_{k=1}^{N_f} p(i|k) m_k \quad (9)$$

where $p_i = Pr(X_i)$ and $m_k = m(A_k)$. This can also be written with P, the vector of the probabilities of the classes:

$$P = KM \quad (10)$$

As $0 \leq p(i|k) \leq 1$ for all (i,k), one can notice from (2) and (3) that for all compatible kernel K,we get the following relation, for all set $X = X_i$:

$$Bel(X) \leq p(X) \leq Pl(X) \quad (11)$$

Thus the probabilities $p_i(X_i)$ of each class X_i are *imprecise probabilities* since they belong to an interval.

3.3 Matrix Representations for Classes

Let N_f be the number of focal sets and N_c the number of classes. One supposes that N_f and N_c are finite. One can still define the gauge matrix G_a of size $N_c \times N_f$ by $G_a(i,j) = 1$ if $X_i \cap A_j \neq \oslash$, and 0 otherwise, for all classes X_i and for all focal sets A_j. Any Markov kernel K compatible with the mass function is zero where G_a is zero. The lines of the transposed matrix G_a^T can be seen as base-2 representations of the focal sets.

One can describe entirely a belief mass by its gauge matrix G_a and its mass vector M.

When the classes are not singletons, the cardinality of a focal set can be defined as the number of classes it meets. This number is obtained by

$$(11...1)G_a = \begin{pmatrix} |A_1| \\ |A_2| \\ \vdots \\ |A_{N_f}| \end{pmatrix}$$

The computation of the belief function, the plausibility function, the commonality and the pignistic probability with matrix products is still possible, as it was shown by Smets [16] and at paragraph 3, for the 2^{N_c} subsets of Ω that are unions of subsets X_i:

$$Bel = G.M \quad (12)$$

$$Pl = G_a.M \tag{13}$$

$$q = S.M \tag{14}$$

$$Bet = \mathbf{BetPfrM}.M \tag{15}$$

where G, G_a, S and **BetPfrM** are all $2^{N_c} \times N_f$-sized matrices defined as in paragraph 3.1.

4 Evidential Markov Chains

4.1 Definition

Let's concider $\Omega = \{a_1, a_2 ... a_N\}$ be the set of the possible random *states* x_t of a system for each time t.

Definition 5. *The probability Pr for each state of the system satisfies the* Markov property *if and only if:*

$$Pr(x_t|x_0, x_1, x_2, x_3, ... x_{t-1}) = Pr(x_t|x_{t-1})$$

The transition matrix *of the system is the $N \times N$ matrix Q of the probabilities of transition from one state at a given time t to another state at the next time $t+1$ defined by:*

$$Q = (q_{ij})_{1 \leq i,j \leq N}$$

where

$$q_{ij} = Pr(x_{t+1} = a_i | x_t = a_j)$$

If a transition matrix exists, the Markov property is satisfied.

Proof. One can write:

$$\begin{aligned} Pr(x_t, x_{t-1} ... x_0) &= Pr(x_t|x_{t-1} ... x_0).Pr(x_{t-1} ... x_0) \\ &= Pr(x_0).\prod_{n=1}^{t} Pr(x_n|x_{n-1}) \end{aligned}$$

by recurrence. So the conditional probability is, by applying Bayes' formula:

$$Pr(x_t|x_{t-1} ... x_0) = \frac{Pr(x_t, x_{t-1} ... x_0)}{Pr(x_{t-1}, x_{t-2} ... x_0)} = Pr(x_t|x_{t-1})$$

If one denotes as P_t the vector of the probabilities of each state, one has the following relation:

$$P_t = QP_{t-1} = Q^t P_0$$

Definition 6. *A* Markov chain *is a triple (Ω, Q, P_0) where $P_0 = Pr(x_0)$ is the* initial probability vector.

Figure 1 shows the scheme (Bayesian Network) of a Markov chain.

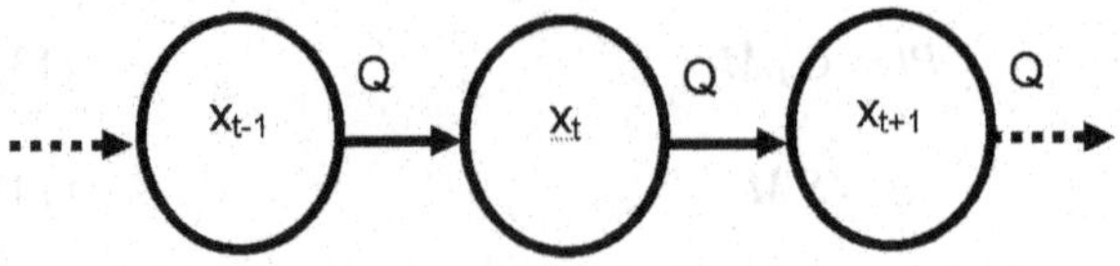

Fig. 1 Scheme of a Markov chain

An Evidential Markov Chain is a classical Markov chain where the random variable representing the possible states of the system is replaced by random (focal) sets [3, 10]:

Definition 7. *Let Ω be a frame of discernment. An* Evidential Markov Chain *(EMC) is a Markov chain $(\mathscr{F}, Q, M_0)$ where $\mathscr{F}$ is a set of focal sets and M_0 is the vector of the initial masses of all the focal sets.*

If the vector of masses at time t is denoted as M_t, one can write the following relation:

$$M_t = QM_{t-1} \tag{16}$$

In the particular case where $\mathscr{F}$ is the set of the N singletons of Ω, the EMC becomes a classical probabilistic Markov chain.

One can verify through equations 12, 13, 14 and 15 that the belief function, the plausibility, the commonality and the pignistic probability of an EMC are Markov chains whose transition matrices are of the form

$$Q' = HQ(H^T H)^{-1} H^T$$

where the matrix H represents G, G_a, S and **BetPfrM** respectively. Of course, if the topology of the focal sets does not satisfy some conditions, the matrix $H^T H$ will not be invertible. But otherwise, a transition matrix exists for these fuzzy measures, so they are Markov chains as showed at paragraph 4.1.

The fact that theses functions are Markov chains is still true when the matrices H are restricted to a subcollection of subsets, for example to singletons. If H is square, it may be invertible, the resulting transition matrix will be:

$$Q' = HQH^{-1}$$

4.2 Conditioning and the Generalized Bayes Theorem

Some classical rules of conditioning have been proposed to take into account a new piece of knowledge in one's beliefs (i.e. a new observation). For example, the famous Jeffrey's rulefor statistical inference is inspired from Bayes formula. Dubois [1] also proposed conbination and conditioning rules.

Our purpose here is not inference, but just to describe a system's behavior. Evidential Markov chains are particular cases of evidential networks [22] since they rely on conditional masses. There have been some works about conditioning in the Dempster-Shafer theory [1, 20].

In probability theory the conditional probability of a subset A given a subset B is the new probability defined on B as probability space, and it is given by the Bayes formula:

$$Pr(A|B) = \frac{Pr(A \cap B)}{Pr(B)} \tag{17}$$

Such relations of conditioning have been proposed for belief masses [13]. Thanks to the probability μ introduced at 2.2, there are four ways to express the conditional probability that the truth is *always / possibly* in A given that it is *always / possibly* in B:

$$\mu(\underline{A}|\underline{B}) = \frac{Bel(A \cap B)}{Bel(B)}$$

$$\mu(\overline{A}|\underline{B}) = \frac{Bel(B) - Bel(B \setminus A)}{Bel(B)}$$

$$\mu(\underline{A}|\overline{B}) = \frac{Pl(B) - Pl(B \setminus A)}{Pl(B)}$$

(Dempster's $Bel(A|B)$ [17])

$$\mu(\overline{A}|\overline{B}) = \frac{Pl(A \cap B)}{Pl(B)}$$

(Dempster's $Pl(A|B)$ [17]).

Dempster also defined the underlying conditional mass [17], which is equal to, in the unnormalized case:

$$m(A \wr B) = \begin{cases} \mu(S \cap B = A) & \text{for } A \subseteq B \\ 0 & \text{otherwise} \end{cases}$$

and, in its normalized version: $m(A|B) = m(A \wr B)/\mu(B^c)$. Smets [15] proposes expressions for the generalization of the Bayes theorem to belief functions (GBT). In this article, the expression proposed for the masses themselves is the Conjunctive Rule of Combination [2], which is exactly the one expressed using $m(A \wr B)$:

$$m_t(A) = \sum_B m(A \wr B) m_{t-1}(B)$$

This is exactly the operation performed in the EMC transition matrix product 16.

Nevertheless, Smets' expressions for the GBT for the belief and the plausibility functions suppose that one of the two frames of discernment is a partition. To be applied to a EMC, the GBT implies that F is a partition of Ω. This is a particular case which is not very interesting since it corresponds to a classical probabilistic case where the belief and the plausibility are all equal to a simple probability.

In conclusion, through its transition matrix, the EMC performs one form of conjunctive rule of combination of Demspter's unnormalized conditional masses.

4.3 Associated Hidden Markov Model

Definition 8. A Hidden Markov Model *(HMM) is a 5-uple* $(\Omega_x, \Omega_y, Q, K, P_0)$ *where* (Ω_x, Q, P_0) *is a Markov chain, and the observation is a random variable y taking values in* Ω_y *and such that the matrix of the emission probabilities (Markov kernel of y given x) is K.*

The internal states x of the Markov chain are not known, except through the knowledge of y. To estimate x_t from an observed sequence y_t (in particular when $|\Omega_y| < |\Omega_x|$), algorithms have been proposed such as the Baum-Welch algorithm and the Viterbi algorithm [11].

To compare a HMM and a EMC (see figures 2 and 3) let's consider a given HMM with transition matrix Q and whose observed varialbe Y is the random focal set. If there exists a compatible EMC with transition matrix Q', the following condition must be satisfied: $(KQ - Q'K)P = 0$ for all probability vector P. If $dim(M) \leq dim(P)$, one solution is

$$Q' = KQK^T(KK^T)^{-1}$$

(if K is not degenerated). Thus it is possible to find one EMC which is compatible with a given HMM, but it is not so easy to find one HMM compatible with a given EMC since $K^T K$ is not invertible in the general case.

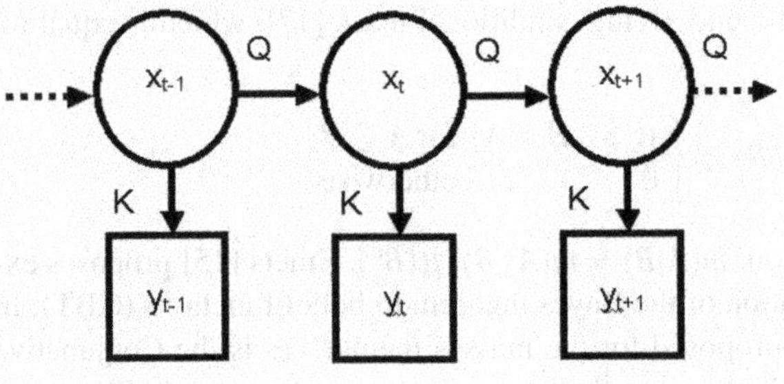

Fig. 2 Scheme of a HMM

Now suppose we have a EMC, and let's consider again the partition of Ω into classes $\mathscr{H} = \{X_i / i \leq i \leq N_c\} \subseteq 2^\Omega$; one assumes that the conditional belief functions $Bel(A_i|X_k)$ are known, and that the mass function on $\mathscr{H}$ is normalized. The GBT can then be applied [15]:

$$\alpha Bel(X_i|A_j) = \prod_{k \neq i} Bel(A_j^c|X_k) - \prod_k Bel(A_j^c|X_k)$$

where α is the normalizing factor:

$$\alpha = 1 - \prod_k Bel(A_j^c|X_k)$$

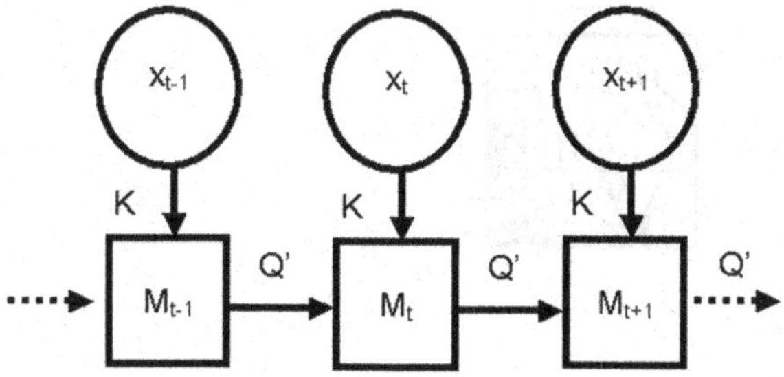

Fig. 3 Scheme of a EMC

Thus, thanks to the Möbius transform, a mass function can be calculated for the classes. If this mass function were a classical probability, the classes X_i could be seen as the hidden internal state of the Markov chain whose observations are the (random) focal sets. Thus, an EMC can be viewed as a generalization of a classical HMM.

In conclusion, we showed that it can be easy to find one EMC compatible with a given HMM. For a given EMC, a family of compatible HMMs exists. When the EMC observations are random focal sets, the EMC internal state can be solved as in a HMM.

5 Applications

The EMC model has been proposed first to achieve image segmentation [7, 3]. It was supposed to be hidden in those cases; this means the mass function could not be observed directly, but through a measurement y such that the conditional probabilities $Pr(y|A)$ are fixed and known (or estimated) for each focal set A. An algorithm derived from the classical HMM identification was proposed [7].

EMCs can be also interesting models for other uncertain systems, particularly if they involve phenomena that are difficult to quantify, like human behaviors. Such modelling can be applied to the forecasting of the future evolution of a system; it can also be useful for simulations in order to measure the performances of other algorithms. Techniques used in the classical statistics, such as Monte-Carlo or Importance Sampling, could be generalized to EMCs.

As an example, an EMC can be used as a simulation model for the tenseness between two countries that could lead to a conflict or a war. This will be detailed here. The decision-making support tool is a situation understanding module.

5.1 Example: Simulation of a Geopolitical Crisis

The simulation of the sequence of events of a crisis is useful to improve tools and algorithms for crisis management, such as the risk assessment module. Its principle in such a situation understanding module for geopolitical crises consists in

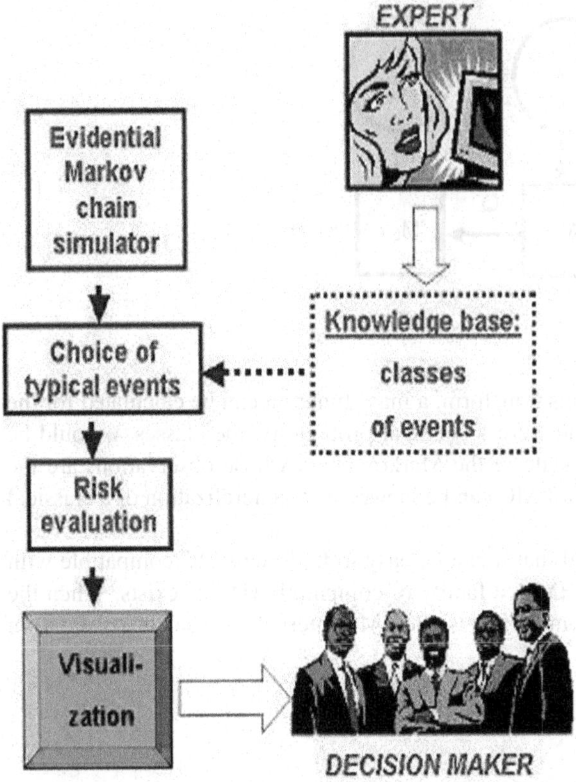

Fig. 4 Scheme of the decision-making support system involving a simulation

receiving events (with any type of sensor, even human observations such as phone calls or texts), then classifying them in the knowledge base, and running the risk measurement tools. The knowledge base is constituted prevously using experts opinions. The simulation consists in replaceing the observation and classification of events by an EMC-based generator, as shown in the diagram at figure 4.

The (political) tenseness is an underlying value that can be estimated only through open sources of information (journal articles, television news...) and indirectly (e.g. through symptomatic events such as demonstrations, declarations, political decisions...). The system should be able to extract a mass function from these events.

5.2 *Crisis Model*

The EMC is the model proposed in this work to describe the behaviour of the crisis situation. The focal sets are overlapping rough estimations of the tenseness, and the tenseness itself is quantized on several values (5 in the example shown figure 5. These intervals of tenseness are the classes $X_i, 1 \leq i \leq 5$.

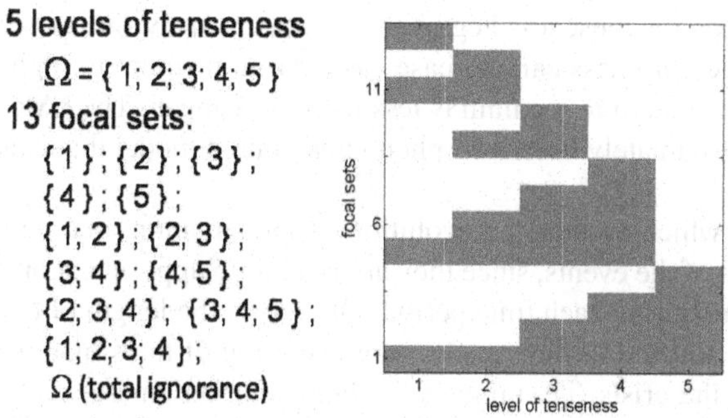

Fig. 5 Example with 13 focal sets over 5 levels of political tenseness between two countries

The observed data are the events. Each type of event corresponds to a gievn mass function which expresses its possible impact on political tenseness, as shown in the example at figure 6.

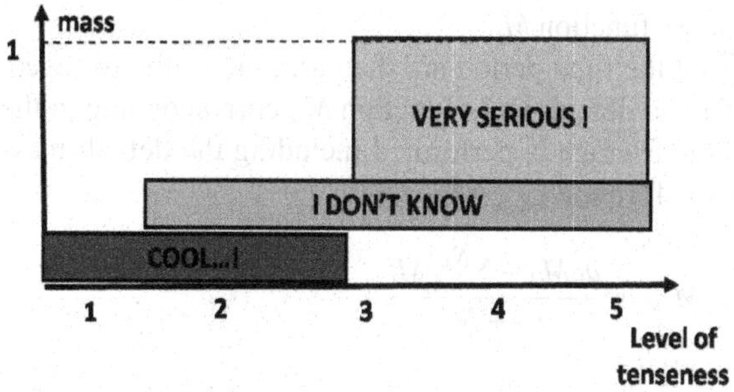

Fig. 6 Example of mass function of an event

A mass function is represented mathematically by a vector of size 13. Each of its coeffficients is the mass of one of the 13 focal sets. The 13×13 transition matrix of the EMC model for the crisis evolution is:

$$Q = \begin{pmatrix} 0.9999 & 0 & 0 & 0 & 0 & 0 & 0 & 0 & 0 & 0 & 0 & 0 & 0 \\ 0 & 0.8 & 0.2 & 0 & 0 & 0 & 0 & 0 & 0 & 0 & 0 & 0 & 0 \\ 0.0001 & 0 & 0.8 & 0 & 0 & 0 & 0 & 0 & 0 & 0 & 0 & 0 & 0 \\ 0 & 0.2 & 0 & 0.8 & 0 & 0 & 0 & 0 & 0 & 0 & 0 & 0 & 0 \\ 0 & 0 & 0 & 0.2 & 0.8 & 0 & 0 & 0 & 0 & 0 & 0 & 0 & 0 \\ 0 & 0 & 0 & 0 & 0.2 & 0.8 & 0 & 0 & 0 & 0 & 0 & 0 & 0 \\ 0 & 0 & 0 & 0 & 0 & 0 & 0.8 & 0 & 0 & 0.2 & 0 & 0 & 0 \\ 0 & 0 & 0 & 0 & 0 & 0 & 0 & 0.8 & 0.2 & 0 & 0 & 0 & 0 \\ 0 & 0 & 0 & 0 & 0 & 0 & 0.2 & 0 & 0.8 & 0 & 0 & 0 & 0 \\ 0 & 0 & 0 & 0 & 0 & 0.2 & 0 & 0 & 0 & 0.8 & 0 & 0 & 0 \\ 0 & 0 & 0 & 0 & 0 & 0 & 0 & 0.2 & 0 & 0 & 0.8 & 0 & 0 \\ 0 & 0 & 0 & 0 & 0 & 0 & 0 & 0 & 0 & 0 & 0 & 0.8 & 0.2 \\ 0 & 0 & 0 & 0 & 0 & 0 & 0 & 0 & 0 & 0 & 0.2 & 0.2 & 0.8 \end{pmatrix}$$

It reflects the fact that when the tenseness begins to rise, it can easily rise more. It shows for example that the tenseness can increase easily from medium to high, but resolution of the crisis from high to medium is less likely to happen. The EMC allows to translate such approximately described phenomena into a model that can be implemented.

In addition to this EMC which models the evolution of the situation, one will perform a kind of integration of the events, since they are random. Suppose that one observes the events occurring during each time period. Of course, the length of the time period is chosen to be adapted to the speed of the evolution of the situation, depending on the nature of the crisis (i.e. minutes for humanitarian emergencies, hours for destructions caused by a natural meteorological phenomenon, weeks for pandemics, months for a political crisis). At each time period t a number N_e of events have been detected and identified. This number N_e is not fixed, and it can be modelled by a random draw. In the simulation, its probability law is uniform on a given interval.

Each one of the events observed during the time period t is classified; it corresponds to type i, which is one of the event types stored in the knowledge base. Thus it receives the corresponding mass function M_i.

All the mass functions during the time period are then arithmetically averaged. If there are no events ($N_e = 0$), the default mass function M_0 corresponding to the non-event case is attributed. The average is performed including the default mass with a weight a_0, as shown in the formula:

$$< M >= \frac{a_0 M_0 + \Sigma_{i=1}^{N_e} M_i}{a_0 + N_e}$$

5.3 Simulation of the Average Mass Function by an EMC

This average mass function $< M >= M_t$ is going to be modelled by the output of an EMC. This is the model used to simulate the crisis, as illustrated on figure 7. The

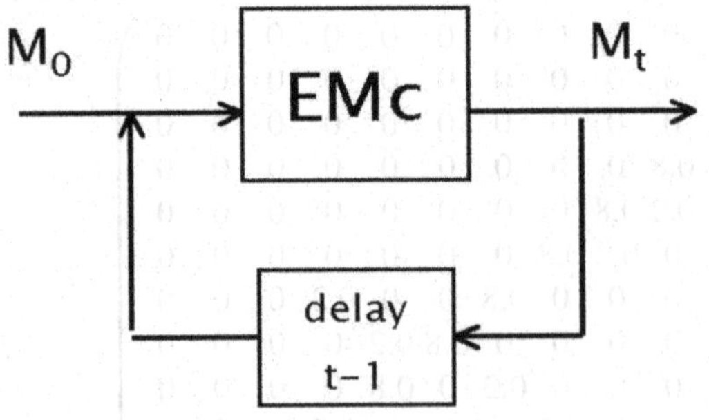

Fig. 7 EMC as a generator of mass functions

mass function obtained at the EMC output at time $t - 1$ is put again at the EMC input to calculate the mass function at the following time t.

5.4 Generation of Events

The output of the EMC described in the previous paragraph represents a reference mass function denoted as M_t. The idea is to choose, more or less randomly, a number of event types (in the knowledge base) whose average mass function $< M >$ is almost equal to the reference mass function. As illustrated at figure 8, the Euclidian distance is calculated between M_t and each one of the mass function vectors M_i belonging to the knowledge base. A subset of the nearest mass functions of the knowledge base is thus defined. Then, the number N_e of simulated events for the given time period is generated randomly; and one event type is picked out of the subset, N_e times. This operation is repeated for each time period t. Then we get two sequences: the sequence of the number N_e for each t, and the sequence for all the event types.

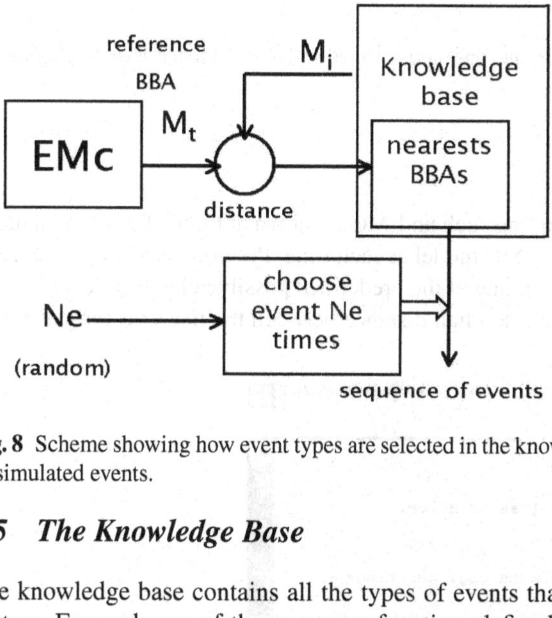

Fig. 8 Scheme showing how event types are selected in the knowledge to generate a sequence of simulated events.

5.5 The Knowledge Base

The knowledge base contains all the types of events that can occur in the studied system. For each one of them, a mass function, defined by an expert, is given. In the present example of geopolitical crisis, we shall use a classification of events that has already been used in the literature [12], called WEIS (World Event Interaction Survey) [8].

Figure 9 illustrates some of these WEIS classes of events. There mass function is represented as imprecise probabilities; that means for each level of political tenseness, the intervals bounded by the minimum and the maximum probabilities (which are equal to the belief and the plausibility functions of this given level of tenseness), are filled with gray colour.

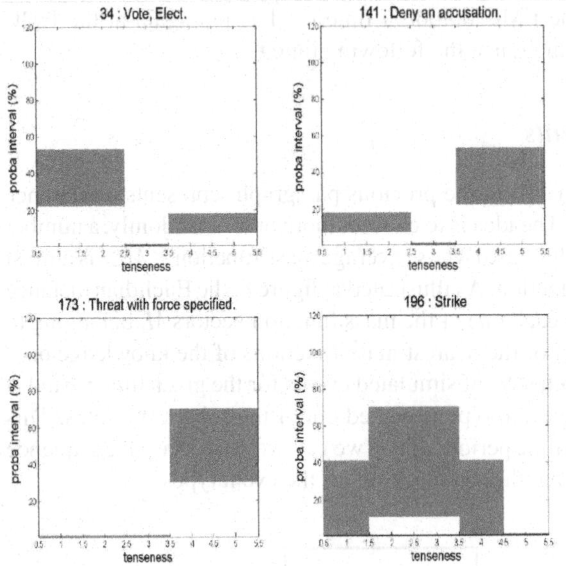

Fig. 9 Scheme showing how event types are selected in the knowledge base to generate a sequence of simulated events.

5.6 *Results*

An example obtained by running such an EMC is shown at figure 10. There, a mass function is generated by the EMC model at each time. Previously, an expert had assigned a mass function to each one of the predefined possible classes of events. The simulator computes then the Euclidian distance between the mass vector generated

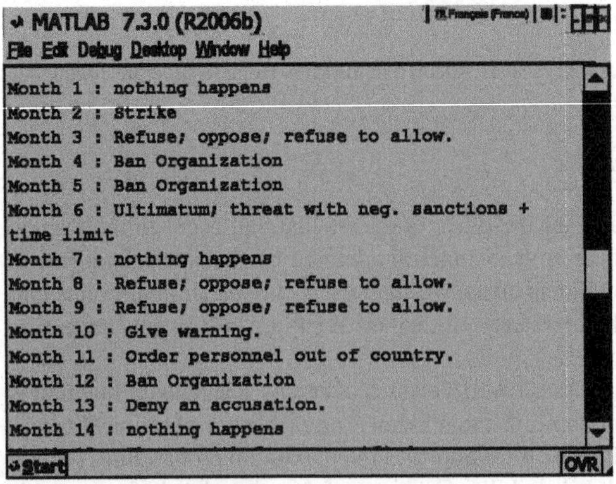

Fig. 10 Random sequence of classes of geopolitical events following an evidential Markov-modelled increase of tenseness

by the EMC and the mass vectors of each class, and it chooses randomly one class amongst the nearest ones.

Thus, one can see that an EMC can be usd to generate the realistic history of an imaginary political crisis. What is displayed is a text: the list of the event descriptions. But the simulator has also provided a sequence of numerical values: the mass function for each generated event.

6 Conclusion

Evidential Markov chains (EMCs) are a generalization of the classical Markov chains. They are Markov chains involving masses on focal sets instead of probabilities on elementary states. They have been proposed in the image segmentation model [3,7]. This paper examines some theoretical aspects of EMCs: it relates them to the Dempster's rules of conditioning and the Smets' Generalized Bayes Theorem; it points out that an EMC is a generalization of a HMM. Some computation tools based on matrix are also proposed.

EMC models have potentially interesting pplications in the field of uncertain systems, particularly those involving human behaviors or imprecise data such as text. An example is given for the simulation of the tenseness between two conflicting countries. The author has developed this study and proposed an algorithm for risk measurement in EMCs [18].

References

1. Dubois, D., Moral, S., Prade, H.: Belief change rules in ordinal and numerical uncertainty theories. In: Dubois, H.P.D. (ed.) Belief Change, pp. 311–392. Kluwer, Dordrecht (1998)
2. Dubois, D., Prade, H.: A set theoretical view of belief functions. Int. J. Gen. Systems 12, 193–226 (1986)
3. Fouque, L., Appriou, A., Pieczynski, W.: An evidential markovian model for data fusion and unsupervised image classification. In: Proc. of 3rd Int. Conf. on Information Fusion, FUSION 2000, Paris, France, pp. YuB4–25–TuB4–31 (2000)
4. Freedman, D.: Markov chains. Holden-Day (1971)
5. Grabisch, M., Murofushi, T., Sugeno, M.: Fuzzy Measures and Integrals. Physica-Verlag (2000)
6. Haenni, R.: Ignoring ignorance is ignorant. Technical report, Center for Junior Research Fellows, University of Konstanz (2003)
7. Lanchantin, P., Pieczynski, W.: Chaînes et arbres de markov évidentiels avec applications à la segmentation des processus non stationnaires. Revue Traitement du Signal 22 (2005)
8. McClelland, C.A.: World event/interaction survey codebook (icpsr 5211) inter-university consortium for political and social research, Ann Arbor (1976)
9. Nuel, G., Prum, B.: Analyse statistique des séquences biologiques. Editions Hermès, Labvoisier, Paris (2007)
10. Pieczynski, W.: Multisensor triplet markov chain and theory of evidence. Int. J. Approximate Reasoning 45, 1–16 (2007)
11. Rabiner, L.R.: A tutorial on hidden markov models and selected applications in speech recognition. Proc. of IEEE 77(2), 257–286 (1989)

12. Schrodt, P.A.: Forecasting conflict in the balkans using hidden markov model (2000)
13. Shafer, G.: A Mathematical Theory of Evidence. Princeton University Press, Princeton (1976)
14. Shafer, G.: Propagating belief functions in qualitative markov trees. Int. J. Approximate Reasoning 1, 349–400 (1987)
15. Smets, P.: Belief functions: the disjunctive rule of combination and the generalized Bayesian theorem. Int. J. Approximate Reasoning 9, 1–35 (1993)
16. Smets, P.: The application of the matrix calculus to belief functions. Int. J. Approximate Reasoning 31, 1–30 (2002)
17. Smets, P., Kennes, R.: The transferable belief model. Artificial Intelligence 66, 191–234 (1994)
18. Soubaras, H.: An evidential measure of risk in evidential markov chains. In: Sossai, C., Chemello, G. (eds.) Symbolic and Qualitative Approaches to Reasoning with Uncertainty - 10th ECSQARU, Verona, Italy, pp. 863–874. Springer, Heidelberg (2009)
19. Soubaras, H., Mattioli, J.: Une approche markovienne pour la prévision du risque. In: Proc. of 7th Congrès int. Pluridisciplinaire Qualité et Sûreté de Fonctionnement, QUALITA 2007, Tanger, Maroc, pp. 64–71 (2007)
20. Xu, H., Smets, P.: Reasoning in evidential networks with conditional belief functions. Int. J. Approximate Reasoning 14, 155–185 (1996)
21. Yaghlane, A.B., Denœux, T., Mellouli, K.: Coarsening approximations of belief functions. In: Benferhat, S., Besnard, P. (eds.) ECSQARU 2001. LNCS (LNAI), vol. 2143, p. 362. Springer, Heidelberg (2001)
22. Yaghlane, B.B., Mellouli, K.: Inference in directed evidential networks based on the transferable belief model. Int. J. Approximate Reasoning 48, 399–418 (2008)

Author Index